Computational Aerodynamic Modeling of Aerospace Vehicles

Computational Aerodynamic Modeling of Aerospace Vehicles

Special Issue Editors

Mehdi Ghoreyshi
Karl Jenkins

MDPI • Basel • Beijing • Wuhan • Barcelona • Belgrade

MDPI

Special Issue Editors
Mehdi Ghoreyshi
United States Air Force Academy
USA

Karl Jenkins
Cranfield University
UK

Editorial Office
MDPI
St. Alban-Anlage 66
4052 Basel, Switzerland

This is a reprint of articles from the Special Issue published online in the open access journal *Aerospace* (ISSN 2226-4310) from 2017 to 2019 (available at: https://www.mdpi.com/journal/aerospace/special_issues/computational_aerodynamic_modeling)

For citation purposes, cite each article independently as indicated on the article page online and as indicated below:

LastName, A.A.; LastName, B.B.; LastName, C.C. Article Title. *Journal Name* **Year**, *Article Number*, Page Range.

ISBN 978-3-03897-610-3 (Pbk)
ISBN 978-3-03897-611-0 (PDF)

Contents

About the Special Issue Editors

Mehdi Ghoreyshi is a senior aerospace engineer at the United States Air Force Academy in Colorado. He is also a visiting lecturer at University of Colorado, Colorado Springs at the Department of Mechanical and Aerospace engineering and serves the President of DANSI Engineering Company. Dr. Ghoreyshi is an active member of numerous NATO applied vehicle technology activities including vortex interaction effects, design of agile NATO vehicles, investigation of shipboard launch and recovery of vehicles, and Reynolds number scaling effects in swept wing flows. Mehdi has been the principal investigator of many projects supported by the U.S. Air Force Academy, U.S. Army, U.S. Navy, and National Academy of Sciences investigating the aerodynamics of ram-air parachutes, airdrop configurations, tiltrotor–obstacle wake interactions, and Dynamic Modeling of Non-linear Databases with Computational Fluid Dynamics (DyMOND-CFD). Mehdi is an active member of the AIAA Applied Aerodynamics Technical Committee and currently is serving as an Associate Editor for Journal of *Aerospace Science and Technology* and a Guest Editor for Journal of *Aerospace*. His research interests include reduced order modeling, system identification, and computational aerodynamic modeling. He is an author of nearly 100 publications in refereed journals and conferences.

Karl Jenkins gained a PhD from the University of Manchester that focused on computational and experimental water waves breaking interacting with coastal structures. This expertise in CFD and High Performance Computing was extended in a post-doctoral position at Cambridge University as the Sir Arthur Marshall Research Fellow, where he studied turbulent combustion using Direct Numerical Simulation. Dr. Jenkins has published over 60 papers and has won the Gaydon prize for the most significant paper contribution at a leading symposium on combustion held in Chicago. He is a member of the United Kingdom Consortium on Turbulent Reacting Flows (UKCTRF). He has been invited to give numerous international and domestic seminars and to participate as a discussion panel member at international HPC DNS/LES conferences in the US. Dr. Jenkins has also worked in industry for Allot and Lomax Consulting Engineers and Davy Distington Ltd., working on various commercial CFD codes and training engineers in their use. He has worked on adaptive parallel grid techniques, and has developed parallel codes for academic use and for blue chip companies such as Rolls Royce plc. He has also been actively involved in one of the Cambridge regional e-Science center projects entitled Grid Technology for Distance CFD.

aerospace

MDPI

Editorial

Special Issue "Computational Aerodynamic Modeling of Aerospace Vehicles"

Mehdi Ghoreyshi

High Performance Computing Research Center, U.S. Air Force Academy, Air Force Academy, CO 80840, USA;
Mehdi.Ghoreyshi@usafa.edu

Received: 3 January 2019; Accepted: 7 January 2019; Published: 8 January 2019

Aerospace, an open access journal operated by MDPI, has published a Special Issue on the Computational Aerodynamic Modeling of Aerospace Vehicles. Dr. Mehdi Ghoreyshi of the United States Air Force Academy, United States and Dr. Karl Jenkins of Cranfield University, United Kingdom served as the Guest Editors. This Special Issue of *Aerospace* contains 13 interesting articles covering a wide range of topics, from fundamental research to real-world applications.

The development of accurate simulations of flows around many aerospace vehicles poses significant challenges for computational methods. This Special Issue presents some recent advances in computational methods for the simulation of complex flows. The research article by El Rafei et al. [1] examines a new computational scheme based on Monotonic Upwind Scheme for Conservation Laws (MUSCL) within the framework of implicit large eddy simulations. The research predictions show the accuracy of the new scheme for refined computational grids. Zingaro and Könözsy [2] present a new adoption of compressible Navier–Stokes equations for predicting two-dimensional unsteady flow inside a viscous micro shock tube. In another article by Teschner et al. [3], the bifurcation properties of the Navier–Stokes equations using characteristics schemes and Riemann Solvers are investigated.

An additional topic of interest covered in this Special Issue is the use of computational tools in aerodynamics and aeroelastic predictions. The problem with these applications is the computational cost involved, particularly if this is viewed as a brute force calculation of a vehicle's aerodynamics and structure responses through its flight envelope. In order to routinely use computational methods in aircraft design, methods based on sampling, model updating, and system identification should be considered. The project report by Zhang et al. [4] demonstrates the use of multi-fidelity aircraft modeling and meshing tools to generate aerodynamic look-up tables for a regional jet-liner. The research article by Ignatyev and Khrabrov [5] presents mathematical models based on neural networks for predicting the unsteady aerodynamic behavior of a transonic cruiser. Silva [6] reviews the application of NASA's AEROM software for reduced-order modeling for the aeroelastic study of different vehicles including the Lockheed Martin N+2 supersonic configuration and KTH's generic wind-tunnel model. Additionally, the article by Berci and Cavallaro [7] demonstrates hybrid reduced-order models for the aeroelastic analysis of flexible subsonic wings. The article by Singh et al. [8] introduces a multi-fidelity computational framework for the analysis of the aerodynamic performance of flight formation. Finally, Ghoreyshi et al. [9] creates reduced-order models to predict the aerodynamic responses of rigid configurations to different wind gust profiles. The results show very good agreement between developed models and simulation data.

The remaining articles show the application of computational methods in simulation of different challenging problems. Satchell et al. [10] shows the numerical results for the simulation of the wake behind a 3D Mach 7 sphere-cone at an angle of attack of five degrees. The article by Aref et al. [11] investigates the propeller–wing aerodynamic interaction effects. Propellers were modeled with fully resolved blade geometries and their effects on the wing pressure and lift distribution are presented for different propeller configurations. In another article by Aref et al. [12], the flow inside a subsonic intake was studied using computational methods. Active and passive flow control methods were studied to improve the intake performance. Finally, the article by Boudreau et al. [13] investigates the use of large eddy simulations in predicting the flow behind a square cylinder at a Reynolds number of 21,400.

The editors of this Special Issue would like to thank each one of these authors for their contributions and for making this Special Issue a success. Additionally, the guest editors would like to thank the reviewers and the *Aerospace* editorial office, in particular Ms. Linghua Ding.

Conflicts of Interest: The author declares no conflict of interest.

References

1. El Rafei, M.; Könözsy, L.; Rana, Z. Investigation of Numerical Dissipation in Classical and Implicit Large Eddy Simulations. *Aerospace* **2017**, *4*, 59. [CrossRef]
2. Zingaro, A.; Könözsy, L. Discontinuous Galerkin Finite Element Investigation on the Fully-Compressible Navier–Stokes Equations for Microscale Shock-Channels. *Aerospace* **2018**, *5*, 16. [CrossRef]
3. Teschner, T.-R.; Könözsy, L.; Jenkins, K.W. Predicting Non-Linear Flow Phenomena through Different Characteristics-Based Schemes. *Aerospace* **2018**, *5*, 22. [CrossRef]
4. Zhang, M.; Jungo, A.; Gastaldi, A.A.; Melin, T. Aircraft Geometry and Meshing with Common Language Schema CPACS for Variable-Fidelity MDO Applications. *Aerospace* **2018**, *5*, 47. [CrossRef]
5. Ignatyev, D.; Khrabrov, A. Experimental Study and Neural Network Modeling of Aerodynamic Characteristics of Canard Aircraft at High Angles of Attack. *Aerospace* **2018**, *5*, 26. [CrossRef]
6. Silva, W.A. AEROM: NASA's Unsteady Aerodynamic and Aeroelastic Reduced-Order Modeling Software. *Aerospace* **2018**, *5*, 41. [CrossRef]
7. Berci, M.; Cavallaro, R. A Hybrid Reduced-Order Model for the Aeroelastic Analysis of Flexible Subsonic Wings—A Parametric Assessment. *Aerospace* **2018**, *5*, 76. [CrossRef]
8. Singh, D.; Antoniadis, A.F.; Tsoutsanis, P.; Shin, H.-S.; Tsourdos, A.; Mathekga, S.; Jenkins, K.W. A Multi-Fidelity Approach for Aerodynamic Performance Computations of Formation Flight. *Aerospace* **2018**, *5*, 66. [CrossRef]
9. Ghoreyshi, M.; Greisz, I.; Jirasek, A.; Satchell, M. Simulation and Modeling of Rigid Aircraft Aerodynamic Responses to Arbitrary Gust Distributions. *Aerospace* **2018**, *5*, 43. [CrossRef]
10. Satchell, M.J.; Layng, J.M.; Greendyke, R.B. Numerical Simulation of Heat Transfer and Chemistry in the Wake behind a Hypersonic Slender Body at Angle of Attack. *Aerospace* **2018**, *5*, 30. [CrossRef]
11. Aref, P.; Ghoreyshi, M.; Jirasek, A.; Satchell, M.J.; Bergeron, K. Computational Study of Propeller–Wing Aerodynamic Interaction. *Aerospace* **2018**, *5*, 79. [CrossRef]
12. Aref, P.; Ghoreyshi, M.; Jirasek, A.; Satchell, M.J. CFD Validation and Flow Control of RAE-M2129 S-Duct Diffuser Using CREATE^{TM}-AV Kestrel Simulation Tools. *Aerospace* **2018**, *5*, 31. [CrossRef]
13. Boudreau, M.; Dumas, G.; Veilleux, J.-C. Assessing the Ability of the DDES Turbulence Modeling Approach to Simulate the Wake of a Bluff Body. *Aerospace* **2017**, *4*, 41. [CrossRef]

aerospace

MDPI

Article

Investigation of Numerical Dissipation in Classical and Implicit Large Eddy Simulations

Moutassem El Rafei, László Könözsy * and Zeeshan Rana

Centre for Computational Engineering Sciences, Cranfield University, Cranfield, Bedfordshire MK43 0AL, UK; moutassem.el-rafei@cranfield.ac.uk or elrafei_moutassem@hotmail.com (M.E.R.); zeeshan.rana@cranfield.ac.uk (Z.R.)
* Correspondence: laszlo.konozsy@cranfield.ac.uk; Tel.: +44-1234-758-278

Received: 21 November 2017; Accepted: 7 December 2017; Published: 11 December 2017

Abstract: The quantitative measure of dissipative properties of different numerical schemes is crucial to computational methods in the field of aerospace applications. Therefore, the objective of the present study is to examine the resolving power of Monotonic Upwind Scheme for Conservation Laws (MUSCL) scheme with three different slope limiters: one second-order and two third-order used within the framework of Implicit Large Eddy Simulations (ILES). The performance of the dynamic Smagorinsky subgrid-scale model used in the classical Large Eddy Simulation (LES) approach is examined. The assessment of these schemes is of significant importance to understand the numerical dissipation that could affect the accuracy of the numerical solution. A modified equation analysis has been employed to the convective term of the fully-compressible Navier–Stokes equations to formulate an analytical expression of truncation error for the second-order upwind scheme. The contribution of second-order partial derivatives in the expression of truncation error showed that the effect of this numerical error could not be neglected compared to the total kinetic energy dissipation rate. Transitions from laminar to turbulent flow are visualized considering the inviscid Taylor–Green Vortex (TGV) test-case. The evolution in time of volumetrically-averaged kinetic energy and kinetic energy dissipation rate have been monitored for all numerical schemes and all grid levels. The dissipation mechanism has been compared to Direct Numerical Simulation (DNS) data found in the literature at different Reynolds numbers. We found that the resolving power and the symmetry breaking property are enhanced with finer grid resolutions. The production of vorticity has been observed in terms of enstrophy and effective viscosity. The instantaneous kinetic energy spectrum has been computed using a three-dimensional Fast Fourier Transform (FFT). All combinations of numerical methods produce a k^{-4} spectrum at $t^* = 4$, and near the dissipation peak, all methods were capable of predicting the $k^{-5/3}$ slope accurately when refining the mesh.

Keywords: large eddy simulation; Taylor–Green vortex; numerical dissipation; modified equation analysis; truncation error; MUSCL; dynamic Smagorinsky subgrid-scale model; kinetic energy dissipation

1. Introduction

The complexity of modelling turbulent flows is perhaps best illustrated by the wide variety of approaches that are still being developed in the turbulence modelling community. The Reynolds-Averaged Navier–Stokes (RANS) approach is the most popular tool used in industry for the study of turbulent flows. RANS is based on the idea of dividing the instantaneous parameters into fluctuations and mean values. The Reynolds stresses that appear in the conservation laws need to be modelled using semi-empirical turbulence models. Direct Numerical Simulation (DNS) is another approach used to study turbulent flows where all the scales of motion are resolved. It should be noted that even using the highest performance computers, it is very difficult to study high Reynolds number flows directly by resolving all the turbulent eddies in space and time. An alternative approach is Large

Eddy Simulation (LES), where the large scales or the energy-containing scales are resolved and the small scales that are characterised by a universal behaviour are modelled. A subgrid-scale tensor must be included to ensure the closure of the system of governing equations. This process reduces the degrees of freedom of the system of equations that must be solved and reduces the computational cost. Implicit Large Eddy Simulation (ILES) is an unconventional LES approach developed by Boris in 1959 [1]. The main idea behind this approach is that no subgrid scale models are used, and the effects of small scales are incorporated in the dissipation of a class of high-order non-oscillatory finite-volume numerical schemes. The latter are characterised by an inherent numerical dissipation that plays the role of an implicit subgrid-scale model that emulates and models the small scales of motion. ILES has not yet received a widespread acceptance in the turbulence modelling community due to the lack of a theoretical basis that proves this approach. In addition to that, pioneers of ILES have worked in a very isolated way unaware of each others' work, which made it difficult to understand the main elements of this approach. Thus, many research groups are using ILES nowadays and are validating it against benchmark test cases, which gives this approach more credibility in many applications.

ILES is an advanced turbulence modelling approach due to its ease of implementation and since it is not based on any explicit Subgrid-Scale (SGS) modelling, which could reduce computational costs. Moreover, the fact that no SGS model is used prevents any modelling errors that affect the accuracy of the numerical solution, in contrast with the explicit large eddy simulation approach where modelling, differentiation and aliasing errors can have impacts on the numerical solution. In addition to that, non-oscillatory finite volume numerical schemes used within the framework of the ILES approach are computationally efficient and parameter free, which means that they do not need to be adapted and modified from one application to another [2]. It should be noted that even if the classical approach to ILES is based on using the inherent dissipation of the convective term as an implicit subgrid-scale model, a recent approach presented an alternative way to perform ILES by a controlled numerical dissipation that is included in the discretisation of the viscous terms through a modified wavenumber used in the evaluation of the second-derivatives in the framework of the finite difference method. The latter approach showed very accurate results compared to DNS data in [3]. This is an indicator of the efficiency of the ILES approach, which is fully independent of modelling of small scales.

Large eddy simulation is becoming widely used in many fundamental research and industrial design applications. Despite this positive situation, LES suffers from weaknesses in its formalism, which make this approach questionable since the mathematical formulation of the LES governing equations is just a model of what is applied in a real LES. In practice, the removal of small scales is carried out by a resolution truncation in space and time along with numerical errors that are not well understood. In LES, a significant part of numerical dissipation is ensured by the subgrid-scale model. However, a truncation error is induced by the mesh resolution and the computational methods being used, and the vast majority of LES studies do not consider this truncation error. The latter should not be neglected since it could overwhelm the subgrid-scale model effect when dissipative numerical methods are used. Sometimes, discretisation and modelling errors cancel each other leading to an increase in the accuracy of the numerical solution, but still, the fact that the truncation error is neglected should be questioned [4]. This conclusion was pointed out in the study carried out by Chow and Moin [5], who showed through a statistical analysis that the truncation error could be comparable and higher than the subgrid-scale error when the grid resolution is equal to the filter width ($\Delta x = \Delta$). Moreover, based on the idea that modelling and truncation errors could cancel each other, some studies pointed out an optimal grid resolution that reduces the numerical errors when the study is not connected to a subgrid-scale model like the Smagorinsky model. The conclusion drawn is that using a grid size less than the filter width allows the control of the truncation error. Unfortunately, this recommendation is rarely followed in the literature [6]. Most LES users are aware of this truncation error, but their conjecture is that it does not affect seriously the results in a wide range of applications. The term "seriously" should be a subject of debate in order to build an awareness of the effect of numerical errors on the accuracy of the solution. ILES suffers as well from some weaknesses that made this approach still be argued about by some scientists. Even if ILES is capable of reproducing the dynamics of Navier–Stokes equations, quantitative studies showed that the numerical dissipation inherent in a class of high-order finite volume numerical schemes could be higher than the subgrid-scale dissipation, which leads to poor results. Another scenario is that the numerical dissipation is smaller than the SGS dissipation yielding good results only in short time integrations. Poor-quality results are obtained in long time integrations due to energy accumulation in high wave numbers [7,8]. Accordingly, there is no clear mechanism to ensure the correct amount of numerical dissipation that should match the SGS dissipation. ILES is often considered as under-resolved DNS; however, ILES terminology should be reserved only for schemes that reproduce the correct amount of numerical dissipation.

Since LES and ILES techniques are often used for modelling mixing processes, therefore the quantitative measure and estimate on the numerical dissipation and dispersion are crucial. Research work on this subject is carried out by other authors, because the quantification of the dissipative properties of different numerical schemes is at the centre of interest in the field of computational physics and engineering sciences. Bonelli et al. [9] carried out a comprehensive investigation of how a high density ratio does affect the near- and intermediate-field of hydrogen jets at high Reynolds numbers. They developed a novel Localized Artificial Diffusivity (LAD) model to take into account all unresolved sub-grid scales and avoid numerical instabilities of the LES approach. In an earlier work, Cook [10] focused on the artificial fluid properties of the LES method in conjunction with compressible turbulent mixing processes dealing with the modified transport coefficients to damp out all high wavenumber modes close to the resolution limit without influencing lower modes. Cook [10] used a tenth-order compact scheme during the numerical investigations. Kawai and Lele [11] simulated jet mixing in supersonic cross-flows with the LES method using an LAD scheme. Their paper devotes particular attention to the analysis of fluid flow physics relying on the computational data extracted from the LES results. De Wiart et al. [12] focused on free and wall-bounded turbulent flows within the framework of a Discontinuous Galerkin (DG)/symmetric interior penalty method-based ILES technique. Aspden et al. [13] carried out a detailed mathematical analysis of the properties of the ILES techniques comparing simulation results against DNS and LES data. The aforementioned contributions made attempts to obtain very accurate results within the framework of LES and ILES methods to gain a deeper insight into the behaviour of the physics of turbulence. The reader can refer to the application of the ILES method in different contexts [14–18], where the quantification of numerical dissipation and dispersion could also be employed to improve the accuracy of the numerical solution.

Due to the above-described reasons, quantifying the numerical dissipation that is inherent in numerical schemes is of great importance to investigate the effect of truncation error on the accuracy of the results. The aim is to examine if the subgrid-scale model is providing the correct amount of dissipation to model the small scales, or otherwise, the contribution of truncation error to the numerical dissipation has a significant effect that could not be neglected, as was done in most large eddy simulation studies. Proving that the truncation error could not be ignored would allow one to work on controlling the effect of numerical dissipation inherent in the scheme in order to predict and provide the correct amount of dissipation needed for modelling the small scales of motion correctly. It should be reminded that the dissipation induced in the numerical methods used within the framework of the ILES approach to act as a subgrid-scale model is mainly related to the convective term of Navier–Stokes equations. Thus, investigating the discretisation error induced in the convective term of the fully-compressible Navier–Stokes equations is a prime objective in this study.

The contribution of truncation error induced in the convective term to the total numerical dissipation of the solution will be evaluated using Modified Equation Analysis (MEA). This approach consists of deriving a modified version of the partial differential equation to which the truncation error of the numerical scheme used to discretize the PDE is added. The MEA approach is based on Taylor series expansions of each component of the discretized convective term. This methodology is inspired by the linear approach introduced in the book of Fletcher in 1988 [19] where the MEA is applied to

1D linear equations. This approach is extended and applied to the three-dimensional Navier–Stokes equations during the course of this study. Since the modified equation analysis needs a substantial amount of algebraic manipulations, the solution-dependent coefficients multiplying the derivatives in the convective term are considered to be frozen as explained in [19]. This approximation works well and gives good results despite the lack of a theoretical basis to prove it. The Taylor–Green Vortex (TGV) is a benchmark test case that matches the aims of this paper. It helps with understanding the transition mechanism for turbulence and small scales' production. Moreover, TGV allows the investigation of the resolving power of the numerical scheme, which is represented by its ability to capture the physical features of the flow. The inviscid Taylor–Green vortex is used within the framework of this study to understand the inherent numerical dissipation of the MUSCL scheme with different slope limiters and the dissipation of the dynamic Smagorinsky subgrid-scale model.

2. Numerical Model and Flow Diagnostics

The dynamics of Taylor–Green Vortex (TGV) are investigated in terms of classical, and implicit large eddy simulation and comparisons with high-fidelity DNS data provided in the study of Brachet et al. [20,21] are performed. The Taylor–Green vortex is considered as a canonical prototype for vortex stretching and small-scale eddies' production. The TGV flow is initialized with solenoidal velocity components represented in the following initial conditions as:

$$u_0 = U_0 \sin(kx) \cos(ky) \cos(kz), \tag{1}$$

$$v_0 = -U_0 \cos(kx) \sin(ky) \cos(kz), \tag{2}$$

$$w_0 = 0. \tag{3}$$

The initial pressure field is given by the solution of a Poisson equation for the given velocity components and could be represented as:

$$p_0 = p_\infty + \frac{\rho_0 U_0^2}{16} \left(2 + \cos\left(2kz\right)\right) \left(\cos\left(2kx\right) + \cos\left(2ky\right)\right), \tag{4}$$

where k represents the wavenumber and a value $k = 1\,\mathrm{m}^{-1}$ is adopted in accordance with the study carried out by Brachet et al. [20]. An ideal gas characterised by a Mach number $M = 0.29$ is considered. For this specified Mach number, compressible effects could be expected since that value falls within the range of mild compressible flows or near incompressible conditions. The initial setup yields the following values for the initial flow parameters:

$$U_0 = 100\frac{\mathrm{m}}{\mathrm{s}}, \quad \gamma = 1.4, \quad \rho_0 = 1.178\frac{\mathrm{kg}}{\mathrm{m}^3}, \quad p_\infty = 101325\frac{\mathrm{N}}{\mathrm{m}^2}. \tag{5}$$

All the results are given in non-dimensional form where for example $t^* = kU_0 t$ is the non-dimensional time and $x^* = kx$ represents a non-dimensional distance. ILES were performed using a fully-compressible explicit finite volume method-based in-house code, which was developed within postgraduate research projects. The Harten–Lax–van Leer–Contact (HLLC) Riemann solver was adopted for the present study, and the MUSCL third-order scheme with three-different slope limiters was employed for spatial discretisation. In addition to this, the ANSYS-FLUENT solver is adopted for the classical LES considering the dynamic Smagorinsky subgrid-scale model and the third-order MUSCL scheme for spatial discretisation. The reason for employing the MUSCL third-order scheme in the in-house code is to be consistent with the FLUENT solver in terms of the order of accuracy. For improving the accuracy of the time-integration, a second-order strong stability preserving Runge–Kutta scheme [22,23] has been employed. In the simulation setup, the box of edge length 2π is considered as the geometry for the problem as shown in Figure 1 where the outer domain is located at $(x, y, z) \in [-\pi, \pi] \times [-\pi, \pi] \times [-\pi, \pi]$. This specific configuration of the Taylor–Green vortex allows triply periodic boundary conditions at the box interfaces.

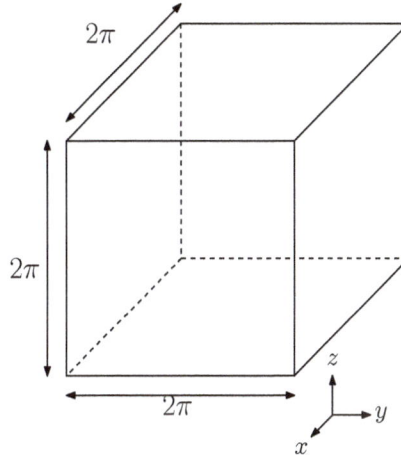

Figure 1. Geometry of the outer computational domain.

The initial problem has eight-fold symmetry, and the computations could be just considered on 1/8 of the domain, which can significantly reduce the computational cost. However, using symmetry conditions at the interfaces prevents the symmetry breaking property, which characterises the numerical scheme. Symmetry breaking means that the flow starts with symmetry and ends up with a non-symmetrical state; thus, if only 1/8 of the domain is studied, the resolving power of the numerical scheme cannot be assessed in a valid and credible way.

A block-structured Cartesian mesh topology was adopted for the discretisation of the computational domain utilized for the ILES and LES approaches. In regards to the ILES, three grid levels where created having 43^3, 64^3 and 96^3 cells. The extrapolation methods employed for the simulations are the second-order MUSCL scheme with Van Albada slope limiter (M2-VA), the third-order MUSCL with Kim and Kim limiter [24] (M3-KK), and the third-order MUSCL scheme with the Drikakis and Zoltak limiter [25] (M3-DD). The second-order strong stability-preserving Runge–Kutta scheme is used for temporal discretisation for the ILES approach. It has been demonstrated that the time discretisation method has only a minor effect on the results obtained using the MUSCL scheme, and for this reason, the previously-mentioned time integration scheme was employed for the simulations since it allows the use of higher Courant–Friedrichs–Lewy (CFL) numbers while preserving the stability [26]. A CFL = 0.8 was employed for all the numerical schemes, which induces equal time steps at each grid level.

In regards to the classical LES, the dynamic Smagorinsky subgrid-scale model is adopted, and the third-order MUSCL scheme (M3) is chosen for spatial discretisation. The latter is built in FLUENT as the sum of upwind and central differencing schemes where an under-relaxation factor is introduced to damp spurious oscillations. The derivation of the MUSCL scheme and the Dynamic Smagorinsky (DSMG) subgrid-scale (SGS) model used in FLUENT are not presented in this study. It should be noted that the simulations were performed using the density-based solver, and the Roe–Riemann solver is considered for the flux evaluation at the cell interfaces. Three grid levels were generated having 64^3, 128^3 and 256^3 cells. One could see that the finest LES grid is much finer than the finest grid used within the ILES approach. The 256^3 mesh will allow the more in depth study of the dynamics of TGV and the investigation of the grid refinement effect on the performance of the SGS model.

As mentioned earlier, the Taylor–Green vortex is a very good test case that allows the study of a numerical scheme's resolving power. The dynamics encountered during the flow evolution in time characterize the behaviour of the numerical scheme being investigated. Hence, several integral quantities have been calculated for the diagnostics of the TGV flow. Nevertheless, some of these quantities are based on the assumption of homogeneous and isotropic turbulence, which is not applied to all phases of Taylor–Green vortex flow evolution. However, investigating those parameters gives a more comprehensive idea about the characteristics of the flow. The volume-averaged kinetic energy can be used as an indicator of the dissipation or the loss of conservation that is related to the numerics being used for the discretisation of the governing equations.

The volume-averaged kinetic energy is defined in an integral form as:

$$E_k = \frac{1}{V} \int_V \frac{1}{2} \vec{u}.\vec{u} dV, \tag{6}$$

where V is the volume of the domain. The kinetic energy could be written in a more compact way as:

$$E_k = \frac{1}{2} \langle |\vec{u}|^2 \rangle, \tag{7}$$

where $\langle . \rangle$ will be used in the rest of this paper to represent the volumetric average of a given quantity. In theory, the kinetic energy should be conserved during the evolution of the flow if the latter is inviscid or for very small viscosity values, since no viscous effects are present to damp the kinetic energy into heat. The assumption of energy conservation holds only when the numerical scheme is able to conserve the kinetic energy or when all the scales of motion could be resolved. Hence, the decay of kinetic energy helps with indicating the onset or the exact time where the flow becomes under-resolved.

The kinetic energy dissipation rate is an important parameter that can be used to quantify the decay of kinetic energy in time and is representative of the slope of the volumetric average kinetic energy development. This parameter is defined as:

$$\epsilon = -\frac{dE_k}{dt}. \tag{8}$$

The production of vorticity could be monitored in terms of enstrophy, which is comparable to the kinetic energy dissipation. The enstrophy should grow to infinity when considering an inviscid flow. Therefore, the behaviour or the evolution of enstrophy can be considered as one of the criteria that are used to assess the resolving power of a numerical scheme and its ability to predict the flow physics accurately. The enstrophy is simply the square of vorticity magnitude and can be expressed as:

$$\langle \omega^2 \rangle = \langle |\vec{\nabla} \times \vec{u}|^2 \rangle. \tag{9}$$

For compressible flows, the dissipation of kinetic energy ϵ is the sum of two other components $\epsilon = \epsilon' + \epsilon''$ given by:

$$\epsilon' = 2\mu \langle \underline{\underline{S}} : \underline{\underline{S}} \rangle \tag{10}$$

$$\epsilon'' = -\langle p \vec{\nabla}.\vec{u} \rangle, \tag{11}$$

where $\underline{\underline{S}}$ is the strain rate tensor. The pressure dilatation-based dissipation ϵ'' is expected to be small in the incompressible limit, and hence, it is neglected during this study. Furthermore, for high Reynolds number flows, the volumetric average of the strain rate tensor product is equal to the enstrophy as derived in [27]. Hence, the dissipation rate could be finally expressed as:

$$\epsilon = \nu_{eff} \langle \omega^2 \rangle, \tag{12}$$

where ν_{eff} is the effective viscosity that represents the viscosity related to the dissipation of kinetic energy, and it is equal to the mean viscous dissipation.

The kinetic energy spectrum is represented for all the numerical methods, as well. Representing the three-dimensional kinetic energy spectra for the Taylor–Green vortex is useful to obtain a deeper understanding of kinetic energy distribution within all the scales of motion numerically resolved. The aim is to investigate whether the TGV presents an inertial subrange or not as predicted by Kolmogorov [28], who found that for homogeneous and isotropic turbulence, the kinetic energy spectrum follows a $k^{-5/3}$ slope in the inertial subrange, where k is the wavenumber. The three-dimensional velocity components are considered for the computation of the energy spectrum, which means that the three-dimensional array of the problem is considered. It should be reminded that the Taylor–Green vortex has an eight-fold symmetry, and some researchers use 1/8 of the domain size to calculate the energy spectra; but for the purpose of this study, all of the domain was considered. A three-dimensional Fast Fourier Transform (FFT) of the velocity components should give a three-dimensional array of amplitudes, which represent the Fourier modes corresponding to wavenumbers (k_x, k_y, k_z). By summation of the FFT square of each velocity component, a three-dimensional array containing the kinetic energy spectrum components is obtained. In the next step, a spherical integration of the three-dimensional array of the spectrum is carried out to obtain a 1D array that represents the kinetic energy spectrum, which is plotted versus the total wave number defined as $k = \sqrt{k_x^2 + k_y^2 + k_z^2}$.

The spherical integral could be expressed as:

$$E(k) = \int_0^{2\pi} \int_0^{\pi} \int_0^{K} A(k, \theta, \phi) \, k^2 \, sin\phi \, dk \, d\theta \, d\phi, \tag{13}$$

where A is the three-dimensional array of Fourier modes' amplitude, $K = \dfrac{2\pi}{L}$ and L is the characteristic length. The TGV is a well-established computational benchmark to investigate the dissipative and dispersive properties of various numerical schemes, and one can find more details about this research area in conjunction with numerical investigations in [29–32].

3. Results and Discussions

3.1. Taylor–Green Vortex Flow Topology

This section is devoted to the discussion of the dynamics of the Taylor–Green vortex. The flow topology is investigated qualitatively on the basis of the results obtained using the third-order MUSCL scheme with the Kim and Kim slope limiter [24] and a grid size of 96^3 cells. The flow is visualized using isosurfaces of a constant Q-criterion (see Figure 2). M3-KK is representative of all the other schemes that will generate similar contours, and the kinetic energy is represented in a dimensionless form in Figure 2. The eight-fold symmetry of the Taylor–Green vortex is clearly visible in Figure 2a, where the initial two-dimensional vorticity field features a symmetry in all planes located at a distance π in the computational domain. No signs of vorticity are predicted at the intersection of the symmetry planes, and the highest vorticity magnitude values were predicted at the centre of the large structures represented by Q-criterion isosurfaces as described in the study carried out by Brachet et al. [20].

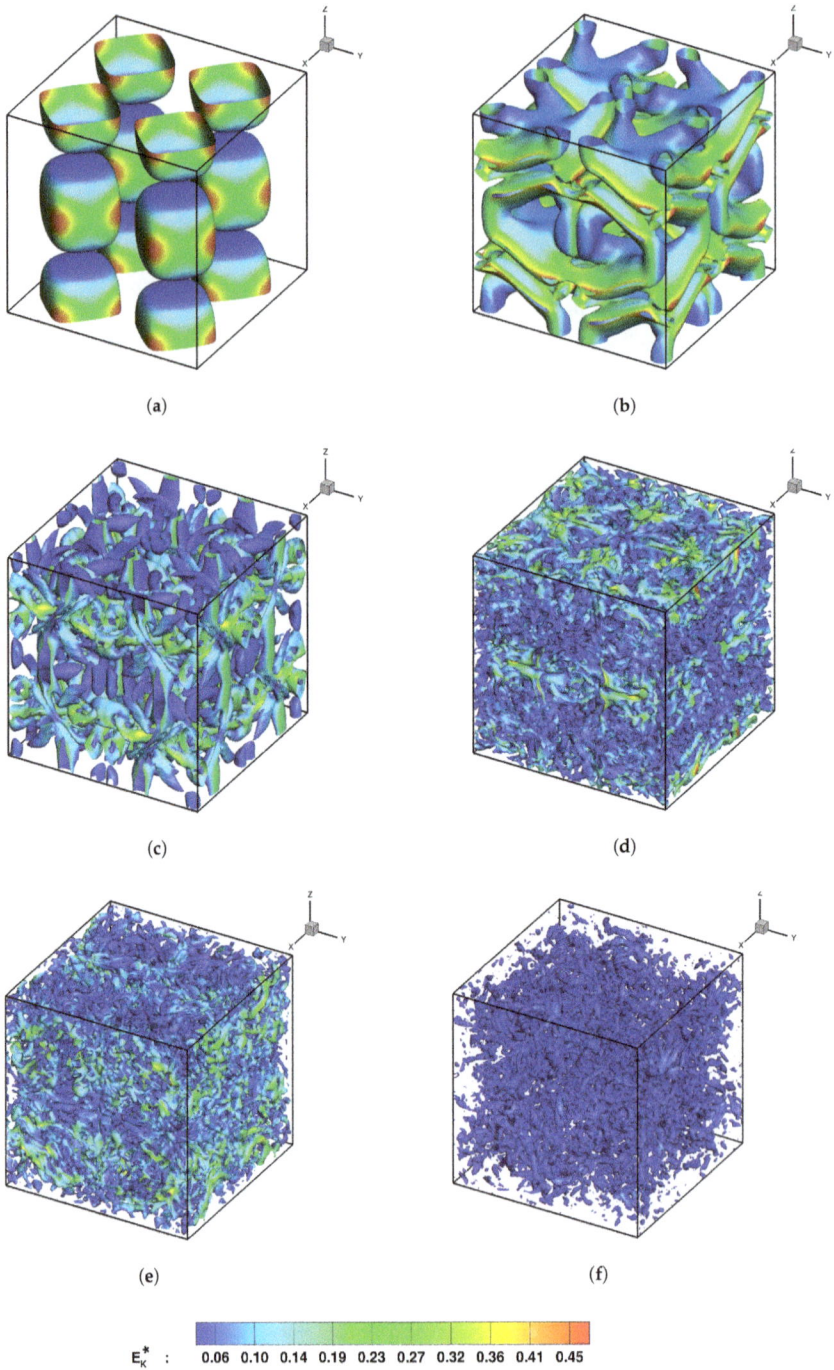

Figure 2. The Taylor–Green Vortex (TGV) flow using isosurfaces of Q-criterion = 0 coloured with the dimensionless kinetic energy obtained through the third-order MUSCL scheme with the Kim and Kim slope limiter (M3-KK) on a 96^3 grid. (a) $t^* = 0$; (b) $t^* = 4$; (c) $t^* = 8$; (d) $t^* = 20$; (e) $t^* = 30$; (f) $t^* = 60$.

When the flow evolves, the vortices begin there descent to the symmetry planes, and vortex sheets are generated, as shown in Figure 2b. At this time stage, the flow starts to become under-resolved, and the kinetic energy can no longer be preserved. As the flow evolves more, the vortex sheet undergoes an instability and loses its coherent structure as presented in Figure 2c. After the disintegration of the vortex sheet due to the instability that was observed at earlier stages, the dynamics of TGV are governed by the interactions of small-scale structures of vorticity that are generated due to vortex stretching characterised by vortex tearing and re-connection. At late stages represented in Figure 2e, the flow becomes extremely disorganized, has no memory of the initial condition that was imposed and the symmetry is no longer maintained. Hence, the symmetry-breaking property that characterizes a numerical scheme is well observed by investigating the TGV flow topology, since the flow starts with a symmetry and ends up non-symmetric. At very late stages, the worm-like vortices fade away (see Figure 2f), and this behaviour is very similar to homogeneously decaying turbulence.

3.2. Effect of Grid Resolution

The dynamics of the Taylor–Green vortex are directly dependent on the resolving power of the numerical scheme and the grid size adopted. In order to assess the effect of grid resolutions on the evolution of TGV flow, simulations were performed using second and third-order MUSCL schemes (M2-VA, M3-KK and M3-DD) considering three levels of grids having sizes of 43^3, 64^3 and 96^3 cells. It should be noted that the second-order strong-stability preserving Runge–Kutta scheme has been utilized for temporal discretisation. The same grid refinement study was performed using the dynamic Smagorinsky subgrid-scale model and third-order MUSCL scheme (DSMG-M3) on 64^3, 128^3 and 256^3 grids. The differences in the flow dynamics during the time evolution of the simulation are of great interest to understand the effect of mesh size on the performance of the numerical scheme or the SGS model. Figure 3 shows the evolution of volumetrically-averaged dimensionless kinetic energy and kinetic energy dissipation rate for all the numerical methods and all grid levels used within this study.

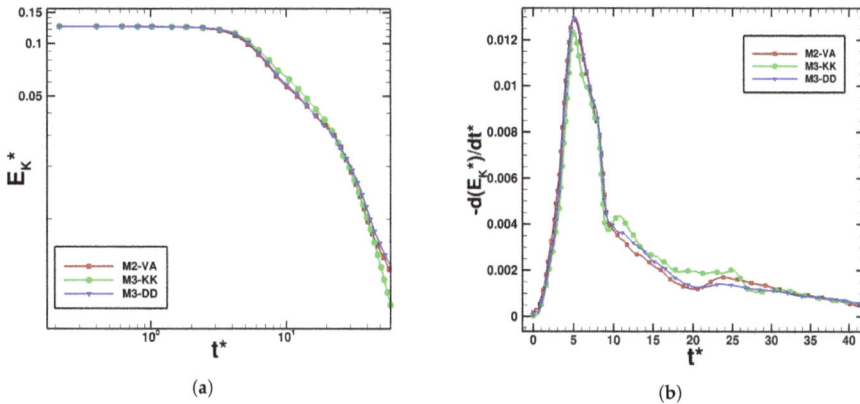

(a)   (b)

Figure 3. *Cont.*

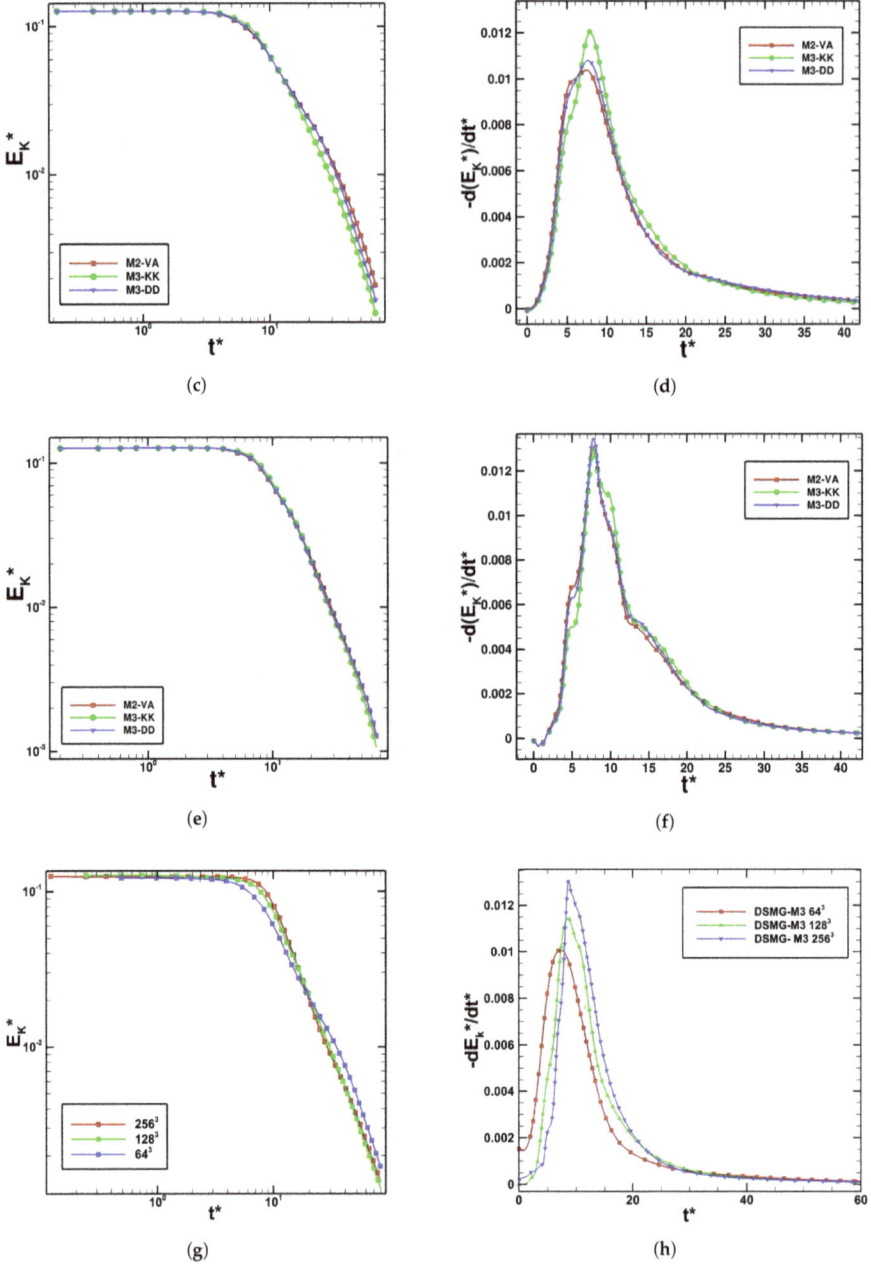

Figure 3. Evolution of the volumetrically-averaged kinetic energy represented in logarithmic scale and kinetic energy dissipation obtained on 43^3, 64^3 and 96^3 grids for the ILES approach and 64^3, 128^3 and 256^3 for the LES approach. (**a**) ILES, E_k^* (43^3); (**b**) ILES, $-dE_K^*/dt^*$ (43^3); (**c**) ILES, E_k^* (64^3); (**d**) ILES, $-dE_K^*/dt^*$ (64^3); (**e**) ILES, E_k^* (96^3); (**f**) ILES, $-dE_K^*/dt^*$ (96^3); (**g**) LES, E_K^*; (**h**) LES, $-dE_K^*/dt^*$.

The kinetic energy is represented in logarithmic scale to understand the trend line that the kinetic energy decay follows. In theory, the kinetic energy for an inviscid flow should be conserved since no viscous effects are present to damp it into heat. However, it is obvious that the latter is decaying during the coarse of the simulations. The decay starts at the end of the laminar stage when the vortex sheet is generated and the flow becomes under-resolved. The kinetic energy dissipation increases due to the instability that the vortex sheet undergoes and to the formation of a smaller and smaller vortical tube. The DNS study performed by Brachet et al. [20,21] predicted that the kinetic energy reaches its dissipation peak at a non-dimensional time level $t^* = 9$. The time level at which the dissipation reaches its highest value is more or less predicted correctly by all the numerical schemes with the exception of DSMG-M3, which underpredicted that time level compared to the other schemes. At a time level higher than $t^* = 9$, the kinetic energy dissipation decreases rapidly and reaches a value of zero at very late time stages, and this behaviour is similar to homogeneously-decaying turbulence. The kinetic energy dissipation rate trend obtained on the 43^3 mesh shows a non-physical behaviour predicted by all the numerical schemes used within the framework of the ILES approach. At late stages, when the flow becomes highly-disorganized ($t^* > 20$), all the numerical schemes predicted an increase in the dissipation of kinetic energy, which created a sort of hump in the trend of $\epsilon = -dE_K^*/dt^*$. In addition to that, the M3-KK scheme predicted an earlier increase in the kinetic energy dissipation at $t^* \sim 12$, which indicates a lower resolving power of that scheme compared to the other methods. Using finer grid resolutions, this non-physical behaviour vanishes, and the decay of kinetic energy at the stages when the flow is fully disorganized is pretty smooth. One could observe that this non-physical trend of kinetic energy dissipation rate was not observed in LES even on the coarsest grid. The dissipation peak represented by DSMG-M3 on a 64^3 mesh appears prior to the peak predicted on finer grids, which indicates that the onset of kinetic energy dissipation is happening earlier on a 64^3 grid. Refining the mesh, the kinetic energy plot represented in Figure 3g is more conserved since the volumetrically-averaged kinetic energy decay starts later on 128^3 and 256^3 grids compared to the coarsest grid, as observed qualitatively. The conclusion that could be drawn from the observation of kinetic energy and the kinetic energy dissipation rate is that the resolving power of the numerical scheme increases using finer grid resolutions. Note that mesh refinement increases the computational cost, which might become quite expensive for very fine grid resolutions. Both the ILES and LES approaches predicted similar dynamics of TGV, which indicates that the numerical dissipation inherent in high-order non-oscillatory numerical schemes could act as an implicit subgrid-scale model in predicting the flow features of homogeneously-decaying turbulence. As mentioned earlier, the kinetic energy $E_K^* = E_K/U_0^2$ is represented in a logarithmic scale in order to understand the trend line of the evolution of kinetic energy in time. It is obvious from Figure 3a,c,e,g that the kinetic energy follows a power law in time. This trend is similar to homogeneous and isotropic turbulence, where Kolmogorov [28] showed that the kinetic energy decay obeys a power law in time as:

$$E_K = (t - t_0)^{-P}, \tag{14}$$

where t_0 represents the onset of kinetic energy decay and P is the power that was evaluated by Kolmogorov and equal to $P = 10/7$. If the length of the largest scales of motion present in the flow are comparable to the grid resolution, the kinetic energy decays following a power law with $P = 2$, as shown in [33]. In this study, the decay exponent P is determined using curve fitting of kinetic energy data between $t^* = 10$ and $t^* = 80$ and values obtained for all numerical methods (see Tables 1 and 2).

The decay exponent predicted by the numerical schemes used within the ILES approach on 64^3 and 96^3 grids falls in the range of the value predicted in [33] ($P = 2$). The value of P is underpredicted by all the numerical methods on the coarsest mesh (43^3 cells), and it was similar to the value of the decay exponent predicted for homogeneous turbulence by Kolmogorov [28]. However, even if the values of P predicted by the coarsest mesh are close to the decay exponent of homogeneous and isotropic turbulence, one should not forget that a non-physical hump was observed in the volumetrically-averaged kinetic energy evolution, which makes the prediction of the decay slope erroneous. Without the presence of this "artificial" hump, the slope of decay would be steeper, and that could explain why the value of P was underpredicted. Refining the mesh, the hump vanishes, and the decay exponent converges to a value similar to the theoretical value provided by Skrbek and Stalp [33]. In regards to LES, it is obvious that the decay exponent obtained on a 64^3 grid is lower than the slope predicted by the medium and fine mesh, which converges to two.

Table 1. Decay exponent and onset of kinetic energy decay obtained using high-order non-oscillatory finite volume numerical schemes.

Grid	43^3			64^3			96^3		
Scheme	M2-VA	M3-KK	M3-DD	M2-VA	M3-KK	M3-DD	M2-VA	M3-KK	M3-DD
Decay Exponent P	1.594	1.843	1.506	1.91	2.183	2.083	2.083	2.197	2.065
Onset of Decay	1.75	2.015	1.845	2.375	2.711	2.618	2.908	3.172	3.026

Table 2. Decay exponent and onset of kinetic energy decay obtained using the dynamic Smagorinsky subgrid-scale model with the MUSCL third-order scheme.

Scheme	DSMG-M3		
Grid	64^3	128^3	256^3
Decay Exponent P	1.588	1.975	1.989
Onset of Decay	1.066	1.875	4.296

The onset is a very important parameter to understand and examine the resolving power of each numerical scheme. The more the kinetic energy is conserved, or in other words, the higher the onset of decay, the better the scheme is in terms of performance. It is obvious from Table 1 that M2-VA starts loosing kinetic energy before M3-DD, and the latter looses energy before M3-KK, which predicts the highest decay onset. Refining the mesh, the kinetic energy is more conserved, and hence, the performance of the numerical scheme is increased. The time at which the kinetic energy starts to decay is shown in Table 2 for the LES. Large differences of decay onset between the coarse, medium and fine grids are clear, and that could explain why the peak in the energy dissipation on the 64^3 grid appears earlier in Figure 3h. Refining the mesh, the onset of decay increases significantly and reaches a value of 4.296 on the 256^3 mesh, which indicates that the kinetic energy is best conserved on the finest grid. Compared to ILES results on a 64^3 grid, the onset of decay predicted by LES is nearly half the values predicted by all numerical schemes (see Table 1), which means that DSMG-M3 looses kinetic energy faster than the high-order non-oscillatory finite volume numerical schemes used for ILES, and the flow becomes under-resolved earlier as predicted by DSMG-M3.

Studies of grid sensitivity of ILES and LES were carried out on all the grids generated and compared to DNS data provided in the study of Brachet et al. [20,21] for Re = 400, 800, 3000 and 5000. It should be noted that explicit data for different Reynolds numbers are not available for the ILES and LES approaches, but the trend of kinetic energy dissipation rate will be used to observe the evolution of kinetic energy dissipation, as will be presented in Figure 4, compared to DNS results, which present the effects of physical viscosity, which damps the kinetic energy into heat. The DNS study showed that at high Reynolds number (Re $\geq$ 3000), the peaks of the kinetic energy dissipation rate remain identical, which means that the dissipation of kinetic energy reaches a Reynolds independent limit. By increasing the Reynolds number above this limit, the dissipation of kinetic energy will have the same trend. This concept is examined in the context of this study where the evolution of kinetic energy dissipation increases the investigated grid resolutions to examine which Reynolds number trends the implicit and explicit numerical viscosities are predicting. Only the results obtained using DSMG-M3 and M3-KK are presented since they are representative of all trends of numerical schemes.

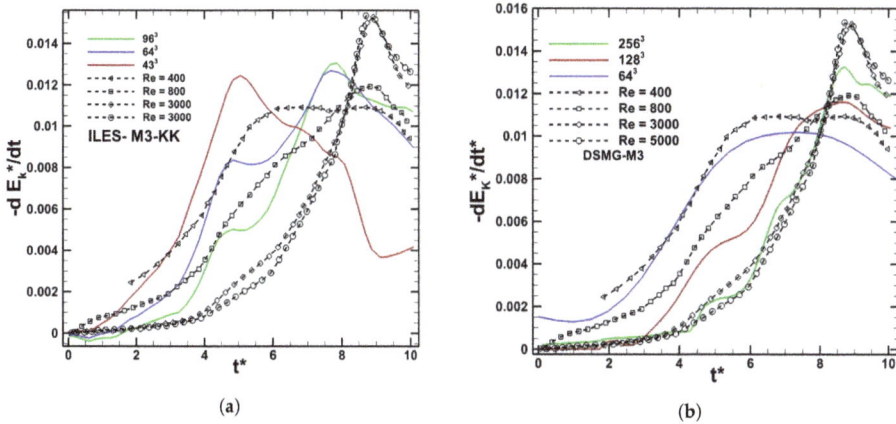

Figure 4. Evolution of the volumetrically-averaged kinetic energy dissipation obtained on all the grid levels and compared to DNS data of Brachet et al. [20,21]. (**a**) ILES, M3-KK; (**b**) LES, DSMG-M3.

Finer grid resolutions correspond to lower kinetic energy dissipation and higher Reynolds number trends. Reynolds-dependent effects are clear from the evolution of the dissipation peak when using coarser grids. The Reynolds number independent effect was clearer on the 256^3 grid used to perform the LES, where the trend of kinetic energy dissipation rate follows the high Reynolds number data, and the peak is reaching the value predicted for Re = 3000 and 5000, respectively. Hence, implicit and explicit subgrid-scale viscosities will predict a sort of Reynolds number independent limit when refining the grid resolution.

The production of vorticity is monitored for all numerical schemes and all grid levels in terms of enstrophy and effective viscosity. The latter is defined as the ratio between the kinetic energy dissipation rate and enstrophy. Figure 5 shows the evolution of the volume average of enstrophy in time and the related development of the volumetrically-averaged effective viscosity obtained using the LES approach. DSMG-M3 results will be representative of all numerical schemes used within the ILES approach, which showed similar trends. In theory, the enstrophy should grow to infinity for an inviscid flow. Thus, this parameter is very important to study the "effective viscosity" that damps the enstrophy and stops it from growing to infinity in time. The enstrophy could be considered, in conclusion, as one of the criteria that could be used to investigate the performance and resolving power of a numerical scheme. The production of vorticity that could be observed at the early laminar stages is due to the stretching of the Taylor–Green vortex where the large-scale vorticity structures are driven towards the symmetry plane. All the numerical schemes predict a decrease in enstrophy at the stage when the flow becomes under-resolved representing the peak of the kinetic energy dissipation rate. The enstrophy decreases due to the dissipation inherent in the numerical scheme or the dissipation of the subgrid-scale model, which could be considered as an implicit or explicit numerical viscosity that is damping the production of vorticity. Using finer grid resolutions, the effective viscosity decreases, and more enstrophy could be produced. This result seems obvious since the enstrophy and effective viscosity are inversely proportional.

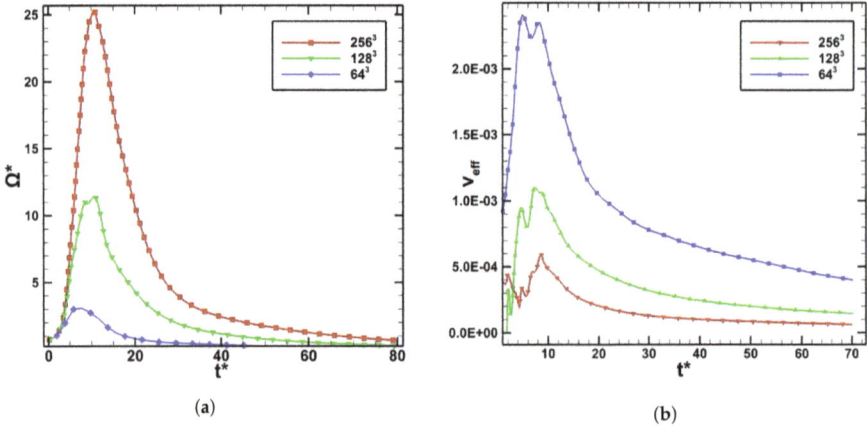

Figure 5. Variation of the volume-averaged enstrophy and the volume-averaged effective viscosity during the course of the simulations performed using DSMG-M3 on three different grid levels. (a) LES, enstrophy; (b) LES, effective viscosity.

The kinetic energy spectra were monitored for all the numerical methods used in this study. The observation of kinetic energy spectra helps with understanding the flow topology and the energy cascade process. Figure 6 presents the kinetic energy spectra obtained at different times using the M3-KK scheme on 64^3 and 96^3, representing the same behaviour as the other schemes used.

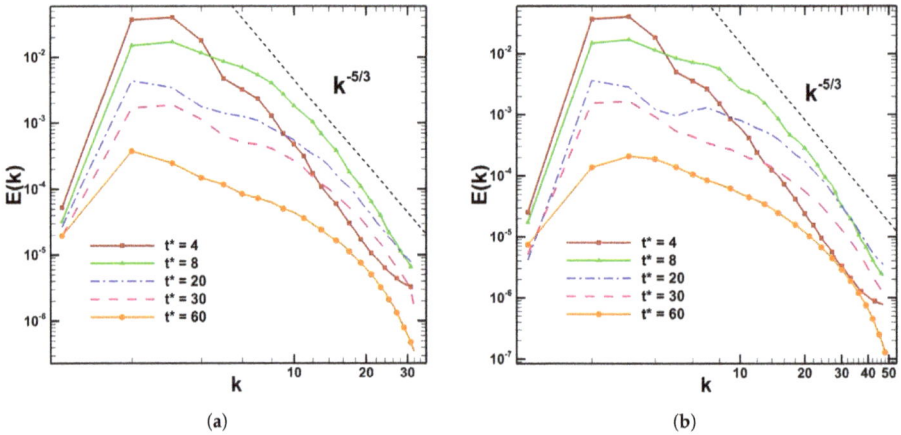

Figure 6. Kinetic energy spectra obtained using the MUSCL scheme with M3-KK slope limiter on 64^3 and 96^3 grids. (a) ILES, M3-KK (64^3 mesh); (b) ILES, M3-KK (96^3 mesh).

A peak in the kinetic energy spectra could be observed for wavenumbers $k = 2 - 4$. The same peak was observed in the study of Drikakis et al. [29], where the authors explain that this peak in the energy spectra represents the imprint of the initial conditions used to initialize the velocity field at $t^* = 0$. At the very early laminar stage ($t^* \sim 4$), all the numerical schemes predicted a k^{-4} spectrum, as has been presented in the DNS study of Brachet et al. and that corresponds to the spectrum usually found for a two-dimensional flow. This result is coherent with the fact that the three-dimensional vortex field of the TGV is initialized by a two-dimensional velocity field, and hence, the two-dimensional character of the flow is explained. As the flow becomes under-resolved, all the numerical methods including the dynamic Smagorinsky subgrid-scale model consistently emulate a $k^{-5/3}$ spectrum as predicted for the decaying turbulence kinetic energy spectrum. It should be noted that the dissipation of kinetic energy and the exchange of energy between the large and small scales is due to numerical dissipation inherent in the schemes being used, which acts as an implicit subgrid-scale model, and the dissipation of the explicit subgrid-scale model used in LES. The fact that all the methods succeeded in predicting the $k^{-5/3}$ spectrum indicates that Taylor–Green vortex flow is characterised by an inertial subrange where the small-scale vortices lose kinetic energy at the grid size level due to numerical dissipation. As the mesh is refined, the $k^{-5/3}$ slope becomes more accurately presented, and one could notice that at very late stages of the flow ($t^* = 60$), the Kolmogorov scale becomes more established. It should be reminded that at this stage of TGV flow, the small-scale worm-like vortices fade-away, similarly to homogeneously-decaying turbulence.

3.3. Modified Equation Analysis

In this section, the effect of the truncation error of the second-order upwind scheme has been investigated, and the discretization of the convective term of the fully-compressible Navier–Stokes equations is carried out within the framework of the LES approach. The modified equation analysis, applied to the convective term of Navier–Stokes equations, yields the following formulation for the truncation error of the second-order upwind scheme as:

$$
\begin{aligned}
\rho\frac{\partial u}{\partial t} + \rho u\frac{\partial u}{\partial x} &+ \rho v\frac{\partial u}{\partial y} + \rho w\frac{\partial u}{\partial z} = \\
&-\frac{1}{2}\rho\Delta t\left(u^2\frac{\partial^2 u}{\partial x^2} + v^2\frac{\partial^2 u}{\partial y^2} + w^2\frac{\partial^2 u}{\partial z^2}\right) - \frac{1}{3}\rho u^3\,(\Delta t)^2\left(1 - \frac{(\Delta x)^2}{u^2\,(\Delta t)^2}\right)\frac{\partial^3 u}{\partial x^3} \\
&-\frac{1}{3}\rho v^3\,(\Delta t)^2\left(1 - \frac{(\Delta y)^2}{v^2\,(\Delta t)^2}\right)\frac{\partial^3 u}{\partial y^3} - \frac{1}{3}\rho w^3\,(\Delta t)^2\left(1 - \frac{(\Delta z)^2}{w^2\,(\Delta t)^2}\right)\frac{\partial^3 u}{\partial z^3} \\
&-\rho u v^2\,(\Delta t)^2\frac{\partial^3 u}{\partial x\partial y^2} - \rho u w^2\,(\Delta t)^2\frac{\partial^3 u}{\partial x\partial z^2} - \rho u^2 v\,(\Delta t)^2\frac{\partial^3 u}{\partial x^2\partial y} \\
&-\rho u^2 w\,(\Delta t)^2\frac{\partial^3 u}{\partial x^2\partial z} - \rho v w^2\,(\Delta t)^2\frac{\partial^3 u}{\partial y\partial z^2} - \rho w v^2\,(\Delta t)^2\frac{\partial^3 u}{\partial z\partial y^2} \\
&-2\rho u v w\,(\Delta t)^2\frac{\partial^3 u}{\partial x\partial y\partial z} - \rho u v\frac{\partial^2 u}{\partial x\partial y} - \rho u w\frac{\partial^2 u}{\partial x\partial z} - \rho v w\frac{\partial^2 u}{\partial y\partial z} \\
&+\mathcal{O}\left[(\Delta t)^3, (\Delta x)^4, (\Delta y)^4, (\Delta z)^4\right].
\end{aligned}
\tag{15}
$$

As explained earlier, the truncation error of the numerical scheme is usually neglected by most of the researchers, who claim that it has a minor effect on the accuracy of the solution, and only the numerical dissipation of the subgrid-scale model is considered. Therefore, this section aims to clarify the idea that the truncation error could have a significant contribution to the numerical dissipation of the solution. The Taylor–Green vortex helps with understanding if the subgrid-scale model dissipation is enough to predict the correct amount of kinetic energy dissipation rate or the truncation error is participating in the dissipation of kinetic energy observed during the course of the TGV simulations. For that purpose, LES performed in ANSYS-FLUENT using the dynamic Smagorinsky subgrid-scale model, second-order upwind scheme for spatial discretisation and first-order upwind scheme for temporal discretisation are considered. The truncation error derived on the right-hand side of Equation (15) is computed using a User-Defined Function (UDF) in ANSYS-FLUENT. One could notice that the truncation error of the second-order upwind scheme has second- and third-order partial derivatives in addition to mixed partial derivatives. Only the effect of second-order partial derivatives, which are considered as the leading-order terms, is investigated in this study. It should be reminded that the second-order partial derivatives that exist in the formulation of the truncation error are present due to the fact that a first-order accurate method is used for temporal discretisation.

Using higher-order temporal methods would induce third- or higher-order partial derivatives in the equation. Figure 7 represents the time evolution of the leading second-order partial derivatives terms in the truncation error formulation of the second-order upwind scheme, without considering the second-order mixed partial derivatives and compared to the kinetic energy dissipation rate obtained using the dynamic Smagorinsky subgrid-scale model and second-order upwind scheme (DSMG-U2) on a 64^3 grid. The expression of the computed terms taken from the truncation error in Equation (15) is:

$$\rho \frac{\partial u}{\partial t} + \rho u \frac{\partial u}{\partial x} + \rho v \frac{\partial u}{\partial y} + \rho w \frac{\partial u}{\partial z} = -\frac{1}{2} \rho \Delta t \left(u^2 \frac{\partial^2 u}{\partial x^2} + v^2 \frac{\partial^2 u}{\partial y^2} + w^2 \frac{\partial^2 u}{\partial z^2} \right) +$$
$$+ \mathcal{O} \left[(\Delta t)^3, (\Delta x)^4, (\Delta y)^4, (\Delta z)^4 \right]. \tag{16}$$

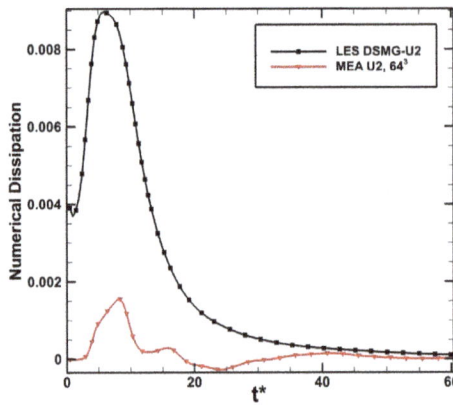

Figure 7. Contribution of second-order partial derivatives in the truncation error of the second-order upwind scheme obtained using Modified Equation Analysis (MEA) on a 64^3 mesh and compared to the kinetic energy dissipation rate obtained using DSMG-U2 on the same grid.

The observed volumetrically-averaged truncation error has a time behaviour similar to the kinetic energy dissipation rate. The trend of the truncation error is characterised by an increase near $t^* = 4$ reaching a peak at about $t^* = 9$ similar to what was predicted by all the numerical schemes and the DNS study of Brachet et al. [20,21]. The truncation error decreases in the stages where the flow becomes disorganized. For $15 < t^* < 40$, negative values of truncation error could be observed, which could be due to a dispersive behaviour of the scheme since the truncation error is derived on the right-hand side in the modified Equation (16), and hence, the negative values will be added to the left-hand side part, which induces a dispersion of the numerical solution. It is obvious from Figure 7 that the truncation error composed of second-order partial derivatives is not negligible compared to the kinetic energy dissipation rate despite the fact that the values of the discretisation error are significantly small compared to the dissipation of kinetic energy. It should be reminded that the dissipation of kinetic energy is due to the dissipation of the subgrid-scale model and the effect of truncation error of the numerical scheme. Hence, the evolution of the kinetic energy dissipation rate represented in Figure 7 contains both effects of the subgrid-scale model and the truncation error of the second-order upwind scheme. Therefore, if the truncation error is subtracted from the kinetic energy dissipation rate, an estimation of the subgrid-scale model dissipation could be obtained, as will be shown in Figure 8.

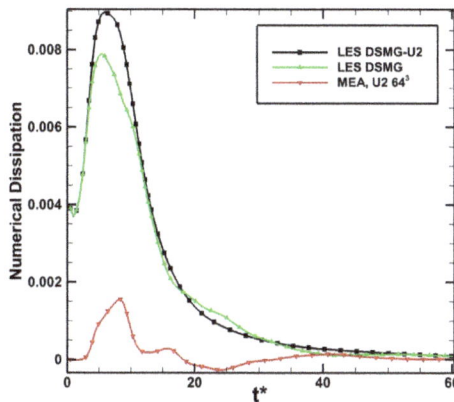

Figure 8. Dissipation of the DSMG model obtained by subtracting the truncation error of the second-order upwind scheme (MEA U2) from the total kinetic energy dissipation rate obtained on a 64^3 grid.

The DSMG model is under-predicting the kinetic energy dissipation since its peak of dissipation is lower than the one predicted by the total dissipation of kinetic energy considering the truncation error of the upwind scheme, as well. At $t^* > 20$, dispersion in the trend of DSMG dissipation is observed due to the negative values of truncation error representing the dispersive behaviour of the scheme. the conclusion that could be drawn is that the truncation error of the numerical scheme could play a significant role in the dissipation of kinetic energy which is added to the subgrid scale model dissipation that is not giving the correct amount of numerical dissipation to model the small scales on its own. That gives motivation for researchers to reconsider their idea of always neglecting the dissipation error that is yielded by the numerical scheme being used.

4. Conclusions

All the numerical methods predicted a decrease in kinetic energy during the course of the simulations performed. The decay is due to the numerical dissipation embedded in the numerical schemes used for ILES and to the dissipation of the subgrid-scale model when classical LES is used. The numerical dissipation of the scheme is beneficial as long as it is lower than the physical dissipation; otherwise, the dissipation could overwhelm the accuracy of the solution. It has been found that increasing the accuracy of the numerical method prolongs the conservation of kinetic energy in time. In effect, the resolving power of the second-order MUSCL scheme with the Van Albada slope limiter was the least among all other numerical schemes. That was expected, since the latter is a second-order method, whereas the other methods have third-order accuracy. In addition to that, the effective viscosity decreases when the mesh is refined and more enstrophy is produced. The conclusion is that the numerical dissipation decreases and more vorticity is produced with finer grid resolutions, which was expected since higher gradients could be handled with finer grids, and hence, the accuracy increases. Nevertheless, all the schemes showed a non-physical behaviour with the 43^3 mesh characterised by an increase in the kinetic energy dissipation at $t^* > 20$, when the flow becomes disorganized. This behaviour vanishes on finer grids, and that decrease in kinetic energy dissipation was not observed in LES since the 43^3 grid was not adopted for the classical LES study. Furthermore, finer grid resolutions corresponded to higher Reynolds number trends compared to DNS data from Brachet et al. [20,21]. Thus, the explicit and implicit subgrid-scale viscosities predict a Re-independent limit when refining the grid resolution, where the peak of kinetic energy dissipation will be the same. The decay exponent falls in the range of the theoretical value predicted in [33] as far as the mesh is refined. The onset of decay showed that M2-VA starts loosing kinetic energy first, then M3-DD, and the least dissipative is the M3-KK scheme. LES results showed that the dissipation of kinetic energy starts earlier than the one predicted by all the schemes used within the framework of ILES on the same grid size. Nonetheless, the kinetic energy starts to decay later when the mesh is refined. Finally, the kinetic energy spectrum was monitored for all schemes and all grid levels in the study of TGV dynamics. A peak in the kinetic energy spectra is present at small wavenumbers, representing the effect of initial conditions used at $t^* = 0$. All the numerical schemes predicted the k^{-4} spectrum at $t^* \sim 4$; the same was pointed out in the study of Brachet et al. [20,21], representing the spectra that are usually observed for two-dimensional turbulence. When the flow becomes under-resolved, all numerical methods succeeded in predicting a $k^{-5/3}$ spectrum, which induces that the Taylor–Green vortex is characterised by an inertial subrange where the kinetic energy dissipates into heat due to numerical dissipation either from the numerical scheme or from the subgrid-scale model.

The analytical formulation of the truncation error of the second-order upwind scheme obtained using the modified equations analysis was investigated. The effect of second-order partial derivatives was examined without considering the influence of mixed derivatives. The study showed that second-order partial derivatives induce a truncation error that is not negligible compared to the total dissipation of kinetic energy. It should be reminded that the kinetic energy dissipation rate contains both the subgrid-scale model and truncation error effects. If the discretisation error is subtracted from the total kinetic energy dissipation, the obtained estimate of the subgrid-scale model dissipation underpredicts the correct amount of kinetic energy dissipation. For $15 < t^* < 40$, negative values of truncation error were observed inducing a dispersive behaviour of the numerical scheme, since the negative values will be added to the solution, which is dispersed.

As a conclusion, the leading-order terms of the truncation error induce a significant dissipation that could not be neglected, as many times done in practice.

Acknowledgments: The present research work was financially supported by the Centre for Computational Engineering Sciences at Cranfield University under Project Code EEB6001R. The authors would like to acknowledge the IT support and the use of the High Performance Computing (HPC) facilities at Cranfield University, U.K. We would like to acknowledge the constructive comments of the reviewers of the *Aerospace* journal.

Author Contributions: Moutassem El Rafei, László Könözsy, and Zeeshan Rana contributed equally to this paper.

Conflicts of Interest: The authors declare no conflict of interest.

Abbreviations

The following abbreviations are used in this manuscript:

CFL	Courant–Friedrichs–Lewy number
DNS	Direct Numerical Simulation
DSMG	Dynamic Smagorinsky model
FFT	Fast Fourier Transform
HLLC	Harten–Lax–van Leer–Contact Riemann solver
ILES	Implicit Large Eddy Simulation
LES	Large Eddy Simulation
MEA	Modified Equation Analysis
M2-VA	Second-order MUSCL scheme with Van Albada slope limiter
M3-DD	Third-order MUSCL scheme with the Drikakis and Zoltak limiter
M3-KK	Third-order MUSCL scheme with the Kim and Kim slope limiter
MUSCL	Monotonic Upwind Scheme for Conservation Laws
PDE	Partial Differential Equation
RANS	Reynolds Averaged Navier–Stokes
SGS	Smagorinsky Subgrid-Scale model
TGV	Taylor–Green Vortex

References

1. Boris, J.P. On large eddy simulation using subgrid subgrid turbulence models. In *Whither Turbulence? Turbulence at the Crossroads. Lecture Notes in Physics*; Lumley, J.L., Ed.; Springer: Berlin/Heidelberg, Germany, 1990; pp. 344–353.
2. Grinstrein, F.F.; Margolin, L.G.; Rider, W.J. *Implicit Large Eddy Simulation—Computing Turbulent Fluid Dynamics*; Cambridge University Press: New York, NY, USA, 2007.
3. Dairy, T.; Lamballais, E.; Laizet, S.; Vassilicos, J.C. Numerical dissipation vs. subgrid scale modelling for large eddy simulation. *J. Comput. Phys.* **2017**, *337*, 252–274.
4. Garnier, E.; Adams, N.; Sagaut, P. *Large Eddy Simulation for Compressible Flows*; Springer: Dordrecht, The Netherlands, 2009.
5. Chow, F.K.; Moin, P.A. Further study of numerical errors in large-eddy simulations. *J. Comput. Phys.* **2003**, *184*, 366–380.
6. Vreman, A.W.; Sandham, N.D.; Luo, K.H. Compressible mixing layer growth and turbulence characteristics. *J. Fluid Mech.* **1996**, *320*, 235–258.
7. Garnier, E.; Mossi, M.; Sagaut, P.; Comte, P.; Deville, M. On the use of shockcapturing schemes for large-eddy simualtion. *J. Comput. Phys.* **1999**, *153*, 273–311.
8. Domaradzki, J.A.; Xiao, Z.; Smolarkiewicz, P.K. Effective eddy viscosities in implicit large eddy simulations of turbulent flows. *Phys. Fluids* **2003**, *15*, 1890–3893.
9. Bonelli, F.; Viggiano, A.; Magi, V. How does a high density ratio affect the near- and intermediate-field of high-Re hydrogen jets. *Int. J. Hydrog. Energy* **2016**, *41*, 15007–15025.
10. Cook, A.W. Artificial fluid properties for Large-Eddy Simulation of compressible turbulent mixing. *Phys. Fluids* **2007**, *19*, 055103, doi:10.1063/1.2728937.
11. Kawai, S.; Lele, S.K. Large-Eddy Simulation of jet mixing in supersonic cross-flows. *AIAA J.* **2010**, *48*, 2063–2083.
12. De Wiart, C.C.; Hillewaert, K.; Bricteux, L.; Winckelmans, G. Implicit LES of free and wall-bounded turbulent flows based on the discontinuous Galerkin/symmetric interior penalty method. *Int. J. Numer. Meth. Fluids* **2015**, *78*, 335–354.
13. Aspden, A.; Nikiforakis, N.; Dalziel, S.; Bell, J. Analysis of implicit LES methods. *Commun. Appl. Math. Comput. Sci.* **2008**, *3*, 103–126.
14. Könözsy, L. Multiphysics CFD Modelling of Incompressible Flows at Low and Moderate Reynolds Numbers. Ph.D. Thesis, Cranfield University, Cranfield, UK, 2012.
15. Tsoutsanis, P.; Kokkinakis, I.W.; Könözsy, L.; Drikakis, D.; Williams, R.J.R.; Youngs, D.L. Comparison of structured- and unstructured-grid, compressible and incompressible methods using the vortex pairing problem. *Comput. Methods Appl. Mech. Eng.* **2015**, *293*, 207–231.
16. Könözsy, L.; Drikakis, D. A unified fractional-step, artificial compressibility and pressure-projection formulation for solving the incompressible Navier–Stokes equations. *Commun. Comput. Phys.* **2014**, *16*, 1135–1180.
17. Rana, Z.A.; Thornber, B.; Drikakis, D. Dynamics of Sonic Hydrogen Jet Injection and Mixing Inside Scramjet Combustor. *Eng. Appl. Comput. Fluid Mech.* **2013**, *7*, 13–39.
18. Sakellariou, K.; Rana, Z.A.; Jenkins, K.W. Optimisation of the surfboard fin shape using computational fluid dynamics and genetic algorithms. *Proc. Inst. Mech. Eng. Part P J. Sports Eng. Technol.* **2017**, *231*, 344–354.
19. Fletcher, C.A.J. *Computational Techniques for Fluid Dynamics 1*, 2nd ed; Springer: Berlin, Germany, 1990.
20. Brachet, M.E. Direct simulation of three-dimensional turbulence in the taylor-green vortex. *Fluid Dyn. Res.* **1991**, *8*, 1–8.
21. Brachet, M.E.; Meiron, D.I.; Orszag, A.; Nickel, B.G.; Morf, R.H.; Frisch, U. Small-scale structure of the taylor-green vortex. *J. Fluid Mech.* **1983**, *130*, 411–452.
22. Spiteriand, R.J.; Ruuth, S.J. A new class of optimal high-order strong-stability-preserving time discretization method. *SIAM J. Numer. Anal.* **2002**, *40*, 469–491.
23. Drikakis, D.; Hahn, M.; Mosedale, A.; Thornber, B. Large eddy simulation using high-resolution and high-order methods. *Philos. Trans. R. Soc. A* **2009**, *367*, 2985–2997.
24. Kim, K.H.; Kim, C. Accurate, efficient and monotonic numerical methods for multi-dimensional compressible flows. Part II: Multi-Dimensional limiting process. *J. Comput. Phys.* **2005**, *208*, 570–615.

25. Zoltak, J.; Drikakis, D. Hybrid upwind methods for the simulation of unsteady shock-wave diffraction over a cylinder. *Comput. Methods Appl. Mech. Eng.* **1998**, *162*, 165–185.

26. Hahn, M. Implicit Large-Eddy Simulation of Low-Speed Separated Flows Using High-Resolution Methods. Ph.D. Thesis, Fluid Mechanics and Computational Science, Cranfield University, Cranfield, UK, 2008.

27. Tennekes, H.; Lumley, J.L. *A First Course in Turbulence*; The MIT Press: Cambridge, MA, USA, 1972.

28. Kolmogorov, A.N. The local structure of turbulence in incompressible viscous fluid for very large Reynolds number. *Proc. USSR Acad. Sci.* **1941**, *30*, 301–305.

29. Drikakis, D.; Fureby, C.; Grinstein, F.; Youngs, D. Simulation of transition and turbulence decay in the Taylor–Green vortex. *J. Turbul.* **2007**, *8*, 1–12.

30. Fauconnier, D.; Langhe, C.D.; Dick, E. Construction of explicit and implicit dynamic finite difference schemes and application to the large-eddy simulation of the Taylor–Green vortex. *J. Comput. Phys.* **2009**, *228*, 8053–8084.

31. Fauconnier, D.; Bogey, C.; Dick, E. On the performance of relaxation filtering for large-eddy simulation. *J. Turbul.* **2013**, *14*, 22–49.

32. Zhou, Y.; Grinstein, F.F.; Wachtor, A.J.; Haines, B.M. Estimating the effective Reynolds number in implicit large-eddy simulation. *Phys. Rev. E* **2014**, *89*, 013303, doi:10.1103/PhysRevE.89.013303.

33. Skrbek, L.; Stalp, S.R. On the decay of homogeneous isotropic turbulence. *Phys. Fluids* **2000**, *12*, 1997–2019.

aerospace

MDPI

Article

Discontinuous Galerkin Finite Element Investigation on the Fully-Compressible Navier–Stokes Equations for Microscale Shock-Channels

Alberto Zingaro and László Könözsy *

Centre for Computational Engineering Sciences, Cranfield University, Cranfield, Bedfordshire MK43 0AL, UK; zingaroalberto@gmail.com
* Correspondence: laszlo.konozsy@cranfield.ac.uk; Tel.: +44-1234-758-278

Received: 24 November 2017; Accepted: 30 January 2018; Published: 3 February 2018

Abstract: Microfluidics is a multidisciplinary area founding applications in several fields such as the aerospace industry. Microelectromechanical systems (MEMS) are mainly adopted for flow control, micropower generation and for life support and environmental control for space applications. Microflows are modeled relying on both a continuum and molecular approach. In this paper, the compressible Navier–Stokes (CNS) equations have been adopted to solve a two-dimensional unsteady flow for a viscous micro shock-channel problem. In microflows context, as for the most gas dynamics applications, the CNS equations are usually discretized in space using finite volume method (FVM). In the present paper, the PDEs are discretized with the nodal discontinuous Galerkin finite element method (DG–FEM) in order to understand how the method performs at microscale level for compressible flows. Validation is performed through a benchmark test problem for microscale applications. The error norms, order of accuracy and computational cost are investigated in a grid refinement study, showing a good agreement and increasing accuracy with reference data as the mesh is refined. The effects of different explicit Runge–Kutta schemes and of different time step sizes have also been studied. We found that the choice of the temporal scheme does not really affect the accuracy of the numerical results.

Keywords: computational fluid dynamics (CFD); microfluidics; numerical methods; gasdynamics; shock-channel; microelectromechanical systems (MEMS); discontinuous Galerkin finite element method (DG–FEM); fluid mechanics

---

1. Introduction

It was 1959 when Richard Feynman gave his famous lecture at the meeting of the American Physical Society at Caltech called "There's Plenty of Room at the Bottom", where he proposed two challenges with a prize of $10,000 each: the first one was to design and build a tiny motor, while the second one was to write the entire Encyclopædia Britannica on the head of a pin.

Nowadays, his speech is considered as the foundation of modern nanotechnology, since he highlighted the possibility to encode a number of pieces of information in very small spaces, hence producing small and compact devices [1]. All those extremely small devices having characteristic length of less than 1 mm but more than 1 micron are called microelectromechanical systems (MEMS) and, as the name suggests, they combine both electrical and mechanical components [2]. MEMS are small devices made of miniaturized structures, sensors, actuators and microeletronics and their components are between 1 and 100 micrometers in size. In recent years, several MEMS have been designed and developed, from small sensors to measure pressure, velocity and temperature, to micro-heat engines and micro-heat pumps and their numerical investigations are indispensable.

From a historical point-of-view, a pioneer experimental work on shock wave propagation in a low-pressure small-scale shock-tube was carried out by Duff (1959) [3], where a non-linear attenuation of the shock wave propagation for a certain diaphragm pressure ratio was observed. Other experimental works were performed by Roshko (1960) [4] and Mirels (1963, 1966) [5,6] confirming the strong attenuation of the shock wave and the acceleration of the contact surface, which propagates behind the shock wave in the classic shock-tube test case. The time interval between the shock wave and the contact surface measured at a certain point—which is also known as flow duration—rarefaction effects and thermal creeping were explained in depth. It is important to note that experimental and numerical studies on shock waves in different fields of engineering sciences attracted researchers over the past seventy years [3–14]. However, researchers paid particular attention to microscale shock waves and rarefied gas dynamics recently, especially for aerospace applications. This is due to the fact that microengines are used in the development of aerospace propulsion systems, because of their reduced size and achievable high power density. One of the greatest difficulty in the design process of microengines is that the fast heat loss results in low efficiency of these microdevices. Therefore, researchers devoted attention to carrying out experimental and numerical works on shock wave propagation and formation in micro shock-tubes and channels for MEMS applications [15–18].

In the aerospace industry, microfluidics is becoming more and more popular having applications mainly in aerodynamics, micropropulsion, micropower generation and in life support and environmental control for space applications. For instance, MEMS can be adopted for flow control problems for both free and wall bounded shear layers flows. In 1998, Smith et al. [15] studied experimentally the control of separated flow on unconventional airfoils using synthetic jet actuators to create a "virtual aerodynamic shaping" of the airfoil in order to modify the airfoil characteristics. Microfluidics is also used through fluidic oscillators in order to produce high-frequency perturbations for example to decrease jet-cavity interaction tones [16]. MEMS-based devices are adopted in the aerospace industry for the sake of turbulent boundary layer control. In fact, the small sizes of those systems (high density devices) allow to study near-wall flow structures [17]. In space applications, micropropulsive devices are designed and developed for miniaturized satellites, mainly used for global positioning systems or to serve generic platforms [18]. A detailed review on the application of microfluidics related devices in the aerospace industrial sector can be found in [18].

The flow behavior at those microscales is in general characterized by a granular nature for liquids and a rarefied behavior for gases; the walls "move", hence the classical no slip boundary conditions adopted in the macro regime fails. In agreement with [19], it is possible to classify the main differences among macrofluidics and microfluidics in the following list: noncontinuum effects, surface-dominated effects, low Reynolds number effects and multiscale and multiphysics effects. Furthermore, it is also observed that the diffusivity effects play an important role at this scales (see, e.g., [20]), especially when compared with the transport effects of the flow. Dealing with gases in micro devices, it is common practice to classify different flow regimes through the dimensionless Knudsen number Kn. Let λ be the mean-free path , which is the average distance traveled by a molecule between two consecutive collisions; denoting with ℓ the characteristic length of the generic problem considered (e.g., the hydraulic diameter for a channel flow problem), the Knudsen number is defined as $Kn = \frac{\lambda}{\ell}$.

Microfluidics can be modeled with two different approaches. The first one is the continuum model, and the flow is considered as a continuous and indivisible matter, while in the molecular model, the fluid is seen as a set of discrete particles. These models are valid in specific flow regimes determined by the Knudsen number and, when Kn increases, the validity of the continuum approach becomes questionable and the molecular approach should be adopted, as briefly sketched in Figure 1.

When the continuum approach is adopted, the fully-compressible Navier–Stokes (CNS) equations must be numerically solved. In the literature, this is usually performed adopting finite volume solvers. In the present work, the authors investigate how the discontinuous Galerkin finite element method (DG–FEM) performs applied to compressible flows at microscale levels in the slip flow regime (low Knudsen number). This method is selected because it takes advantages from the classical finite element method (FEM) and the finite volume method (FVM) since discontinuous polynomial functions are used and a numerical flux is defined among cells to reconstruct the solution. To verify the DG–FEM code, due to a lack of experimental data in microfluidics, the Zeitoun's test case [21] is adopted, which consists of a mini viscous shock channel problem numerically solved. In particular, they adopted the following models: the CNS equations in a FVM context, DSMC (Direct Simulation Monte Carlo) for the Boltzmann equation and the kinetic model BGKS (Bhatnagar–Gross–Krook with Shakhov equilibrium distribution function) model. In our work, the open source MATLAB code—developed by Hesthaven and Warburton [22]—has been adopted, modified and further improved.

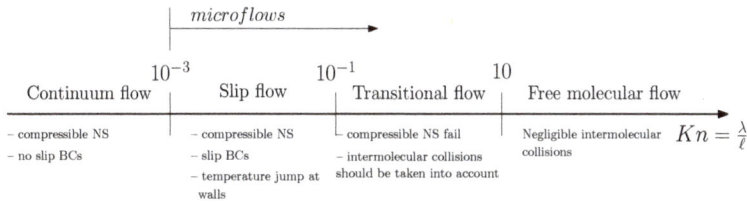

Figure 1. Different flow regimes in function of the Knudsen number *Kn*.

2. Mathematical Formulation and Solution Methodology

2.1. Compressible Navier–Stokes Equations

Consider a generic domain $\Omega \subset \mathbb{R}^d$ being $d = 1,2,3$ the dimension, provided with a sufficiently regular boundary $\partial \Omega \subset \mathbb{R}^{d-1}$ oriented by outward pointing normal unit vector $\hat{\mathbf{n}}$. On a two-dimensional ($d = 2$) cartesian reference system characterized by unit vectors **i** and **j**, the position vector is $\mathbf{x} = x\mathbf{i} + y\mathbf{j}$. Consider a gas with vector velocity field $\mathbf{u} = u\mathbf{i} + v\mathbf{j}$, density ρ, pressure p and total energy E. All the properties considered are both space and time dependent, e.g., $u = u(x,y,t)$. The fully-compressible set of governing equations made of the continuity equation, Navier–Stokes momentum equations and energy conservation form a set of m partial differential equations which can be written in a vectorial form as

$$\frac{\partial \mathbf{w}}{\partial t} + \frac{\partial \mathbf{f}_c}{\partial x} + \frac{\partial \mathbf{g}_c}{\partial y} = \frac{\partial \mathbf{f}_v}{\partial x} + \frac{\partial \mathbf{g}_v}{\partial y}. \tag{1}$$

In the equation above, $\mathbf{w}(\mathbf{x},t) = (\rho, \rho u, \rho v, E)^T$ is the vector of conserved variables; $\mathbf{f}_c(\mathbf{w}) = (\rho u, \rho u^2 + p, \rho uv, (E + p)u)^T$ and $\mathbf{g}_c(\mathbf{w}) = (\rho v, \rho uv, \rho v^2 + p, (E + p)v)^T$ are the convective fluxes in the x and y directions, respectively; and $\mathbf{f}_v(\mathbf{w}, \nabla \otimes \mathbf{w}) = (0, \tau_{xx}, \tau_{xy}, \tau_{xx}u + \tau_{xy}v)^T$ and $\mathbf{g}_v(\mathbf{w}, \nabla \otimes \mathbf{w}) = (0, \tau_{xy}, \tau_{yy}, \tau_{xy}u + \tau_{yy}v)^T$ are the viscous fluxes in the x and y directions, respectively. The terms τ_{ij} are the entries of the second-order viscous stress tensor τ. The latter is related to the velocity field according to the Navier–Stokes hypothesis for Newtonian, isotropic, viscous fluid through the following formulation: $\tau = 2\mu\mathbb{S} - \frac{2}{3}\mu(\nabla \cdot \mathbf{u})\mathbb{I}$, being μ the dynamic viscosity, $\mathbb{S} = \frac{1}{2}[(\nabla \otimes \mathbf{u}) + (\nabla \otimes \mathbf{u})^T]$ the strain rate tensor and $\mathbb{I}$ the identity matrix. The viscous stress tensor entries are

$$\tau_{xx} = 2\mu\frac{\partial u}{\partial x} - \frac{2}{3}\mu\left(\frac{\partial u}{\partial x} + \frac{\partial v}{\partial y}\right), \tag{2}$$

$$\tau_{xy} = \tau_{yx} = \mu\left(\frac{\partial u}{\partial y} + \frac{\partial v}{\partial x}\right), \tag{3}$$

$$\tau_{yy} = 2\mu\frac{\partial v}{\partial y} - \frac{2}{3}\mu\left(\frac{\partial u}{\partial x} + \frac{\partial v}{\partial y}\right). \tag{4}$$

The total energy is linked to the other fluid properties through the following equation of state (EOS) for a calorically ideal gas: $E = \frac{p}{\gamma-1} + \frac{1}{2}\rho|\mathbf{u}|^2$, where γ is the specific heat capacity ratio and

$|\mathbf{u}| = \sqrt{u^2 + v^2}$. Consider the compressible Navier–Stokes equations written in compact form (1) where the viscous fluxes are taken on the on the left hand side by

$$\frac{\partial \mathbf{w}}{\partial t} + \frac{\partial}{\partial x}(\mathbf{f}_c - \mathbf{f}_v) + \frac{\partial}{\partial y}(\mathbf{g}_c - \mathbf{g}_v) = 0, \tag{5}$$

if one defines $\mathbf{F}(\mathbf{w})$ as a $m \times d$ matrix having as columns the differences among the convective and viscous flux vectors, respectively, in the x and y direction

$$\mathbf{F} = [\mathbf{f}_c - \mathbf{f}_v | \mathbf{g}_c - \mathbf{g}_v] = \begin{bmatrix} \rho u & \rho v \\ \rho u^2 + p - \tau_{xx} & \rho u v - \tau_{xy} \\ \rho u v - \tau_{xy} & \rho v^2 + p - \tau_{yy} \\ (E+p)u - (\tau_{xx}u + \tau_{xy}v) & (E+p)v - (\tau_{xy}u + \tau_{yy}v) \end{bmatrix}, \tag{6}$$

Equation (5) can be expressed as

$$\frac{\partial \mathbf{w}}{\partial t} + \nabla \cdot \mathbf{F} = 0. \tag{7}$$

In particular, $\mathbf{w}(\mathbf{x}, t) : \mathbb{R}^d \times [0, T] \to \mathbb{R}^m$ and $\mathbf{F}(\mathbf{w}, \nabla \otimes \mathbf{w}) : \mathbb{R}^m \times [0, T] \to \mathbb{R}^m \times \mathbb{R}^d$.

2.2. Discontinuous Galerkin Finite Element Method (DG–FEM) Formulation

The physical domain is approximated by the computational domain Ω_h which consists of an unstructured grid made of K geometry conforming non-overlapping elements D_k, with $k = 1, \ldots, K$. A non-negative integer N is introduced for each element k and let $\mathbb{P}_N$ be the space of polynomials of global degree less than or equal to N. The following discontinuous finite element approximation space is introduced [23]:

$$V_h = \{\mathbf{v} \in (L^2(\Omega_h))^m \quad : \quad w|_k \in (\mathbb{P}_N(k))^m, \quad \forall k \in \Omega_h \}, \tag{8}$$

being $L^2(\Omega_h)$ the Hilbert space of square integrable functions on Ω_h. Using DG–FEM, the vector of conserved variables $\mathbf{w}(\mathbf{x}, t)$ is approximated by a function $\mathbf{w}_h(\mathbf{x}, t)$, which is the direct sum of K local polynomial solution $\mathbf{w}_h^k(\mathbf{x}, t)$ by

$$\mathbf{w}(\mathbf{x}, t) \simeq \mathbf{w}_h(\mathbf{x}, t) = \bigoplus_{k=1}^{K} \mathbf{w}_h^k(\mathbf{x}, t). \tag{9}$$

Analogously, one has

$$\mathbf{f}_c \simeq \mathbf{f}_{c_h} = \mathbf{f}_c(\mathbf{w}_h), \quad \mathbf{g}_c \simeq \mathbf{g}_{c_h} = \mathbf{g}_c(\mathbf{w}_h), \quad \mathbf{f}_v \simeq \mathbf{f}_{v_h} = \mathbf{f}_v(\mathbf{w}_h, \nabla \otimes \mathbf{w}_h), \quad \mathbf{g}_v \simeq \mathbf{g}_{v_h} = \mathbf{g}_v(\mathbf{w}_h, \nabla \otimes \mathbf{w}_h), \tag{10}$$

which means that $\mathbf{F} \simeq \mathbf{F}_h = \mathbf{F}(\mathbf{w}_h, \nabla \otimes \mathbf{w}_h)$. The local solution is expressed as a polynomial of order N through a nodal representation as

$$\mathbf{w}_h^k(\mathbf{x}, t) = \sum_{i=1}^{N_p} \mathbf{w}_h^k(\mathbf{x}_i, t) \ell_i^k(\mathbf{x}), \tag{11}$$

being ℓ_i^k the multidimensional interpolating Lagrange polynomial defined by grid points $\mathbf{x}_i$ on the element D_k and N_p the number of terms within the expansion which is related to the order of polynomial N through the relation $N_p = \frac{(N+1)(N+2)}{2}$. From this perspective, recalling Equation (7), the residual is formed as

$$\mathcal{R}_h(\mathbf{x}, t) = \frac{\partial \mathbf{w}_h}{\partial t} + \nabla \cdot \mathbf{F}_h. \tag{12}$$

The residual can vanish requiring that it is orthogonal to all test functions $\phi_h(\mathbf{x}) \in V_h$ on all the K grid elements

$$\int_{D_k} \mathcal{R}_h(\mathbf{x}, t) \phi_h(\mathbf{x}) d\Omega = 0 \implies \int_{D_k} \left(\phi_h \frac{\partial \mathbf{w}_h}{\partial t} + \phi_h (\nabla \cdot \mathbf{F}_h) \right) d\Omega = 0. \tag{13}$$

Using the Gauss' theorem, it can be easily shown that the latter reduces to

$$\int_{D_k} \left(\phi_h \frac{\partial \mathbf{w}_h}{\partial t} - \nabla \phi_h \cdot \mathbf{F}_h \right) d\Omega = - \oint_{\partial D_k} \phi_h \mathbf{F}_h \cdot \hat{n} d\Gamma. \tag{14}$$

From the RHS of the last equation, one can observe that the solution at the element interfaces is multiply defined, thus, it is possible to refer to a solution $\mathbf{F}_h^*$ to be determined. Reconsidering the flux vectors of the matrix $\mathbf{F}$, and considering that the normal vector is defined as $\hat{n} = \hat{n}_x \mathbf{i} + \hat{n}_y \mathbf{j}$, one has

$$\nabla \phi_h \cdot \mathbf{F}_h = (\mathbf{f}_{c_h} - \mathbf{f}_{v_h}) \frac{\partial \phi_h}{\partial x} + (\mathbf{g}_{c_h} - \mathbf{g}_{v_h}) \frac{\partial \phi_h}{\partial y}, \tag{15}$$

$$\mathbf{F}^*_h \cdot \hat{n} = (\hat{n}_x (\mathbf{f}_{c_h} - \mathbf{f}_{v_h}) + \hat{n}_y (\mathbf{g}_{c_h} - \mathbf{g}_{v_h}))^*, \tag{16}$$

which gives the following weak form:

$$\text{find } \mathbf{w}_h \in V_h: \quad \int_{D_k} \left(\phi_h \frac{\partial \mathbf{w}_h}{\partial t} - (\mathbf{f}_{c_h} - \mathbf{f}_{v_h}) \frac{\partial \phi_h}{\partial x} - (\mathbf{g}_{c_h} - \mathbf{g}_{v_h}) \frac{\partial \phi_h}{\partial y} \right) d\Omega =$$

$$= - \oint_{\partial D_k} (\hat{n}_x (\mathbf{f}_{c_h} - \mathbf{f}_{v_h}) + \hat{n}_y (\mathbf{g}_{c_h} - \mathbf{g}_{v_h}))^* \phi_h d\Gamma, \quad \forall \phi_h \in V_h. \tag{17}$$

The numerical flux indicated with the superscript '*' is computed through the local Lax–Friedrich flux as

$$(\hat{n}_x (\mathbf{f}_{c_h} - \mathbf{f}_{v_h}) + \hat{n}_y (\mathbf{g}_{c_h} - \mathbf{g}_{v_h}))^* = \hat{n}_x \{\{ \mathbf{f}_{c_h} - \mathbf{f}_{v_h} \}\} + \hat{n}_y \{\{ \mathbf{g}_{c_h} - \mathbf{g}_{v_h} \}\} + \frac{\hat{\lambda}}{2} [\![\mathbf{w}_h]\!], \tag{18}$$

where $\hat{\lambda}$ in general represents the local maximum of the directional flux Jacobian and an approximate local maximum linearized acoustic wave speed can be given [22] by

$$\hat{\lambda} = \max_{s \in [u_h^-, u_h^+]} \left(|\mathbf{u}(s)| + \sqrt{\frac{\gamma p(s)}{\rho(s)}} \right). \tag{19}$$

Note that, even if this flux has a dissipative nature, hence strong shock wave in supersonic regime can have a smeared trend, it gives accurate results for subsonic and weakly supersonic flows. For a generic quantity, the superscripts "$-$" and "$+$" here indicate, respectively, an interior and exterior information, i.e., if the quantity is taken at the internal or external side of the face of the element considered. The symbols $\{\{ \cdot \}\}$ and $[\![\cdot]\!]$ are the average and the jump along a normal $\hat{n}$, which are defined, for a generic vector $\mathbf{v}$, as $\{\{ \mathbf{v} \}\} = \frac{\mathbf{v}^- + \mathbf{v}^+}{2}$ and $[\![\mathbf{v}]\!] = \hat{n}^- \cdot \mathbf{v}^- + \hat{n}^+ \cdot \mathbf{v}^+$.

2.3. Temporal Integration Schemes

Considering the semi-discrete problem, written in the form of a system of ordinary differential equations (ODEs), the corresponding initial-value problem when initial conditions are given at time $t = t_0$ is

$$\begin{cases} \dfrac{d\mathbf{w}_h}{dt} = \mathcal{L}_h(\mathbf{w}_h, t), \\ \mathbf{w}_h(t_0) = \mathbf{w}_h^0. \end{cases} \tag{20}$$

$\mathcal{L}(\cdot)$ is the elliptic operator. Since the flow is strongly characterized by flow discontinuities, the strong stability-preserving Runge–Kutta (SSP–RK) schemes are adopted because they do not introduce spurious oscillations. Referring to Gottlieb et al. [24], the optimal second-order, two-stage and third-order three-stage SSP–RK schemes are expressed as

$$2^{nd}\text{-order, two-stage SSP–RK}: \begin{cases} \mathbf{v}^{(1)} = \mathbf{w}_h^n + \Delta t \mathcal{L}_h(\mathbf{w}_h^n, t^n), \\ \mathbf{w}_h^{n+1} = \mathbf{v}^{(2)} = \dfrac{1}{2}\left(\mathbf{w}_h^n + \mathbf{v}^{(1)} + \Delta t \mathcal{L}_h(\mathbf{v}^{(1)}, t^n + \Delta t)\right). \end{cases} \tag{21}$$

$$3^{rd}\text{-order, three-stage SSP–RK}: \begin{cases} \mathbf{v}^{(1)} = \mathbf{w}_h^n + \Delta t \mathcal{L}_h(\mathbf{w}_h^n, t^n), \\ \mathbf{v}^{(2)} = \dfrac{1}{4}\left(3\mathbf{w}_h^n + \mathbf{v}^{(1)} + \Delta t \mathcal{L}_h(\mathbf{v}^{(1)}, t^n + \Delta t)\right), \\ \mathbf{w}_h^{n+1} = \mathbf{v}^{(3)} = \dfrac{1}{3}\left(\mathbf{w}_h^n + 2\mathbf{v}^{(2)} + 2\Delta t \mathcal{L}_h(\mathbf{v}^{(2)}, t^n + \dfrac{1}{2}\Delta t)\right). \end{cases} \tag{22}$$

Gottlieb et al. [24] showed that it is not possible to design a fourth-order, four-stage SSP–RK where all the coefficients are positive. The classical fourth-order four-stage explicit RK method (ERK4) might be adopted, however the main disadvantage of this approach is its high computational effort since for each time step, four arrays must be stored in the memory. A valid alternative to this method, is given by the low storage explicit Runge–Kutta (LSERK) scheme, firstly introduced in 1994 in [25]. The fourth-order LSERK is defined by

$$4^{th}\text{-order LSERK}: \begin{cases} \mathbf{p}^{(0)} = \mathbf{w}_h^n, \\ \text{for } i \in [1, \dots, 5]: \begin{cases} \mathbf{k}^i = a_i \mathbf{k}^{(i-1)} + \Delta t \mathcal{L}_h(\mathbf{p}^{(i-1)}, t_n + c_i \Delta t), \\ \mathbf{p}^{(i)} = \mathbf{p}^{(i-1)} + b_i \mathbf{k}^i, \end{cases} \\ \mathbf{w}_h^{n+1} = \mathbf{p}^{(5)}. \end{cases} \tag{23}$$

The coefficients a_i, b_i and c_i are listed in Table 1. As the formula above shows, different from the classical ERK4, in this case, only one additional storage level is required. However, the LSERK requires five stages instead of four.

Table 1. Coefficients a_i, b_i and c_i used for the low storage five-stage fourth-order explicit Runge–Kutta method.

i	a_i	b_i	c_i
1	0	$\dfrac{1,432,997,174,477}{9,575,080,441,755}$	0
2	$-\dfrac{567,301,805,773}{1,357,537,059,087}$	$\dfrac{5,161,836,677,717}{13,612,068,292,357}$	$\dfrac{1,432,997,174,477}{9,575,080,441,755}$
3	$-\dfrac{2,404,267,990,393}{2,016,746,695,238}$	$\dfrac{1,720,146,321,549}{2,090,206,949,498}$	$\dfrac{2,526,269,341,429}{6,820,363,962,896}$
4	$-\dfrac{3,550,918,686,646}{2,091,501,179,385}$	$\dfrac{3,134,564,353,537}{4,481,467,310,338}$	$\dfrac{2,006,345,519,317}{3,224,310,063,776}$
5	$-\dfrac{1,275,806,237,668}{842,570,457,699}$	$\dfrac{2,277,821,191,437}{14,882,151,754,819}$	$\dfrac{2,802,321,613,138}{2,924,317,926,251}$

The time step size Δt that ensures a stable solution is computed [22] as

$$\Delta t = \frac{1}{2} \min \left(\frac{1}{(N+1)^2 \frac{|\mathbf{u}| + |a|}{\vartheta} + (N+1)^4 \frac{\mu}{\vartheta^2}} \right), \tag{24}$$

where a is the local speed of sound, which, using the ideal gas law, reads $a = \sqrt{\gamma R T} = \sqrt{\gamma \frac{p}{\rho}}$, while the geometrical factor ϑ is computed as $\vartheta = \frac{2}{F_{scale_{(i,k)}}}$, where $\mathbf{F}_{scale}$ is a matrix having dimension $N_{faces} \times K$ and its entries are the ratio of surface to volume Jacobian of face i on element k. From Equation (24), one can observe that, with very high order polynomials ($N \gg 1$), this time step restriction becomes impracticable; furthermore, the time step decreases as the dynamic viscosity μ increases, hence for highly viscous fluids, this expression for the time step might be unfeasible.

2.4. Slope Limiting Procedure

Due to strong flow discontinuities, the solution might be affected by spurious unphysical oscillations, hence, slope limiters are added to the existing code in order to properly model and catch the large gradients in the flow field. In particular, van Albada type slope limiter suitable for DG–FEM, throughly described by Tu and Aliabadi in [26], are adopted.

2.5. Benchmark Test Problem with Its Initial and Boundary Conditions

The viscous shock wave propagation is studied in a microchannel characterized by characteristic length (hydraulic diameter) equal to $H = 2.5$ mm. The viscous shock channel problem of Zeitoun et al. [21] is characterized by a driver and a driven chamber, quantities referred to these states are denoted with subscripts 4 and 1 and summarized in Table 2. The driver chamber is characterized by a higher pressure and density. The gas used in both chambers is Argon (Ar) and the main fluid properties of this gas, considered in standard condition, are listed in Table 2c. A sketch of the microchannel in the cartesian reference system (x, y) is given in Figure 2. Geometric information, initial conditions and the main flow properties of the argon are given in Table 2.

Table 2. Zeitoun's test case: geometric information and output time for the simulation (a); flow properties in the driver and driven chambers (b); and Argon properties in standard condition (c).

(a)	
characteristic length H (mm)	2.5
size of the domain (mm)	$32H \times 2H$
diaphragm position x_d (mm)	29.60
Output time T_f (µs)	80

(b)	Left Driver	Right Driven
state	4	1
Gas	Ar	Ar
ρ (kg/m^3)	8.43×10^{-3}	7.08×10^{-4}
u (m/s)	0	0
v (m/s)	0	0
p (Pa)	525.98	44.2

(c)	
specific gas constant R (J/(kg·K))	208.0
specific heat ratio γ (–)	1.67
thermal conductivity κ (W/(m·K))	0.0172
Sutherland's reference viscosity μ_0 (kg/(m·s))	2.125×10^{-5}
Sutherland's reference temperature T_0 (K)	273.15
Sutherland's temperature C (K)	144.4

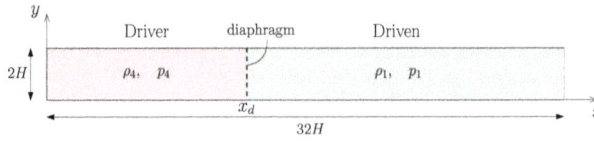

Figure 2. The sketch of the viscous shock-channel of Zeitoun et al. [21] for microfluidic applications.

Since the continuum approach is adopted, the rarefaction effects are usually taken into account imposing at wall the following conditions:

- velocity slip boundary condition;
- temperature jump boundary condition.

The small area where thermodynamic disequilibria occur is called Knudsen layer, having thickness of order of the mean free path λ. A generic form of the slip boundary condition is proposed by Maxwell. Let u_{slip} be the fictitious velocity required to predict the velocity profile out of the layer, the slip velocity can be expressed as

$$u_{slip} = u_f - u_{wall} = \frac{2 - \sigma_u}{\sigma_u} \lambda \frac{\partial u_f}{\partial n}\bigg|_{wall} + \frac{3}{4} \frac{\lambda}{k_2} \sqrt{\frac{R}{T}} \frac{\partial T}{\partial s}\bigg|_{wall}, \tag{25}$$

being u_f the fluid velocity, n and s the normal and parallel directions to the wall, and σ_u the tangential momentum accommodation coefficient which denotes the fractions of molecules absorbed by the walls due to the wall roughness, condensation and evaporation processes [27]. For microchannels, accurate values of σ_u are in the range 0.8–1.0 [28]. For the temperature jump [21], a condition is imposed by

$$T_s - T_{wall} = \frac{2 - \sigma_T}{\sigma_T} \frac{2\gamma}{\gamma + 1} \frac{\lambda}{Pr} \frac{\partial T}{\partial n}\bigg|_{wall}, \tag{26}$$

where T_s is the temperature that must be computed at wall that takes into account the gas rarefied conditions, T_{wall} is the reference wall temperature, Pr is the Prandtl number and σ_T is the thermal accommodation coefficient. In the literature, different empirical and semi-analytical expressions are available for λ and they are based on the way the force exerted among molecules is defined. In this work, the inverse power law (IPL) model is used, firstly introduced in 1978 by Bird in [29]. The model is based on a description of the mean free path based on the repulsive part of the force. It defines λ as

$$\lambda = k_2 \frac{\mu}{\rho \sqrt{RT}}, \tag{27}$$

where k_2 is a coefficient which varies according to the model taken into account. According to the Maxwell Molecules (MM) model, this constant is equal to $k_2 = \sqrt{\frac{\pi}{2}}$. Hence, if the temperature variation at walls is neglected, the slip boundary conditions becomes

$$u_{slip} = u_f - u_{wall} = \frac{2 - \sigma_u}{\sigma_u} \sqrt{\frac{\pi}{2}} \frac{\mu}{\rho \sqrt{RT}} \frac{\partial u_f}{\partial n}\bigg|_{wall}, \tag{28}$$

and the temperature jump

$$T_s - T_{wall} = \frac{2 - \sigma_T}{\sigma_T} \frac{2\gamma}{\gamma + 1} \sqrt{\frac{\pi}{2}} \frac{\mu}{\rho \sqrt{RT}} \frac{1}{Pr} \frac{\partial T}{\partial n}\bigg|_{wall}. \tag{29}$$

The slip boundary condition and temperature jump at wall are conditions desired for high Knudsen number regimes, whereas rarefied condition of the gas becomes predominant. However, since the present work focuses on the investigation of low Knudsen number regimes, the no-slip boundary condition case is used as initial approach.

The shock channel problem consists of two chambers at high (on the left, denoted with number 4) and low (on the right, denoted with number 1) pressure separated by a diaphragm in a known position x_d. When the diaphragm is instantaneously removed, i.e., when $t > 0$, due to the initial pressure difference, a combination of different wave patterns arises. The flow considered is originally at rest ($\mathbf{u} = 0$) and the initial conditions of the problem are given by

$$\rho(x,y,0) = \begin{cases} \rho_4 & \text{if } x < x_d, \\ \rho_1 & \text{if } x \geq x_d, \end{cases} \tag{30}$$

$$u(x,y,0) = 0, \tag{31}$$

$$v(x,y,0) = 0, \tag{32}$$

$$p(x,y,0) = \begin{cases} p_4 & \text{if } x < x_d, \\ p_1 & \text{if } x \geq x_d. \end{cases} \tag{33}$$

Numerical values of the quantities above are summarized in Table 2b.

3. Results and Discussion

To verify the MATLAB code and to validate the numerical results achieved, the benchmark test problem of Zeitoun et al. [21] on the investigation of viscous shock waves are considered, because this is one of the most frequently used benchmark problem for microscale applications. In their work, the viscous shock channel problem is solved at micro scales adopting three different approaches: compressible Navier–Stokes (CNS) equations with slip and temperature jump BCs using the CARBUR solver, the statistical Direct Simulation Monte Carlo (DSMC) method for the Boltzmann equation and the kinetic model Bhatnagar–Gross–Krook with the Shakhov equilibrium distribution function (BGKS).

3.1. Grid Convergence Study

The validation of the numerical results achieved with DG–FEM is performed through a grid convergence study using four grid levels. The grid levels are indicated with the index i, respectively, equal to 4, 3, 2 and 1. Let Δx and Δy be the mesh widths, respectively, in the x and y directions and N_x and N_y the number of grid points. The mesh widths in both directions have same length. A constant refinement ratio $R = \frac{\Delta x_{i+1}}{\Delta x_i} = \frac{\Delta y_{i+1}}{\Delta y_i}$ among all grid levels equal to 2 is considered. The required data for the mesh refinement study are listed in Table 3, whereas the quantity h is the dimensionless grid spacing which is the ratio among the grid spacing of the i-th grid level considered and the grid spacing of the finer mesh, defined as $h_i = \frac{\Delta x_i}{\Delta x_1} = \frac{\Delta y_i}{\Delta y_1}$, $i = 1, \ldots, 4$.

Table 3. Grid levels adopted in the mesh refinement study.

Mesh	Level i	N_x	N_y	Δx (mm)	Δy (mm)	h (–)	$1/h$ (–)
Coarse	4	97	7	8.33×10^{-1}	8.33×10^{-1}	8	0.125
Medium	3	193	13	4.17×10^{-1}	4.17×10^{-1}	4	0.25
Fine	2	385	25	2.08×10^{-1}	2.08×10^{-1}	2	0.5
Finer	1	769	49	1.04×10^{-1}	1.04×10^{-1}	1	1

The simulations are performed setting the Knudsen number equal to 0.05 and the order of polynomials is kept equal to 1 for all the grid levels. The results are considered at the output time $T_f = 80$ μs: before this time, the acoustic waves are propagating in the channel without considering the reflection at lateral walls.

Figure 3a,b shows, respectively, the dimensionless density ρ/ρ_1 and temperature T/T_1 extracted at the centerline of the channel ($y = H$) plotted against the dimensionless x/H coordinate. Qualitatively speaking, referring to the density profile in Figure 3a, one can see that the accuracy of the numerical results achieved increases when the mesh used is finer; in particular, with the coarse and medium mesh, the profile is very diffusive yielding to an incorrect and inaccurate representation of the discontinuities in the flow field. In fact, the density in the rarefaction wave region is over predicted and, as a result, the positions of the contact wave and of the shock wave are imprecise. The real flow physics is matched when the fine and finer mesh are considered, since less numerical diffusion can be observed and, as a result, the density jumps are properly caught. Analogous considerations can be done for the dimensionless temperature profile shown in Figure 3b. Firstly, one can observe that the accuracy quickly increases as the mesh is refined, in fact, for instance, the results achieved adopting the coarse mesh do not match at all the jumps in the flow field observed by Zeitoun et al. [21]. Furthermore, considering the finer grid level, it is observed that the position of the contact wave is properly achieved using DG–FEM, however the whole jump in temperature is slightly bigger than the one observed in the reference data (the relative error observed is approximately 10.6%). This produces a small under prediction of the shock wave position (relative error equal to 3.8%).

The numerical results are validate also in terms of streamwise velocity u (Figure 3c) in $x = 25H$, which represents the position immediately before the shock wave (pre-shock state). The velocity is made dimensionless using the speed of sound in the driven chamber defined as $a_1 = \sqrt{\gamma R T_1}$ and plotted against the dimensionless y/H coordinate (for the first half of the channel's height). The velocity profile obtained with the coarse and medium mesh under predicts the outcomes given by Zeitoun et al. [21], while bigger values are achieved adopting the fine and finer mesh; however, when $y/H \approx 0.3$, the profiles obtained overcome the reference data. The numerical results obtained for the stream wise velocity are quite accurate even if they are studied in the no slip fluid flow regime. However, as the Knudsen number increases, the slip condition becomes mandatory [19].

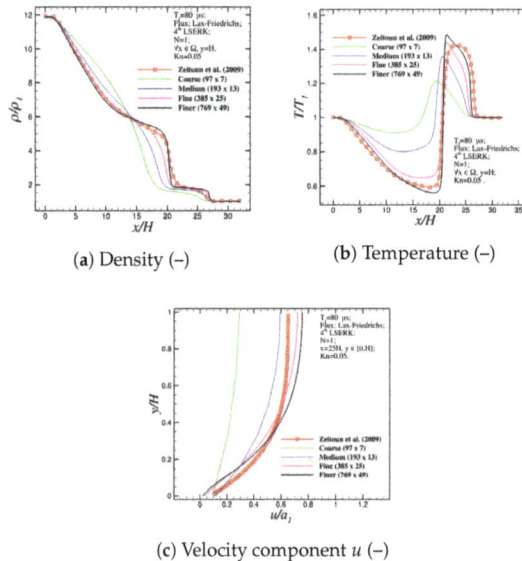

(a) Density (–)

(b) Temperature (–)

(c) Velocity component u (–)

Figure 3. Dimensionless density (**a**); temperature (**b**) profiles in the centerline of the microchannel; and stream wise velocity u (**c**) in the cross section $x = 25H$ using four grid levels at the final time $T_f = 80$ µs with $Kn = 0.05$ in comparison with reference data of Zeitoun et al. [21].

To confirm that the numerical results achieved grid convergence, a simulation on an additional grid level 5—which is the finest mesh—is performed to compare the results between the finer and the finest mesh (see Figure 4). The refinement ratio is kept equal to 2, so the finest grid level is characterized by $N_x = 1536$ and $N_y = 96$ grid points in the x and y directions, respectively, and mesh widths $\Delta x = \Delta y = 0.52 \times 10^{-1}$ mm. Figure 4 shows that the mesh further refinement does not improve significantly the already obtained accuracy of the numerical simulation results, which means that the further refinement of the mesh compared to the grid level 4 could increase the computational cost without significant further improvement of the achieved accuracy.

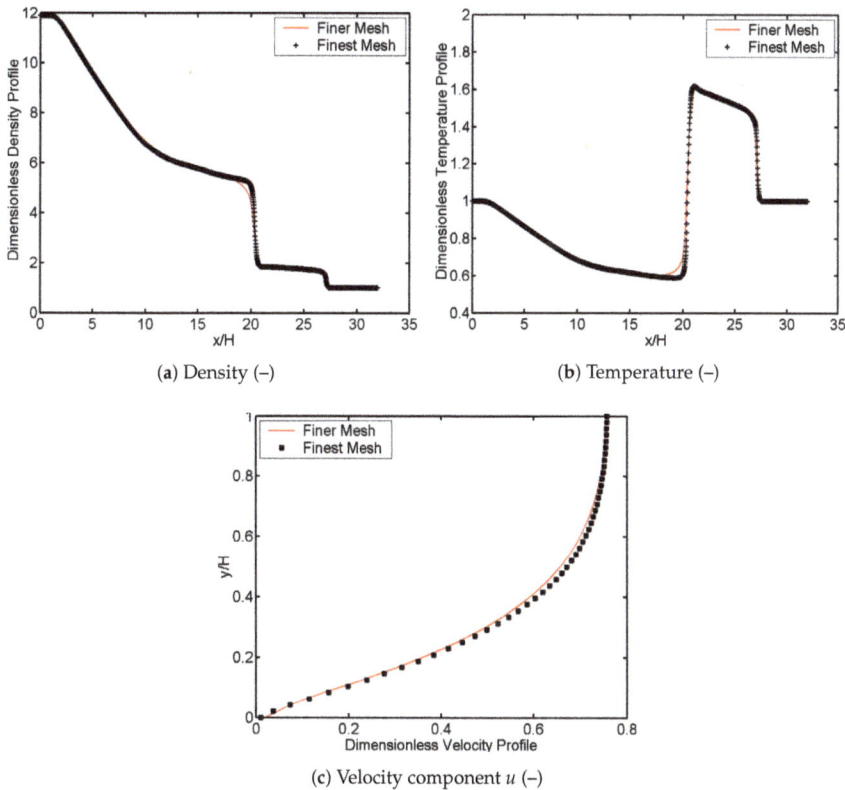

(a) Density (–)

(b) Temperature (–)

(c) Velocity component u (–)

Figure 4. Dimensionless density (**a**); temperature profiles (**b**) in the centerline of the microchannel; and stream wise velocity u (**c**) in the cross section $x = 25H$ on the finer and the finest grid levels at the final time $T_f = 80$ μs with $Kn = 0.05$.

For the sake of a quantitative analysis, the L^0, L^1 and L^2 norms of the absolute error between the numerical results and the reference data given by Zeitoun et al. [21] are computed. Figure 5 shows the logarithmic plots of the L^0 (Figure 5a), L^1 (Figure 5b) and L^2 (Figure 5c) norms of the absolute error between the results achieved with DG–FEM and reference data given by [21] against the inverse of the dimensionless grid spacing h. Regarding the density and the temperature, the three plots clearly show that as the mesh is refined, the error drops in terms of all the norms considered. The same trend can be observed for the velocity profile, however, going from the fine to the finer grid level, the error increases after $log(1/h) = 0.5$. The reason is that the present investigation focuses on the low Knudsen number flow regime, where rarefaction effects start to become important but still not dominant. Rarefaction effects are taken into account in a continuum approach—i.e., using compressible Navier–Stokes equations—imposing wall slip boundary conditions producing hence a different velocity profile [19]. The slip boundary condition becomes a mandatory requirement as the Knudsen number increases, which would yield to more physical and accurate results as confirmed by other authors in [19,21] when other continuum based numerical approaches were employed as well. This behavior is also met in the qualitative discussion above, since it is seen that the velocity profile achieved through the finer mesh overcomes the reference data when $y/H \approx 0.3$. Furthermore, the lowest error norms are observed for the stream wise velocity profile extracted in $x = 25H$. For the sake of a complete analysis, the error norms are also summarized in Table 4.

Table 4. L^0, L^1 and L^2 norms of the absolute error between the results achieved with DG–FEM and reference data in [21]. Results presented for density (a), temperature (b) in the centerline of the microchannel, and stream wise velocity (c) in the cross section $x = 25H$. Results obtained at the final time $T_f = 80$ µs with $Kn = 0.05$.

(a) Density				
$\rho(x, H)$				
	Coarse	Medium	Fine	Finer
$\|e_{abs}\|_0$ (–)	3.15963	2.72327	1.70752	0.81022
$\|e_{abs}\|_1$ (–)	0.94709	0.56817	0.26207	0.17318
$\|e_{abs}\|_2$ (–)	0.21601	0.15045	0.08903	0.04143
(b) Temperature				
$T(x, H)$				
	Coarse	Medium	Fine	Finer
$\|e_{abs}\|_0$ (–)	1.19697	1.08770	0.66427	0.44463
$\|e_{abs}\|_1$ (–)	0.34085	0.29573	0.21112	0.13444
$\|e_{abs}\|_2$ (–)	0.07237	0.06797	0.04617	0.02911
(c) Velocity Component u				
$u(25H, y)$				
	Coarse	Medium	Fine	Finer
$\|e_{abs}\|_0$	0.37619	0.10962	0.08258	0.10261
$\|e_{abs}\|_1$	0.28596	0.07559	0.04886	0.06401
$\|e_{abs}\|_2$	0.03477	0.00921	0.00615	0.00804

A convergence test is performed in order to understand if the formal order of accuracy matches (or not) the observed order of accuracy. Hence, within this approach, one can understand if the discretization error is reduced at the expected rate [30]. The formal order of accuracy can be achieved from a truncation error analysis, and, in the FEM approach, it is equal to N. In particular, in this case, $N = 1$ since first-order polynomials are considered. The observed order of accuracy $\mathcal{P}$ is computed from the numerical outputs on systematically refined grids [30]. The observed order of accuracy is based on the trend of the error. Consider two generic grid levels i and $i + 1$, being the i-th level the finer among them, and let $e_{abs}^{(i)}$ and $e_{abs}^{(i+1)}$ be the absolute errors for these grid levels. The observed order of accuracy, based on the L^p norms of the errors is defined by

$$\mathcal{P} = \frac{\ln\left(\dfrac{\|e_{abs}^{(i+1)}\|_p}{\|e_{abs}^{(i)}\|_p}\right)}{\ln(R)}, \quad \text{with} \quad p = 0, 1, 2. \tag{34}$$

In this work, different grid levels are taken into account along with different observed orders of accuracy for each flow property and for each norms. Figure 6 shows three logarithmic plots of the observed order of accuracy adopting the L^0, L^1 and L^2 norms of the absolute error between the results achieved with DG–FEM and reference data given by [21] against the dimensionless grid spacing h.

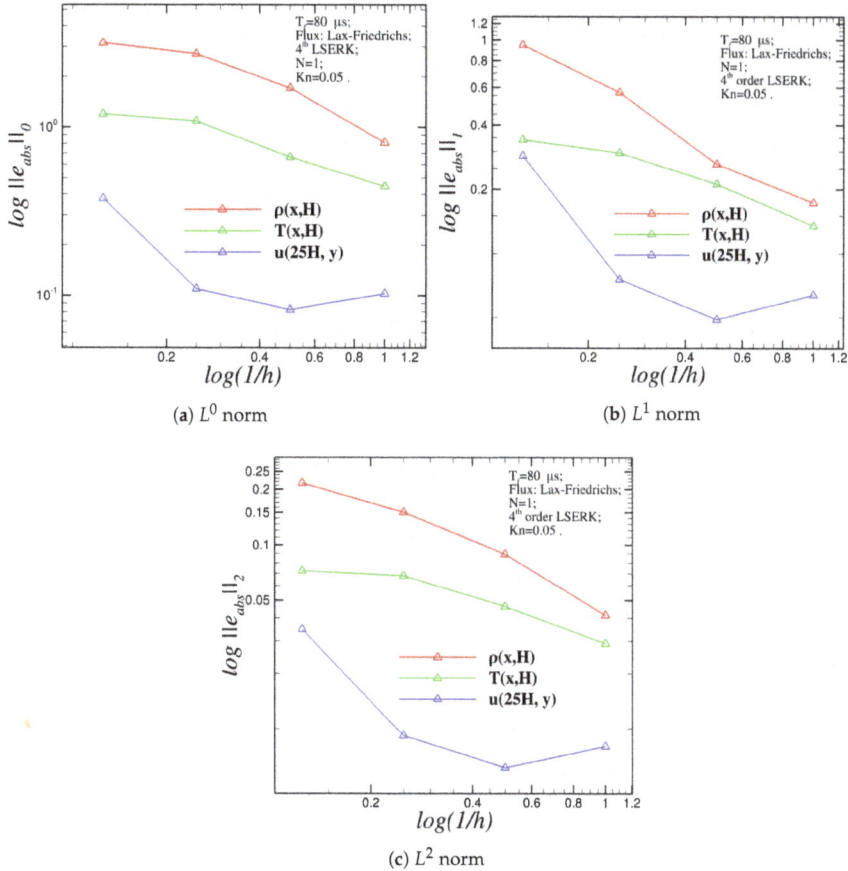

(a) L^0 norm

(b) L^1 norm

(c) L^2 norm

Figure 5. L^0, L^1 and L^2 norms of the absolute error between the results achieved with DG–FEM and reference data in [21] against the inverse of the dimensionless grid spacing h. Results presented for density and temperature in the centerline of the microchannel and stream wise velocity in the cross section $x = 25H$. Results obtained at the final time $T_f = 80$ μs with $Kn = 0.05$.

The simulations are performed using the first-order polynomial representation $N = 1$ and, for a sake of clarity, the observed order of accuracy for each quantity, for each norm, is also listed in Table 5. Figure 6a shows that, regarding the density at the centerline of the microchannel, for the L^0 and L^2 norms, the observed order of accuracy increases as the mesh is refined, reaching the values 1.076 and 1.104, respectively, which are higher than the theoretical order $N = 1$. The same trend is not shared by the L^1 norm, which exhibits a maximum between the medium and fine grid level. Concerning the temperature profile (Figure 6b), all the values are below the theoretical order $N = 1$ and the observed order increases as the mesh is refined for the L^1 and L^2 norms, while the L^0 norm shows a maximum between the medium and fine mesh. The stream wise dimensionless velocity profile in Figure 6c shows a different trend, which means that the observed order of accuracy $\mathcal{P}$ decreases as the mesh is refined, as also confirmed by the previous discussions about Figures 3c and 5.

(a) Density

(b) Temperature

(c) Velocity component *u*

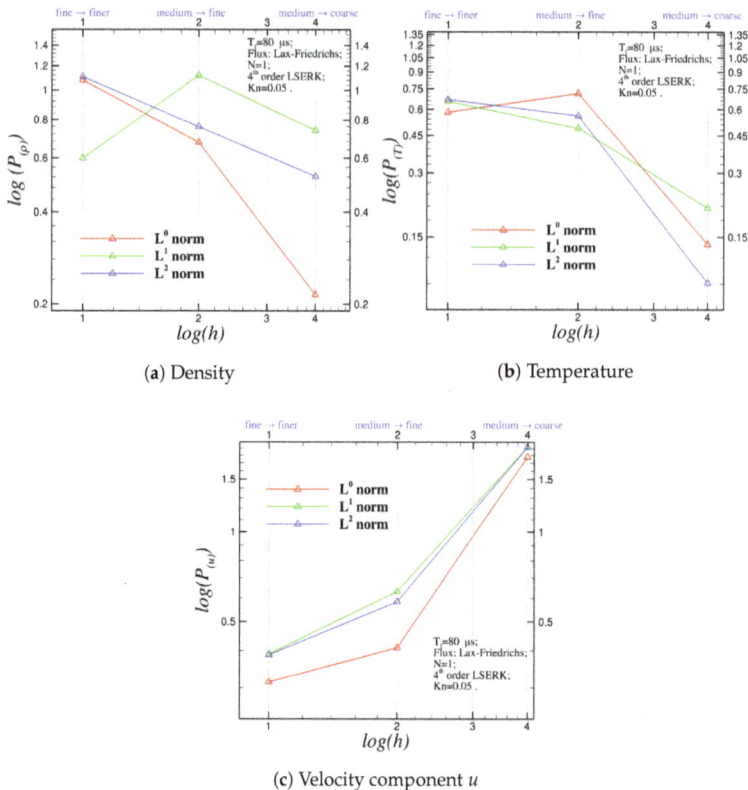

Figure 6. Logarithmic plots of the observed order of accuracy $\mathcal{P}$ using the L^0, L^1 and L^2 norms of the absolute error between the results achieved with DG–FEM and reference data in [21] against the dimensionless grid spacing h. Results presented for: density (**a**); temperature (**b**) in the centerline of the microchannel; and streamwise velocity (**c**) in the cross section $x = 25H$. Results obtained at the final time $T_f = 80$ μs with $Kn = 0.05$.

Table 5. Observed order of accuracy $\mathcal{P}$ for density, temperature and u velocity based on L^0, L^1 and L^2 norms of the absolute error between numerical solution obtained with DG–FEM and numerical data in [21].

Grid Level (from → to)	$\rho(x,H)$			$T(x,H)$			$u(25H,y)$		
	$1 \to 2$	$2 \to 3$	$3 \to 4$	$1 \to 2$	$2 \to 3$	$3 \to 4$	$1 \to 2$	$2 \to 3$	$3 \to 4$
$\mathcal{P}_0$ (–)	1.076	0.673	0.214	0.579	0.711	0.138	0.313	0.409	1.779
$\mathcal{P}_1$ (–)	0.598	1.116	0.737	0.651	0.486	0.205	0.390	0.629	1.920
$\mathcal{P}_2$ (–)	1.104	0.757	0.522	0.666	0.558	0.091	0.387	0.582	1.917

3.2. Effect of Different Time Integration Schemes

The effect of different time integration schemes is investigated. The following explicit Runge–Kutta schemes are taken into account: second-order, two-stage SSP–RK, third-order, three-stage SSP–RK and fourth-order LSERK. The limiting procedure is applied for each stage of the methods. The investigation is performed adopting both the medium and fine mesh and Figure 7 shows the results obtained for density and streamwise velocity. Referring to Figure 7a,b, where the density profile is plotted, respectively, adopting the medium and fine mesh with the three RK schemes presented, no big differences are observed. On the right of the figures, some zooms are shown: a unique trend is not observed, with the medium mesh the accuracy of the second-order, two-stage SSP–RK is always between the other two schemes. When the fine mesh is used, the differences among the schemes become even smaller and the third-order, three-stage SSP–RK seems to hold an average trend among the other methods. Broadly speaking, the same trend can be observed for the stream wise velocity profiles in Figure 7c,d , where the medium and fine grids are used, respectively. When the medium mesh is used, the third-order, three-stage SSP–RK scheme gives the most accurate profile, while using the fine mesh the results achieved are very similar. In particular, the second-order, two-stage SSP–RK is more accurate in the first part of the profile and the fourth-order LSERK in the second.

Of course, one can see that, due to the small differences among the outputs obtained, a qualitative analysis cannot determine correctly which scheme yields to the most accurate results. Hence, as previously done, the L^0, L^1 and L^2 norms of the absolute error are computed and listed in Table 6 for the medium and fine mesh. On the one hand, regarding the medium mesh, it is possible to observe that the third-order, three-stage SSP–RK scheme is the least accurate, since the lowest error norms are gotten. This behavior is met for all norms, for all the physical quantities considered. On the other hand, when the fine mesh is adopted, a unique trend is not observed: the fourth-order LSERK scheme is more accurate for temperature and density (except for the L^0 norm), while the second-order, two-stage SSP–RK is the most accurate for the stream wise velocity.

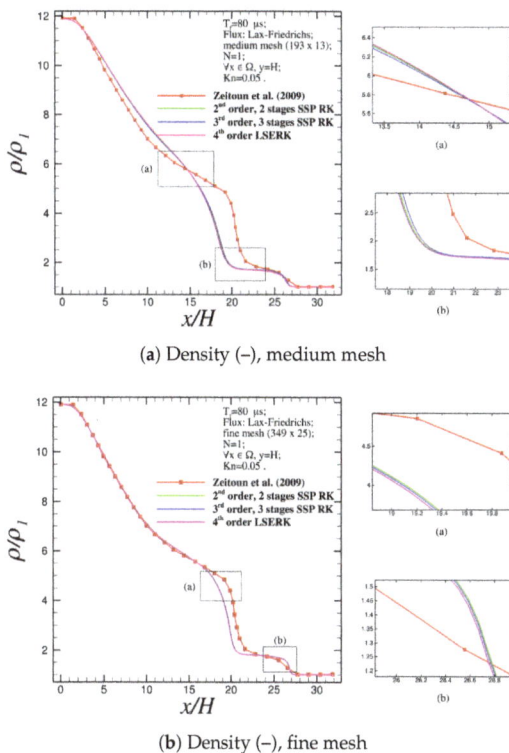

(a) Density (–), medium mesh

(b) Density (–), fine mesh

Figure 7. *Cont.*

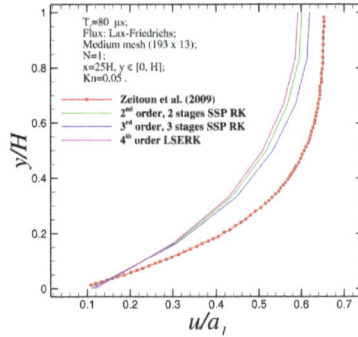

(c) Velocity component u (–), medium mesh

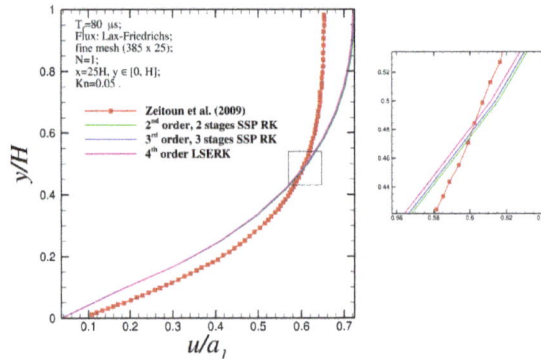

(d) Velocity component u (–), fine mesh

Figure 7. Dimensionless density profile in the centerline of the microchannel (**a**,**b**); and streamwise velocity u in the cross section $x = 25H$ (**c**,**d**) using the: medium (**a**,**c**); and fine (**b**,**d**) mesh at the final time $T_f = 80$ μs with $Kn = 0.05$ and $N = 1$. Comparison between different explicit Runge–Kutta schemes: second-order, two-stage SSP–RK, third-order, three-stage SSP–RK, and fourth-order LSERK. Validation with reference data of Zeitoun et al. [21].

To understand exactly how those time integration schemes perform, the relative difference among the previous error norms is compared. In particular, the relative difference among different temporal schemes using the L^p norms are defined by

$$\epsilon_{3/2} = 100 \cdot \left| \frac{||e_{abs}||_p^{III} - ||e_{abs}||_p^{II}}{||e_{abs}||_p^{II}} \right|, \quad \text{for } p = 0, 1, 2, \tag{35}$$

$$\epsilon_{4/3} = 100 \cdot \left| \frac{||e_{abs}||_p^{IV} - ||e_{abs}||_p^{III}}{||e_{abs}||_p^{III}} \right|, \quad \text{for } p = 0, 1, 2, \tag{36}$$

where the superscripts $'II'$, $'III'$ and $'IV'$, respectively, indicate the error norms computed using the second-order, two-stage SSP–RK, third-order, three-stage SSP–RK and fourth-order LSERK. These quantities are collected, in percentage, in Table 7. From the table, one can see that the relative difference is bigger when the medium mesh is adopted and this trend is emphasized when the stream wise velocity is taken into account. However, generally speaking, when the fine mesh is considered, the relative differences has as order of magnitude 1%, which can be considered a negligible outcome. For this reason, it is possible to conclude that when a fine (or even a finer) mesh is considered, the choice of the temporal scheme do not really affect the accuracy of the numerical results.

Table 6. L^0, L^1 and L^2 norms of the absolute error between the results achieved with DG–FEM and reference data in [21]. Comparison between different explicit RK schemes using the medium and fine mesh. Results presented for density (**a**); temperature (**b**) in the centerline of the microchannel; and stream wise velocity (**c**) in the cross section $x = 25H$. Results obtained at the final time $T_f = 80$ µs with $Kn = 0.05$ and $N = 1$.

(a) Density (–)

	Medium			Fine		
	$\|e_{abs}\|_0$	$\|e_{abs}\|_1$	$\|e_{abs}\|_2$	$\|e_{abs}\|_0$	$\|e_{abs}\|_1$	$\|e_{abs}\|_2$
2nd–order, 2 stage SSP–RK	2.7233	0.5682	0.1505	1.7075	0.2621	0.0890
3rd–order, 3 stage SSP–RK	2.6024	0.5333	0.1442	1.6846	0.2566	0.0874
4th–order LSERK	2.6870	0.5563	0.1484	1.6689	0.2530	0.0862

(b) Temperature (–)

	Medium			Fine		
	$\|e_{abs}\|_0$	$\|e_{abs}\|_1$	$\|e_{abs}\|_2$	$\|e_{abs}\|_0$	$\|e_{abs}\|_1$	$\|e_{abs}\|_2$
2nd–order, 2 stage SSP–RK	1.0877	0.2957	0.0680	0.6643	0.2111	0.0462
3rd–order, 3 stage SSP–RK	1.0480	0.2817	0.0658	0.6465	0.2086	0.0454
4th–order LSERK	1.0756	0.2912	0.0673	0.6348	0.2070	0.0448

(c) Velocity Component u (–)

	Medium			Fine		
	$\|e_{abs}\|_0$	$\|e_{abs}\|_1$	$\|e_{abs}\|_2$	$\|e_{abs}\|_0$	$\|e_{abs}\|_1$	$\|e_{abs}\|_2$
2nd–order, 2 stage SSP–RK	0.1096	0.0756	0.0092	0.0826	0.0489	0.00615
3rd–order, 3 stage SSP–RK	0.0903	0.0585	0.0072	0.0833	0.0494	0.00622
4th–order LSERK	0.1019	0.0692	0.0084	0.0834	0.0496	0.00625

Table 7. Relative difference among L^0, L^1 and L^2 norms of the absolute error achieved using second-order, two-stage SSP–RK, third-order, three-stage SSP–RK and fourth-order LSERK scheme. Results obtained using the medium and fine mesh density and temperature in the microchannel's centerline and stream wise velocity in the cross section $x = 25H$. Results obtained at the final time $T_f = 80$ µs with $Kn = 0.05$ and $N = 1$.

		Medium Mesh			Fine Mesh		
		L^0	L^1	L^2	L^0	L^1	L^2
$\rho(x,H)$	$\epsilon_{3/2}$ (%)	4.43947	6.14220	4.18605	1.34233	2.07547	1.88136
	$\epsilon_{4/3}$ (%)	3.25085	4.31277	2.91262	0.93030	1.39436	1.27821
$T(x,H)$	$\epsilon_{3/2}$ (%)	3.64473	4.73038	3.15965	2.67499	1.17253	1.76798
	$\epsilon_{4/3}$ (%)	2.63372	3.35658	2.23801	1.81116	0.79209	1.17433
$u(25H,y)$	$\epsilon_{3/2}$ (%)	17.66372	22.60352	21.77167	0.90700	1.09284	1.11081
	$\epsilon_{4/3}$ (%)	12.92501	18.20015	17.17119	0.93030	1.39436	1.27821

3.3. Effect of Different Time Step Sizes

The effect of different time step sizes Δt on the accuracy of the numerical results is investigated. The time step size is computed using Equation (24) to get a stable solution according to [22]. The following time step sizes are considered: $4\Delta t$, $2\Delta t$, Δt, $\Delta t/2$ and $\Delta t/4$. The numerical simulations are performed using the fine mesh with second-order, two-stage SSP–RK scheme. Changing the time step size still gives stable and bounded solutions and relevant changes in terms of accuracy are not observed. The L^0, L^1 and L^2 norms of the absolute error among reference data and numerical solutions are computed and it is observed that they share same precision up to the fourth digit. As for the temporal integration schemes study, the relative difference among the error norms is computed as:

$$\epsilon_{j+1/j} = 100 \cdot \left| \frac{||e_{abs}||_p^{(j+1)} - ||e_{abs}||_p^{(j)}}{||e_{abs}||_p^{(j)}} \right|, \quad \text{for } j = 1, \ldots, 4 \quad \text{and } p = 0, 1, 2. \tag{37}$$

The index j indicates a time step size level, in particular $j = 1$ corresponds to $4\Delta t$ and $j = 5$ to $\Delta t/4$. Table 8 shows these relative differences and it can be observed that the order of magnitude, in percentage, is between 10^{-7} and 10^{-3}. For this reason, it is possible to conclude that time step sizes do not affect the accuracy of the numerical solution.

Table 8. Relative difference among L^0, L^1 and L^2 norms of the absolute error achieved using different time step sizes. Results obtained with fine mesh and second-order, two-stage SSP–RK scheme and shown for density and temperature in the centerline of the microchannel and stream wise velocity in the cross section $x = 25H$. Results obtained at the final time $T_f = 80$ μs with $Kn = 0.05$ and $N = 1$.

		$\epsilon_{5/4}$ (%)	$\epsilon_{4/3}$ (%)	$\epsilon_{3/2}$ (%)	$\epsilon_{2/1}$ (%)
		$\frac{\Delta t}{4} \rightarrow \frac{\Delta t}{2}$	$\frac{\Delta t}{2} \rightarrow \Delta t$	$\Delta t \rightarrow 2\Delta t$	$2\Delta t \rightarrow 4\Delta t$
	L^0	2.8756×10^{-4}	1.6312×10^{-3}	1.4415×10^{-5}	1.2184×10^{-4}
$\rho(x, H)$	L^1	2.8834×10^{-5}	1.6530×10^{-4}	4.3485×10^{-7}	6.1623×10^{-5}
	L^2	1.4868×10^{-5}	8.4653×10^{-5}	6.8205×10^{-7}	9.2102×10^{-6}
	L^0	3.0293×10^{-5}	3.8387×10^{-4}	1.9821×10^{-4}	1.0443×10^{-5}
$T(x, H)$	L^1	3.0859×10^{-6}	3.9217×10^{-5}	6.3271×10^{-6}	6.3682×10^{-6}
	L^2	1.3106×10^{-6}	1.6615×10^{-5}	7.3049×10^{-6}	1.0169×10^{-6}
	L^0	1.2996×10^{-4}	1.6198×10^{-3}	1.2461×10^{-3}	1.0256×10^{-4}
$u(25H, y)$	L^1	5.6303×10^{-5}	7.0370×10^{-4}	5.4373×10^{-4}	4.4575×10^{-5}
	L^2	7.6356×10^{-6}	9.5516×10^{-5}	7.3887×10^{-5}	6.0460×10^{-6}

4. Conclusions

The two-dimensional unsteady fully-compressible Navier–Stokes equations for microfluidic problems are numerically solved adopting the nodal discontinuous Galerkin finite element method (DG–FEM) in space and different explicit Runge–Kutta (RK) schemes for the temporal integrations. The equations are solved at microscale level considering a miniaturized version of the shock-channel problem as test case. Unstructured meshes are generated and a MATLAB code developed by Hesthaven and Warburton [22] is adopted, modified and further improved. The numerical results are validated with reference data provided by Zeitoun et al. [21] where CNS equations in a FVM context, DSMC for the Boltzmann equation and the kinetic model BGKS (Bhatnagar–Gross–Krook with Shakhov equilibrium distribution function) models are adopted. A mesh refinement study is performed using four grid levels. The study showed that, as the mesh is refined, more accurate results are achieved. This trend is always confirmed for the density and temperature profiles, while for the streamwise velocity profile, the error slightly increases from the fine to the finer mesh. In fact, when the finer grid is adopted, it is observed that the method overestimates the velocity profile. A slip boundary condition implementation could improve this result. The observed order of accuracy based on different error norms for different flow properties is also computed and it is shown that, when the first-order of polynomial $N = 1$ is adopted (theoretical order of one), the following results are achieved: for the density and temperature, as the mesh is refined, the observed order of accuracy increases reaching (and overcoming for the density) the theoretical order; a different trend is achieved for the velocity profile, i.e., the order is higher as the mesh is coarser. The reason behind this behavior is the same as the previous one for the error norms. An additional grid level is then considered, and it is seen that the mesh further refinement does not improve significantly the already obtained accuracy. In addition to this, the effect of different temporal schemes (explicit Runge–Kutta) is investigated considering second-order, two-stage SSP–RK; third-order, three-stage SSP–RK; and fourth-order LSERK. A qualitative and quantitative analysis is done computing error norms and their relative differences, however no big differences among the schemes are observed as the mesh is refined and it is concluded that the choice of the temporal scheme does not really affect the accuracy of the numerical results, especially when fine meshes are used. Finally, different time step sizes have also been considered, and the solution remains stable and the accuracy unaffected.

Acknowledgments: The present research work was financially supported by the Centre for Computational Engineering Sciences at Cranfield University under project code EEB6001R. We would also like to acknowledge the constructive comments of the reviewers of the Aerospace journal.

Author Contributions: Alberto Zingaro and László Könözsy contributed equally to this paper.

Conflicts of Interest: The authors declare no conflict of interest.

Abbreviations

The following abbreviations are used in this manuscript:

BC	Boundary Condition
BGKS	Bhatnagar–Gross–Krook with Shakhov equilibrium distribution function
CFD	Computational Fluid Dynamics
CNS	Compressible Navier–Stokes
DG–FEM	Discontinuous Galerkin Finite Element Method
DSMC	Direct Simulation Monte Carlo
ERK4	Fourth-Order Four-Stage Explicit Runge–Kutta
FEM	Finite Element Method
FVM	Finite Volume Method
LSERK	Low Storage Explicit Runge–Kutta
MEMS	Microelectromechanical Systems
ODE	Ordinary Differential Equation
PDE	Partial Differential Equation
RK	Runge–Kutta
SSP–RK	Strong Stability–Preserving Runge–Kutta

References

1. Gribbin, J.; Gribbin, M. *Richard Feynman: A Life in Science*; Plume: New York, NY, USA, 1998.
2. Gad-el-Hak, M. Use of Continuum and Molecular Approaches in Microfluidics. In Proceedings of the 3rd Theoretical Fluid Mechanics Meeting, St. Louis, MO, USA, 24–26 June 2002.
3. Duff, R.E. Shock-Tube Performance at Low Initial Pressure. *Phys. Fluids* **1959**, *2*, 207–216.
4. Roshko, A. On Flow Duration in Low-Pressure Shock Tubes. *Phys. Fluids* **1960**, *3*, 835–842.
5. Mirels, H. Test Time in Low-Pressure Shock Tubes. *Phys. Fluids* **1963**, *6*, 1201–1214.
6. Mirels, H. Correlation Formulas for Laminar Shock Tube Boundary Layer. *Phys. Fluids* **1966**, *9*, 1265–1272.
7. Pant, J.C. Some Aspects of Unsteady Curved Shock Waves. *Int. J. Eng. Sci.* **1969**, *7*, 235–245.
8. Huilgol, R.R. Growth of Plane Shock Waves in Materials with Memory. *Int. J. Eng. Sci.* **1973**, *11*, 75–86.
9. Ting, T.C.T. Intrinsic Description of 3-Dimensional Shock Waves in Nonlinear Elastic Fluids. *Int. J. Eng. Sci.* **1981**, *19*, 629–638.

10. Shankar, R. On Growth and Propagation of Shock Waves in Radiation-Magneto Gas Dynamics. *Int. J. Eng. Sci.* **1989**, *27*, 1315–1323.

11. Suliciu, I. On Modelling Phase Transitions by Means of Rate-Type Constitutive Equations. Shock Wave Structure. *Int. J. Eng. Sci.* **1990**, *28*, 829–841.

12. Shi, Z.; Reese, J.M.; Chandler, H.W. The Application of a Shock Wave Model to Some Industrial Bubbly Fluid Flows. *Int. J. Eng. Sci.* **2000**, *38*, 1617–1638.

13. Bhardwaj, D. Formation of Shock Waves in Reactive Magnetogasdynamic Flow. *Int. J. Eng. Sci.* **2000**, *38*, 1197–1206.

14. Zel'dovich, Y.B.; Raizer, Y.P. *Physics of Shock Waves and High-Temperature Hydrodynamic Phenomena*; Hayes, W.D., Probstein, R.F., Eds.; Dover Publications, Inc.: Mineola, NY, USA, 2002.

15. Smith, D.; Amitay, M.; Kibens, V.; Parekh, D.; Glezer, A. Modification of Lifting Body Aerodynamics Using Synthetic Jet Actuators. In Proceedings of the 36th AIAA Aerospace Sciences Meeting and Exhibit, Reno, NV, USA, 12–15 January 1998; Paper No.: 98-0209.

16. Raman, G.; Raghu, S.; Bencic, T.J. Cavity Resonance Suppression Using Miniature Fluidic Oscillators. In Proceedings of the 5th AIAA/CEAS Aeroacoustics Conference, Seattle, WA, USA, 10–12 May 1999; Paper No.: 99-1900.

17. Gad-el-Hak, M. The Fluid Mechanics of Microdevices—The Freeman Scholar Lecture. *J. Fluids Eng.* **1999**, *121*, 5–33.

18. Sushanta, K.M.; Suman, C. *Microfluidics and Nanofluidics Handbook: Fabrication, Implementation and Applications*; CRC Press: Boca Raton, FL, USA, 2011.

19. Karniadakis, G.; Beskok, A.; Aluru, N. *Microflows and Nanoflows: Fundamentals and Simulation*; Springer: New York, NY, USA, 2005.

20. Brouillette, M. Shock Waves at microscales. *Shock Waves* **2003**, *13*, 3–12.

21. Zeitoun, D.E.; Burtschell, Y.; Graur, I.A.; Ivanov, M.S.; Kudryavtsev, A.N.; Bondar, Y. A. Numerical Simulation of Shock Wave Propagation in Microchannels Using Continuum and Kinetic Approaches. *Shock Waves* **2009**, *19*, 307–316.

22. Hesthaven, J.S.; Warburton, T. *Nodal Discontinuous Galerkin Methods: Alghorims, Analysis, and Applications*; Springer: New York, NY, USA, 2008.

23. Peraire, J.; Nguyen, N.C.; Cockburn, B. A Hybridizable Discontinuous Galerkin Method for the Compressible Euler and Navier-Stokes Equations. In Proceedings of the 48th AIAA Aerospace Sciences Meeting Including the New Horizons Forum and Aerospace Exposition, Orlando, FL, USA, 4–7 January 2010; pp. 1–11.

24. Gottlieb, S.; Shu, W.; Tadmor, E. Strong Stability Preserving High Order Time Discretization Methods. *SIAM Rev.* **2001**, *43*, 89–112.

25. Carpenter, M.H.; Kennedy, C. *Fourth-Order 2N–Storage Runge–Kutta Schemes*; NASA Report TM 109112; NASA Langley Research Center, Hampton, VA, USA, 1994.

26. Tu, S.; Aliabadi, S. A Slope Limiting Procedure in Discontinuous Galerkin Finite Element Method for Gasdynamics Application. *Int. J. Numer. Anal. Model.* **2005**, *2*, 163–178.

27. Kennard, E.H. *Kinetic Theory of Gases*; McGraw–Hill: New York, NY, USA, 1938.

28. Kandlikar, S.G.; Garimella, S.; Li, D.; Colin, S.; King, M.R. *Heat Transfer and Fluid Flow in Minichannels and Microchannels*; Elsevier: Oxford, UK, 2006.

29. Bird, G.A. Monte Carlo simulation of gas flows. *Annu. Rev. Fluid. Mech.* **1978**, *10*, 11–31.

30. Veluri, S.P. Code Verification and Numerical Accuracy Assessment for Finite Volume CFD Codes. Ph.D. Thesis, Virginia Polytechnic Institute, Blacksburg, VA, USA, 2010.

aerospace

MDPI

Article

Predicting Non-Linear Flow Phenomena through Different Characteristics-Based Schemes

Tom-Robin Teschner, László Könözsy * and Karl W. Jenkins

Centre for Computational Engineering Sciences, Cranfield University, Cranfield, Bedfordshire MK43 0AL, UK;
t.teschner@cranfield.ac.uk (T.-R.T.); k.w.jenkins@cranfield.ac.uk (K.W.J.)
* Correspondence: laszlo.konozsy@cranfield.ac.uk; Tel.: +44-1234-758-278

Received: 18 November 2017; Accepted: 18 February 2018; Published: 24 February 2018

Abstract: The present work investigates the bifurcation properties of the Navier–Stokes equations using characteristics-based schemes and Riemann solvers to test their suitability to predict non-linear flow phenomena encountered in aerospace applications. We make use of a single- and multi-directional characteristics-based scheme and Rusanov's Riemann solver to treat the convective term through a Godunov-type method. We use the Artificial Compressibility (AC) method and a unified Fractional-Step, Artificial Compressibility with Pressure-Projection (FSAC-PP) method for all considered schemes in a channel with a sudden expansion which provides highly non-linear flow features at low Reynolds numbers that produces a non-symmetrical flow field. Using the AC method, our results show that the multi-directional characteristics-based scheme is capable of predicting these phenomena while the single-directional counterpart does not predict the correct flow field. Both schemes and also Riemann solver approaches produce accurate results when the FSAC-PP method is used, showing that the incompressible method plays a dominant role in determining the behaviour of the flow. This also means that it is not just the numerical interpolation scheme which is responsible for the overall accuracy. Furthermore, we show that the FSAC-PP method provides faster convergence and higher level of accuracy, making it a prime candidate for aerospace applications.

Keywords: characteristics-based scheme; multi-directional; Riemann solver; Godunov method; bifurcation

---

1. Introduction

Despite the advances made in the field of computational fluid dynamics over the past decades, predicting flow patterns around aerodynamic shapes remains a challenge for aerospace applications. The flow around a wing can have a transonic behaviour which, at high angles of attack, may be supplemented by flow separation, strong crossflow gradients as well as a hysteresis in the lift slope [1,2]. Traub [3] highlighted further that at low Reynolds number flows, laminar separation bubbles exist which have an inherently unsteady behaviour. These separation bubbles may reattach to the wing or transition into a fully turbulent flow, depending on the pressure gradient. Panaras [4] reviewed recent published studies on Reynolds-averaged Navier–Stokes (RANS) and large-eddy simulation (LES) results and pointed out that the turbulent kinetic energy in regions of reversed flow is low and can be considered to be almost laminar. Thus, it is important to capture the transition between laminar and turbulent flows as flow separation and reattachment ultimately have a strong influence on predicting the lift slope, stall angle and hysteresis. Furthermore, since RANS models are commonly based on the linear Boussinesq assumption, results may provide only a moderate accuracy while models based on non-linear theories may be more efficient in predicting non-isotropic flow behaviour.

From the considerations given above, it is important to realize that the non-linear term in the Navier–Stokes equations is the most sensitive part and an accurate numerical description is mandatory to capture those flow features. The literature on numerical schemes is vast and this article does not intend to give a comprehensive overview—interested readers are referred to reference [5,6]—however a brief overview is given. Initially the second-order central scheme was used and later augmented by artificial dissipation. This was necessary to account for the lost dissipation on the small scales due to an under-resolved computational grid, essentially mimicking the same approach RANS turbulence model take in order to provide turbulent, or physical, dissipation. The central scheme lacks transportiveness and so is not suitable to predict flows containing discontinuities. A range of high-resolution schemes were developed to address this issue, of which the monotonic upwind scheme for conservation laws (MUSCL), total variation diminishing (TVD), essential non-oscillatory (ENO) and weighted essential non-oscillatory (WENO) schemes are the most widely used schemes nowadays. These schemes all have the same goal; to capture the correct behaviour of the non-linear term. However, their closure is based on numerical considerations. In order to enhance these scheme and to provide highly-accurate physical values for the time integration procedure, two possible routes are available. The first one is commonly employed in the compressible community where an approximate Riemann solver (RS) is used. In-fact, by including the local Riemann problem, the values obtained through the numerical scheme are modified based on the local eigenstructure of the system of equations, thus presenting a hybrid approach to couple numerical and physical fluxes. The second approach is to take a characteristics-based (CB) scheme which shares many similarities of the local Riemann problem, although being rather different in the details. Here, the characteristics of the flow are found at each computational node where the primitive variables are updated along the compatibility equations, which are unique to each node. The CB scheme found widespread use in the 1960s and onwards for the calculation of compressible flows [7].

With the introduction of the Artificial Compressibility (AC) method of Chorin [8], which was the first hyperbolic method available in the framework on incompressible flows, it became possible to apply both the local Riemann problem as well as CB schemes to incompressible flows. Drikakis et al. [9] were among the first to introduce the CB scheme in conjunction with the AC method in the finite volume method, which was followed by a development in the finite element framework by Zienkiewicz and Codina [10] and Zienkiewicz et al. [11]. The CB scheme of Drikakis [9] was developed for one-dimensional flows but can be extended to higher dimensions by applying the same equations to any coordinate direction in space. Thus, the approach is termed the single-directional characteristics-based, or SCB, scheme. The scheme was later revised by Neofytou [12] and tested by Su et al. [13], concluding that, although the revised scheme being more mathematical rigorous, there was little difference in the results using the original and revised SCB scheme.

A vastly different approach was taken by Razavi et al. [14] who proposed to use a multi-directional characteristics-based, or MCB, scheme. It was extended to handle curve-linear coordinate systems [15] as well as unstructured grids [16]. Improved boundary conditions in combination with the MCB scheme was proposed by Hashemi and Zamzamian [17] for the far-field and Zamzamian and Hashemi [18] for solid boundary conditions. Fathollahi and Zamzamian [19] investigated the influence of the number of compatibility equations used at each node. They showed that increasing the number of compatibility equation did not affect the accuracy of the solution while increasing the convergence rate. The scheme has been tested for heat transfer and turbulent flows by Razavi and Adibi [20] and Razavi and Hanifi [21], respectively. In the MCB scheme, instead of tracking characteristic lines through each computational node as done in the SCB scheme, characteristic surfaces are multiplied into the governing equations and their solution is found by means of the system's eigenvalues and geometrical considerations. This creates a characteristic cone along which the compatibility equations are emanating. This allows the scheme to capture any anisotropic properties of the flow and therefore lends itself to capture non-linear flow features.

In the present work, our aim is to show the applicability of the MCB scheme to predict non-linear flow behaviours at low reynolds numbers. As a comparison, we also employ the SCB scheme and apply Rusanov's RS which provides a hybrid multi-directional, Godunov-type treatment for the non-linear term of the Navier–Stokes equations. We apply the scheme to a channel with a sudden expansion. This geometry was found to be particularly suitable to predict the onset of bifurcation and subsequently the reattachment point of the flow. Both phenomena are underlying features that are important to calculate turbulent flows and thus highly relevant to the discussion in the context of aerospace applications.

We use the AC method in order to apply the CB scheme and RS, which require a hyperbolic system of equations. Furthermore, we also employ a recent method developed by Könözsy [22], later introduced by Könözsy and Drikakis [23], which unifies Chorin's AC method with Chorin's and Temam's Fractional-Step, Pressure projection (FS-PP) method [24,25] which is termed the Fractional-Step, Artificial Compressibility with Pressure-Projection, or FSAC-PP method (see Section 2.2). It has been tested for classical benchmark cases for incompressible flows [22], multi-species and variable density flows in micro-channels [26], trapping and positioning of cryogenic propellants [27], forced separated flows over a backward facing step [28] and the vortex pairing problem [29]. The FSAC-PP method was originally introduced using the SCB scheme and extended by Teschner et al. [30] to the MCB scheme. Smith et al. [31] removed the CB approach and used different Riemann solvers instead to obtain the inviscid fluxes. The FSAC-PP has shown to have superior convergence properties over the AC and FS-PP method, especially in low Reynolds number flows, while showing a similar level of accuracy.

The remaining structure of this article is as follows. In Section 2, we give numerical details on the AC and FSAC-PP method. Section 3 summarises the SCB and MCB scheme used in this study while the computational setup is presented in Section 4. We present the results for the sudden expansion geometry in Section 5 and highlight final remarks in Section 6.

2. Methodology

In this section, we briefly review the concepts of the AC and FSAC-PP method, which are employed throughout this study. Details can be found in [8] about the AC method and in [22,23] about the FSAC-PP approach.

2.1. The Artificial Compressibility Method

The absence of a thermodynamic functional relationship between the pressure and density has led to the notion of an Artificial Compressibility approach. Here, the density is replaced by the pressure in the time derivative of the continuity equation. Since it could be difficult to give a clear relationship between the two state variables, a numerical constant is introduced, β, which is a user-defined convergence parameter [8]. Thus, the continuity equation loses any physical meaning, however, it provides a mechanism to predict the pressure in the momentum equation. Once a steady state is obtained, the time derivative vanishes to zero and so a divergence-free velocity field is obtained. The governing equations with the extensions of the AC method thus become

$$\frac{1}{\beta}\frac{\partial p}{\partial \tau} + \nabla \cdot \mathbf{U} = 0, \tag{1}$$

$$\frac{\partial \mathbf{U}}{\partial \tau} + (\mathbf{U} \cdot \nabla)\,\mathbf{U} = -\frac{1}{\rho}\nabla p + \nu \nabla^2 \mathbf{U}, \tag{2}$$

where ρ and ν are the density and viscosity, respectively. The equations are iterated in pseudo-time due to the non-physical time derivative of pressure. In order to advance the system of Equations (1) and (2) in real time, a dual-time stepping procedure needs to be employed for which a real time derivative is added to the momentum equation. The time-step has to satisfy the following condition as

$$\Delta \tau = \min \left[\frac{\mathrm{CFL}\,\Delta \mathbf{x}}{|\mathbf{U}| + \sqrt{\mathbf{U}^2 + \beta}}, \frac{\mathrm{CFL}\,\Delta \mathbf{x}^2}{4\nu} \right]. \tag{3}$$

2.2. The Unified Fractional-Step, Artificial Compressibility and Pressure Projection Method

The FSAC-PP method unifies both the AC and FS-PP method into a single framework. In the first Fractional-Step, the continuity equation of the AC method is used along with the momentum equation of the FS-PP method, in which the pressure gradient is dropped. The governing equations of the numerical method can be written in a semi-discretised form by

$$\frac{1}{\beta}\frac{\partial p}{\partial \tau} + \nabla \cdot \mathbf{U} = 0, \tag{4}$$

$$\frac{\mathbf{U}^* - \mathbf{U}^{(n)}}{\Delta \tau} + (\mathbf{U} \cdot \nabla)\mathbf{U} = \nu \nabla^2 \mathbf{U}. \tag{5}$$

From Equations (4) and (5), we obtain an intermediate velocity field denoted by $\mathbf{U}^*$ which is used in the second Fractional-Step of the numerical procedure to recover the pressure field as

$$\frac{\mathbf{U}^{(n+1)} - \mathbf{U}^*}{\Delta t} = -\frac{1}{\rho}\nabla p^{(n+1)}. \tag{6}$$

The Helmholtz-Hodge decomposition requires the velocity field at time level $(n+1)$ to become divergence-free. Thus, by taking the divergence of Equation (6), the following Poisson equation for the pressure is obtained by

$$\nabla^2 p^{(n+1)} = \frac{\rho}{\Delta t}\nabla \cdot \mathbf{U}^*. \tag{7}$$

With an updated pressure value, we can recover the velocity at the next time level from Equation (6) as

$$\mathbf{U}^{n+1} = \mathbf{U}^* - \frac{\Delta t}{\rho}\nabla p^{(n+1)}. \tag{8}$$

Equations (5)–(8) are consistent with the FS-PP method of Chorin [24] and Temam [25] while Equation (4) provides an initial pressure field for the Poisson equation through the perturbed continuity equation of the AC method. Thus, the hyperbolic and elliptic features of the AC and FS-PP method are coupled through the Fractional-Step procedure which allows for a characteristics-based treatment of the convective fluxes while the pressure is stabilised through the elliptic Poisson equation.

3. Characteristic-Based Schemes for Incompressible Flows

While the FSAC-PP method was originally introduced using the SCB scheme [9], Teschner et al. [30] showed the extension to a multi-directional characteristics-based (MCB) scheme, which is based on the derivation given by Razavi et al. [14] for the AC method.

3.1. A Single-Directional Closure

The SCB scheme splits the governing equations into one-dimensional equations and assumes that the SCB scheme is equally applicable for any direction. The primitive variables are recovered in the context of the hyperbolic-type AC method as

$$\tilde{u}_{i+\frac{1}{2}}^n = \tilde{x}R + \tilde{y}(u_0\tilde{y} - v_0\tilde{x}), \tag{9}$$

$$\tilde{v}_{i+\frac{1}{2}}^n = \tilde{y}R + \tilde{x}(u_0\tilde{y} - v_0\tilde{x}), \tag{10}$$

$$\tilde{p}_{i+\frac{1}{2}}^n = \frac{1}{2\sqrt{\lambda_0^2 + \beta}}(\lambda_1 k_2 - \lambda_2 k_1), \tag{11}$$

where

$$R = \frac{1}{2\sqrt{\lambda_0^2 + \beta}}[(p_1 - p_2) + \tilde{x}(\lambda_1 u_1 - \lambda_2 u_2) + \tilde{y}(\lambda_1 v_1 - \lambda_2 v_2)], \tag{12}$$

$$k_1 = p_1 + \lambda_1(u_1\tilde{x} + v_1\tilde{y}), \tag{13}$$

$$k_2 = p_2 + \lambda_2(u_2\tilde{x} + v_2\tilde{y}). \tag{14}$$

The values of $\tilde{x}$ and $\tilde{y}$ are set to $\tilde{x} = 1$, $\tilde{y} = 0$ when considering the x-direction and $\tilde{x} = 0$, $\tilde{y} = 1$ for the y-direction, respectively. The eigenvalues are found as

$$\lambda_0 = u\tilde{x} + v\tilde{y}, \tag{15}$$

$$\lambda_1 = \lambda_0 + \sqrt{\lambda_0^2 + \beta}, \tag{16}$$

$$\lambda_2 = \lambda_0 - \sqrt{\lambda_0^2 + \beta}. \tag{17}$$

The primitive variables with indices $j = 0, 1, 2$ are found through Godunov's RS

$$\mathbf{U}_{j,i+\frac{1}{2}} = \frac{1}{2}[(1 + \text{sign}\lambda_j)\mathbf{U}_{i+\frac{1}{2}}^L + (1 - \text{sign}\lambda_j)\mathbf{U}_{i+\frac{1}{2}}^R], \tag{18}$$

where the intercell flux values $\mathbf{U}_{i+1/2}^L$ and $\mathbf{U}_{i+1/2}^R$ are found through a third-order interpolation procedure, see Section 4. In the FSAC-PP method, the characteristic pressure is set to zero as it is dropped in the momentum equation. From Equations (9) and (10) it can be seen, however, that the pressure provided by the continuity equation enters the characteristic velocity field which enforces a stronger coupling between velocity and pressure. More details on the SCB scheme can be found in [9,22,23].

3.2. A Multi-Directional Closure

In contrast to the SCB scheme, the multi-directional approach applies a characteristic surface to the governing equations. For a two-dimensional flow, the governing equations of the AC method are

$$\frac{1}{\beta}\frac{dp}{df}\frac{\partial f}{\partial \tau} + \frac{du}{df}\frac{\partial f}{\partial x} + \frac{dv}{df}\frac{\partial f}{\partial y} = 0, \tag{19}$$

$$\frac{du}{df}\frac{\partial f}{\partial \tau} + u\frac{du}{df}\frac{\partial f}{\partial x} + v\frac{du}{df}\frac{\partial f}{\partial y} + \frac{1}{\rho}\frac{dp}{df}\frac{\partial f}{\partial x} = 0, \tag{20}$$

$$\frac{dv}{df}\frac{\partial f}{\partial \tau} + u\frac{dv}{df}\frac{\partial f}{\partial x} + v\frac{dv}{df}\frac{\partial f}{\partial y} + \frac{1}{\rho}\frac{dp}{df}\frac{\partial f}{\partial y} = 0. \tag{21}$$

Multiplying Equations (19)–(21) by df and introducing the shorthand notation $\Psi = f_\tau + uf_x + vf_y$, the equations reduce to

$$\frac{f_\tau}{\beta}dp + f_x du + f_y dv = 0, \tag{22}$$

$$\Psi du + \frac{f_x}{\rho}dp = 0, \tag{23}$$

$$\Psi dv + \frac{f_y}{\rho}dp = 0. \tag{24}$$

The determinant of the coefficient matrix corresponding to Equations (22)–(24) yields to solution in the form of $\Psi_1 = 0$ and $\Psi_2 = \beta/(\rho n_\tau)$, where $n_\tau = f_\tau$ is the component of the unit normal vector and use has been made of the fact that the normal direction corresponds to the derivative of the characteristic surface in the form of $\mathbf{n} = \nabla f$. In-fact, it can be shown that $\Psi_{1,2}$ corresponds to two independent characteristic surfaces [7,32–34] which are called the stream- (Ψ_1) and wave- (Ψ_2) surface in accordance with the literature on the compressible MCB scheme. Thus, inserting $\Psi_{1,2}$ into Equations (23) and (24) yields two sets of compatibility equations along with Equation (22). Rusanov [7] showed that a mix of compatibility equations on Ψ_1 and Ψ_2 are necessarily needed to obtain the characteristic values on the cell interfaces. For the incompressible version of the MCB scheme, however, it is sufficient to only use the compatibility equations corresponding to Ψ_2. This can be verified by computing the rank of the coefficient matrix of Equations (22)–(24) for Ψ_2 which will result in a non-rank deficient matrix. Thus, three compatibility equations are found over four wave directions as

$$\tilde{u}_{i+\frac{1}{2}}^{(n)} = \frac{1}{\beta(f_{\tau,1} + f_{\tau,2})} \left[(p_1 - p_2)f_{\tau,1}f_{\tau,2} + (u_1 f_{\tau,2} + u_2 f_{\tau,1})\beta \right], \tag{25}$$

$$\tilde{v}_{i+\frac{1}{2}}^{(n)} = \frac{1}{\beta(f_{\tau,1} + f_{\tau,2})} \left[(p_3 - p_4)f_{\tau,3}f_{\tau,4} + (v_3 f_{\tau,4} + v_4 f_{\tau,3})\beta \right], \tag{26}$$

$$\tilde{p}_{i+\frac{1}{2}}^{(n)} = \frac{1}{4} \sum_{j=1}^{4} (-1)^{j+1} p_j + (-1)^j \frac{\beta}{f_{\tau,j}} (\tilde{\mathbf{U}} - \mathbf{U}_j)^{\|}. \tag{27}$$

The characteristic pressure is an average over all four wave directions and $(\tilde{\mathbf{U}} - \mathbf{U}_j)^{\|}$ indicates that the velocity components, which are aligned with the wave directions, are to be used. The normal vector is

$$n_{\tau,j} = f_{\tau,j} = \frac{-(u_j \cos \phi_j + v_j \sin \phi_j) \pm \sqrt{(u_j \cos \phi_j + v_j \sin \phi_j)^2 + 4\beta/\rho}}{2}, \tag{28}$$

where $j = 1, 2, 3, 4$ indices correspond to the intersection of the characteristic surfaces with the time level (n), which is illustrated in Figure 1, and the primitive variables at those locations (wave directions) are found through interpolation. Razavi et al. [14] introduced two versions of the MCB scheme, a first- and second-order scheme, where primitive variables at $j = 1, 2, 3, 4$ are set to the reconstructed intercell variables at $i + \frac{1}{2}$ for the first-order scheme and are interpolated along the characteristic surfaces in the second-order scheme. Since the SCB scheme classifies as a first-order CB scheme under this definition, the first-order MCB scheme is used in the remaining investigation to compare both methods.

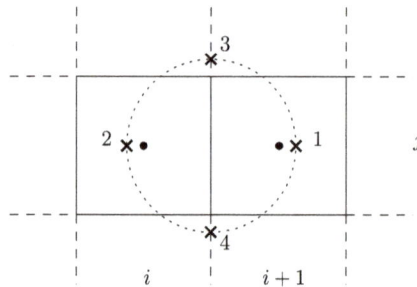

Figure 1. Intersection of the characteristic surfaces with time level (n).

The SCB scheme was introduced using Godunov's RS, Equation (18), which provided the transportiveness property of the SCB scheme. The MCB scheme, as introduced by Razavi et al. [14], did not feature any RS treatment. Thus, in order to ensure the transportiveness of the MCB scheme, a numerical interpolation scheme needs to be selected which provides this property. Alternatively, the MCB scheme can be supplemented with a RS to remove the necessity to provide transportiveness through the numerical interpolation scheme and make it a property of the MCB scheme itself. Another benefit of RS-based approaches is that different RS provide different level of numerical dissipation which may become important at high Reynolds number flows. This combination allows to use a relative cheap but high-order interpolation scheme for increased accuracy at reduced computational cost, where we make the properties traditionally linked to the numerical scheme now a property of the CB framework. Furthermore, the inclusion of the RS promotes a multi-directional Godunov-type characteristic framework for incompressible flows which has so far only been tested and introduced by

Teschner et al. [30] for a simple channel geometry. Therefore, we make use of Rusanov's approximate RS [35] in the form of

$$\mathbf{F}(\tilde{\mathbf{U}})_{i+1/2} = \frac{1}{2}\left[\mathbf{F}(\tilde{\mathbf{U}})_R + \mathbf{F}(\tilde{\mathbf{U}})_L\right] - \frac{S^+}{2}\left(\tilde{\mathbf{U}}_R - \tilde{\mathbf{U}}_L\right), \tag{29}$$

where $\mathbf{F}(\tilde{\mathbf{U}})_{i+1/2}$ is the inviscid flux at the cell interface from which the primitive variables are found directly through flux differentiation. We follow the approach of Davis [36] to obtain the signal velocity S^+ as

$$S^+ = \max\left\{|A_L^-|, |A_R^-|, |A_L^+|, |A_R^+|\right\}, \tag{30}$$

with

$$A_L^- = \mathbf{U}_L - \sqrt{\mathbf{U}_L^2 + \beta/\rho}, \tag{31}$$

$$A_R^- = \mathbf{U}_R - \sqrt{\mathbf{U}_R^2 + \beta/\rho}, \tag{32}$$

$$A_L^+ = \mathbf{U}_L + \sqrt{\mathbf{U}_L^2 + \beta/\rho}, \tag{33}$$

$$A_R^+ = \mathbf{U}_R + \sqrt{\mathbf{U}_R^2 + \beta/\rho}. \tag{34}$$

Using this closure, we have extended the MCB scheme into the Godunov framework and will demonstrate that even a simple approximate RS has favourable influence on the accuracy of the overall procedure.

4. Computational Setup and Numerical Schemes

The intercell values of the inviscid fluxes are obtained through a third-order interpolation polynomial as

$$\mathbf{U}_{i+1/2}^L = \frac{1}{6}(5\mathbf{U}_i^n - \mathbf{U}_{i-1}^n + 2\mathbf{U}_{i+1}^n), \tag{35}$$

$$\mathbf{U}_{i+1/2}^R = \frac{1}{6}(5\mathbf{U}_{i+1}^n - \mathbf{U}_{i+2}^n + 2\mathbf{U}_i^n). \tag{36}$$

The viscous fluxes, on the other hand, use a second-order reconstruction scheme which, for Cartesian coordinates, defaults to a finite difference discretization in the form of

$$\frac{\partial^2 \phi}{\partial x^2} = \left[\frac{\phi_{i-1} + 2\phi_i - \phi_{i+1}}{(\Delta x)^2}\right] \tag{37}$$

The time integration is carried out by a third-order Runge–Kutta TVD scheme in which each stage is updated by the preceding one as

$$\mathbf{U}^* = \mathbf{U}^{(n)} + \Delta\tau\text{RHS}(\mathbf{U}^{(n)}),$$

$$\mathbf{U}^{**} = \frac{3}{4}\mathbf{U}^{(n)} + \frac{1}{4}\mathbf{U}^* + \frac{1}{4}\Delta\tau\text{RHS}(\mathbf{U}^*),$$

$$\mathbf{U}^{(n+1)} = \frac{1}{3}\mathbf{U}^{(n)} + \frac{2}{3}\mathbf{U}^{**} + \frac{2}{3}\Delta\tau\text{RHS}(\mathbf{U}^{**}), \tag{38}$$

where

$$\text{RHS}(\mathbf{U}) = -(\mathbf{U}\cdot\nabla)\mathbf{U} - \frac{\alpha}{\rho}\nabla p + \nu\nabla^2\mathbf{U}. \tag{39}$$

Here, we set $\alpha = 1$ for the AC method and $\alpha = 0$ for the FSAC-PP method. The numerical convergence parameter is set to $\beta = 1.0$, the CFL number for the explicit Runge–Kutta integration scheme is kept at CFL $= 0.8$ while the SOR algorithm applied to the Poisson solver, Equation (7), uses an under-relaxation factor of $\omega = 0.7$ with a total of ten Poisson iteration per time-step. Könözsy [22] found that the Poisson equation can be over-relaxed to gain further computational speed-up and suggested to use $\omega = 1.7$ instead. Unlike the FS-PP method, in which the Poisson solver is the most computational expensive part to obtain a correct pressure field, the FSAC-PP method benefits from an initial prediction of the pressure field through the continuity equation of the AC method which substantially reduces the number of iteration needed for the Poisson equation. It was found that of the order of ten iterations is sufficient to speed up the convergence rate while keeping the same order of accuracy as obtained with the FS-PP method. The residuals are calculated based on

$$R_U = \max \left(|\nabla \cdot \mathbf{U}^{(n+1)} - \nabla \cdot \mathbf{U}^{(n)}| \right), \tag{40}$$

where R_U is normalised based on the residual at the first iteration and compared against the convergence criterion of $\varepsilon = 10^{-12}$ to ensure that the residuals have reduced by 12 orders of magnitude based on the divergence free constrain.

5. Results and Discussion

To investigate the performance of the MCB scheme using the Rusanov RS against the classical, SCB scheme, we simulate the flow inside a suddenly expanding channel with an expansion ratio of 3:1, see Figure 2. We have performed a grid convergence study and calculated the grid convergence index (GCI) based on the reattachment length according to Roache [37]. Due to the Cartesian nature of the solver, the mesh cannot be locally refined, especially near walls where a good resolution is required. This, however, creates a challenging test case for both the AC and FSAC-PP method as strong velocity gradients are present close to the wall which also amplifies any differences between the two methods. Table 1 shows the results obtained for the grid dependency study where we give the reattachment length at the upper and lower wall, as well as its corresponding GCI value and the time and iteration it took to compute the results. The Cartesian mesh elements were divided by a factor of two for subsequent mesh levels, ensuring a refinement ratio of four in all cases. We also provide the expected value based on the Richardson extrapolation which can also be found in [37]. At a Reynolds number of Re = 30, for which the GCI study has been carried out, the expected reattachment length as given by Oliveira [38] is $X_{r1}, X_{r2} = 3.080$. Looking at the results, we can see that the FSAC-PP method converges towards that results for the finest level, while the AC method fails to predict the results correctly. The FSAC-PP method also shows a lower level of GCI which is of the order of 10% while the AC method produces GCI values which are close to double of that result. In terms of relative change, the reattachment length changes less than 5% from the medium to the finest mesh level for both AC and FSAC-PP method. Although the AC method overpredicts the reattachment, in this case using no CB scheme and no RS, we will see later that the results are slightly better with other schemes. Hence, focusing on the FSAC-PP method, it would be desired to use the finest mesh level for all subsequent simulations. The computational time, however, becomes prohibitive in the case of the AC method if the finest mesh level was to be used, where a single simulation would last for more than 32 h which would cause excessive simulation times as several simulations at different Reynolds numbers are required. Furthermore, for very low Reynolds numbers, the AC method requires even more computational time [23,27]. Thus, we chose the medium mesh with 22,400 elements as the next closest mesh for which results in a reasonable amount of time can be obtained.

To ensure that the convergence criterion of 10^{-12} is sufficient, we furthermore carried out a convergence study where we have checked the influence of the convergence parameter from 10^{-6} to 10^{-12} for Equation (40). To judge convergence, we check by how much the reattachment length changes on the upper and lower wall. The results are summarised in Table 2. We can see that the chosen convergence threshold at 10^{-12} is a sufficient indicator for convergence. The results for the FSAC-PP method do not change from 10^{-11} while the results for the AC method do only differ for the fourth significant figure. Thus, we haven chosen 10^{-12} as a convergence threshold for all subsequent simulations.

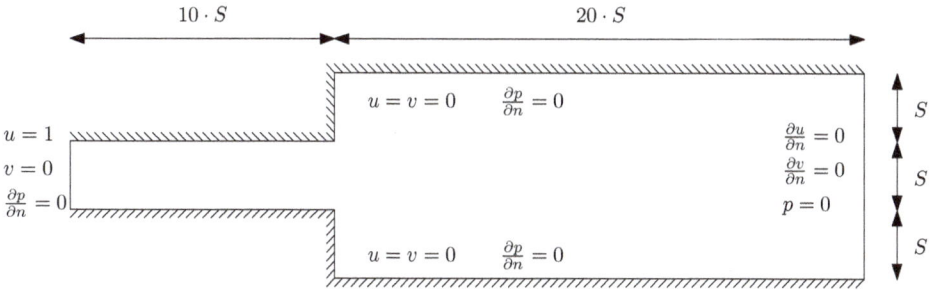

Figure 2. Geometry of the channel with a sudden expansion. Boundary conditions are given for the inflow, outflow and solid (wall) boundary condition.

Table 1. Grid dependency study for the AC and FSAC-PP method at a Reynolds number of Re = 30. The expected value of the reattachment length is $X_{r1}, X_{r2} = 3.080$ as given by Oliveira [38].

Method	Cells	X_{r1}	X_{r2}	GCI (r1)	GCI (r2)	Time (h)	Iterations
AC	1400	3.65448	3.65448	-	-	0.006	40,051
	5600	3.92843	3.92843	0.3269	0.3269	0.13	222,404
	22,400	4.2131	4.2131	0.3167	0.3167	2.73	1,221,133
	89,600	4.40196	4.40196	0.2011	0.2011	32.2	2,768,606
	Richardson Extrapolation	4.4146	4.4146	-	-	-	-
FSAC-PP	1400	2.78324	2.83458	-	-	0.003	11,534
	5600	2.85779	2.87495	0.1223	0.0658	0.019	22,408
	22,400	2.9544	2.96133	0.1533	0.1367	0.22	58,896
	89,600	3.03597	3.03854	0.1259	0.1191	3.19	177,801
	Richardson Extrapolation	3.0414	3.0437	-	-	-	-

Table 2. Convergence study for different level of convergence at a Reynolds number of Re = 30 based on the reattachment length X_{r1} and X_{r2} at the upper and lower wall.

Method		10^{-6}	10^{-7}	10^{-8}	10^{-9}	10^{-10}	10^{-11}	10^{-12}
AC	X_{r1}	3.28289	3.89382	4.00063	4.16895	4.20950	4.21217	4.21310
	X_{r2}	3.28289	3.89382	4.00063	4.16895	4.20950	4.21217	4.21310
FSAC-PP	X_{r1}	3.05442	2.97130	2.96041	2.96123	2.96134	2.96133	2.96133
	X_{r2}	3.04702	2.96438	2.95348	2.95430	2.95441	2.95440	2.95440

Figure 3 shows contours of the u-velocity component while Figure 4 presents the velocity profile at $x/S = 5$, where S indicates the step height, see Figure 2. We show results here obtained with the AC and FSAC-PP method for two Reynolds number. The first Reynolds number is Re = 34.6, based on the average inlet velocity and a characteristic lengthscale of unity, which is below the critical Reynolds number of $\text{Re}_{\text{crit}} \approx 54$ [38,39] and where a symmetrical flow pattern prevails. The second Reynolds number of Re = 80 is above the critical one and exhibits a breaking in the symmetry. All velocity profiles are compared against experimental data of Fearn et al. [39].

Figure 4a shows the AC method at a sub-critical Reynolds number. Not using any characteristics or RS overpredicts the flow slightly which is similar to the MCB scheme by itself. Both the standard non-linear treatment in conjunction with the Rusanov RS, as well as the hybrid Rusanov MCB scheme show an under-prediction of the peak velocity. A similar observation can be made for the SCB scheme. Since the SCB scheme has been proposed together with Godunov's RS, we can see that all RS-based approaches are more dissipative than a non RS-based approach. Figure 4b shows the AC method with the same schemes above the critical Reynolds number. Here, only the non-CB and MCB approach were capable of predicting the bifurcation to a non-symmetric flow behaviour. All RS-based approaches do not predict accurately the onset of bifurcation which may be explained by the inherent numerical dissipation to the Riemann solvers themselves.

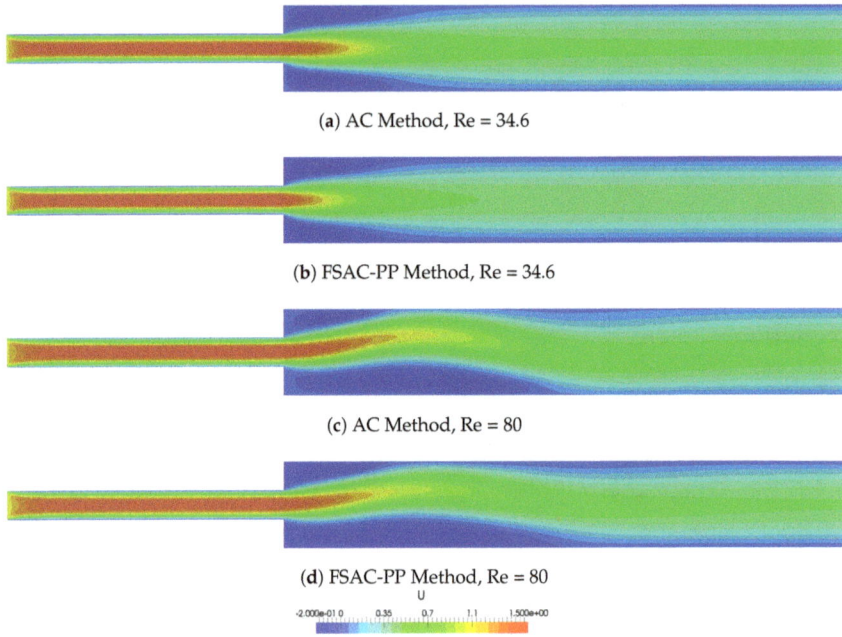

(**a**) AC Method, Re = 34.6

(**b**) FSAC-PP Method, Re = 34.6

(**c**) AC Method, Re = 80

(**d**) FSAC-PP Method, Re = 80

Figure 3. Contour plots of the axial velocity component using a non-CB treatment.

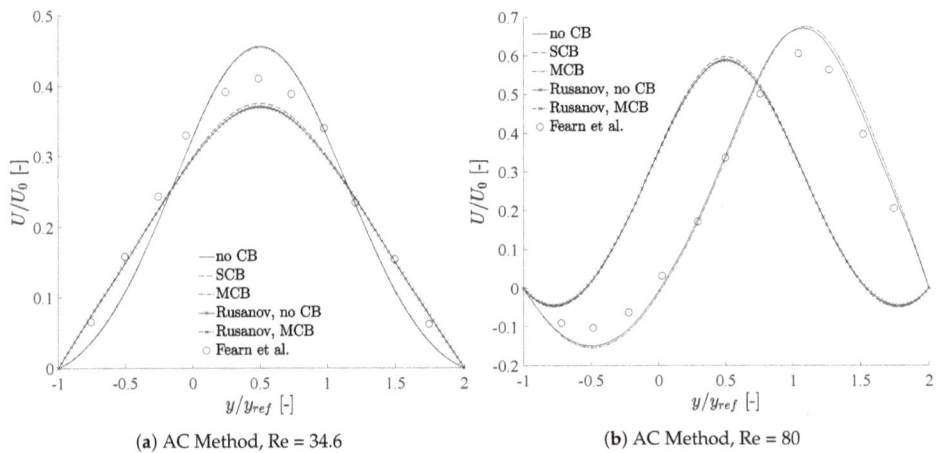

(**a**) AC Method, Re = 34.6

(**b**) AC Method, Re = 80

Figure 4. *Cont.*

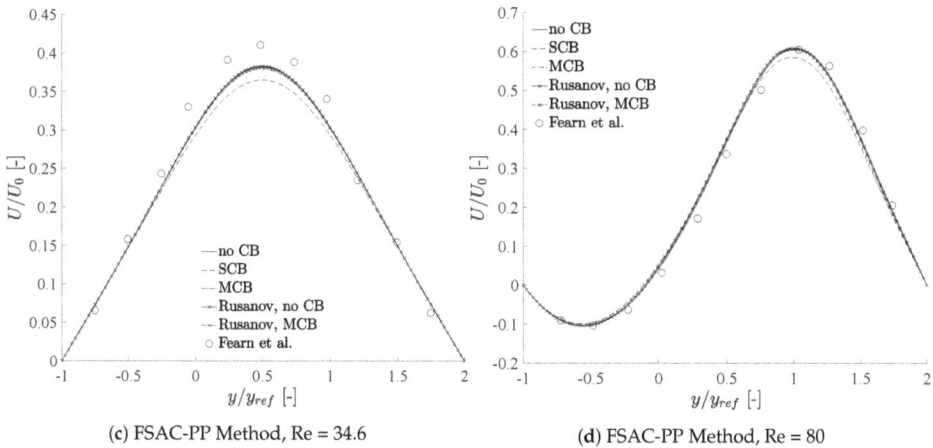

(c) FSAC-PP Method, Re = 34.6

(d) FSAC-PP Method, Re = 80

Figure 4. Velocity profiles at $x/S = 5$ for the AC (**top**) and FSAC-PP (**bottom**) method using different combinations of CB schemes and the Rusanov Riemann solver at Re = 34.6 (**left**) and Re = 80 (**right**).

The picture is fundamentally different for the FSAC-PP method. At Re = 34.6, Figure 4c, the results are independent of the method, except for the SCB scheme, which under-predicts the peak velocity slightly more than the other schemes. In Figure 4d we see the bifurcated state and all schemes, including the RS-based approaches, do predict the velocity profile correctly. We see here that not just the numerical scheme, but also the incompressible method used to perform the calculations makes a difference. This highlights that not just the numerical scheme that is used makes a difference, but also the incompressible closure assumption of a given method can have a significant influence. It should be kept in mind that the AC method is purely hyperbolic while the FSAC-PP method has a mixed hyperbolic-elliptic behaviour and it is not surprising that both method perform differently. However, it is surprising that the FSAC-PP method does predict the bifurcation correctly, regardless of the numerical scheme, while the AC method is highly depending on the numerical scheme. This is against the common believe that only the numerical scheme influences the accuracy of the solution while the numerical methods may only differ in the number of iterations. We make a case here and show that different mathematical behaviours (hyperbolic and elliptic) do have a significant effect on the final result. Since the current study is investigating the non-linear term of the Navier–Stokes equations, this is an important finding and it shows that for studies where the non-linear term is of importance, such as stall prediction around wings, it is worth considering not just a suitable scheme for the convective term but also a suitable incompressible method. The mixed hyperbolic-elliptic behaviour of the FSAC-PP method means that any pressure disturbances in the flow will instantaneously propagate through the domain using the elliptic Poisson solver. The purely hyperbolic nature of the AC method, on the other hand, means that pressure disturbances travel at a fixed signal velocity, determined by the local eigenstructure of the system. The onset of bifurcation does critically depend on the pressure boundary value at the convex corners which, in turn, are influenced by the recirculation area upstream. The FSAC-PP method allows that any change in the recirculation area will influence the upstream convex corner points and vice versa. The AC method, on the other hand, lags behind due to the finite information propagation speed and thus faces difficulties in predicting the onset of bifurcation. In-fact, the same argumentation was used by Teschner et al. [40] where a novel incompressible method was devised based on a parabolic pressure transport equation. It was argued that the pressure treatment plays a crucial role for incompressible flows and that a parabolic behaviour of the pressure matches the physical expected behaviour more closely. Thus, the elliptic part of the FSAC-PP method plays a crucial part in the current discussion as the pressure disturbances are handled in a different way as in the AC method. The outcome has a rather drastic impact, as can be seen from the results, where the bifurcation is predicted by the FSAC-PP method but not by the AC method. At the same time, however, we need to point out that the bifurcation phenomena can be predicted using the AC method, see for example Drikakis [41]. We do not claim that it is not capable of predicting the non-linear behaviour correctly, however, the AC method may be more prone to different modelling approaches, for example, it may become necessary to use stretched grids near the wall to capture the correct flow physics. The FSAC-PP method does not show the same modelling issues which can be attributed, in parts, to the Poisson solver. Not only does it provide an elliptic behaviour which advantages have been discussed above, but also it provides a stabilisation and smoothing procedure for the pressure. A stabilised pressure field, in turn, will provide a more realistic velocity field. Alternative approaches, using a fully hyperbolic compressible (Euler/Navier–Stokes) or incompressible (AC) solver are, for example, to use artificial dissipation, to stabilise the velocity field. The FSAC-PP method contains the stabilising feature by default which is another indicator as to why the result show not such a drastic difference as the AC method and furthermore predict the correct results regardless of the numerical scheme employed. In essence this is what we aim to demonstrate in this study, that we can take classical properties of numerical scheme, such as transportiveness and stability, and make them properties of the CB scheme as well as the numerical method. In this way, we provide a robust framework which is capable of treating non-linear and anisotropic flow behaviours correctly while modelling errors due to different reconstruction schemes are removed.

The discussion above is also given quantitatively in Tables 3 and 4. Focusing on the L_1 norm, we can see the order of magnitude is similar for both methods and Reynolds numbers, where the correct flow was predicted and is between 2% and 4%. The inability of the AC method under the current set up to predict the bifurcation results in rather large errors while the FSAC-PP method shows similar errors for the bifurcated and non-bifurcated state. We can also see that the application of the Rusanov RS produces larger errors for the FSAC-PP methods and also fails to predict the bifurcation for the AC method, while the non Rusanov RS-based approached predicted the bifurcation. This can be explained by the numerical dissipation inherent to the Rusanov RS. It may seem as a disadvantage for the present results, however, its advantages become only apparent at high Reynolds numbers, where the added dissipation acts as physical dissipation where a non-dissipative interpolation scheme is used. For the present study, where laminar flows are concerned, it is not expected to see a large difference in the result. Rather, it is worth mentioning that the added dissipation does not contaminated the results significantly for low Reynolds numbers in the case of the FSAC-PP method, while the effect is felt rather strongly in the case of the AC method, where the bifurcation is no predicted. Thus, compared to the AC method, we show that our hybrid Rusanov scheme in conjunction with the MCB scheme does predict the bifurcation correctly and that the gains are significant. The numerical results could be improved by using a less dissipative RS like the Harten, Lax, van Leer (HLL) or HLL-Contact (HLLC) RS, however, Smith et al. [31] highlighted that the signal speed prediction becomes problematic. Thus, we accept the numerical dissipation by the Rusanov RS but gain conservativeness through the RS which allows to use a relative simple numerical interpolation scheme. At the same time, once real aerospace applications are of interest, the numerical dissipation provided by the Rusanov RS is sufficient to produce stable results. This, by itself, could be regarded as an alternative Implicit Large Eddy Simulation (ILES) approach where we do not take the numerical dissipation from the interpolation scheme but rather from the Riemann solver directly.

Figure 5 shows the bifurcation diagram for the AC (Figure 5a) and FSAC-PP (Figure 5b) method. Here, we have defined DX as the difference between the upper and lower reattachment length and show results obtained in the range of $10 \leq \text{Re} \leq 100$. We further show numerical results by Oliveira et al. [38] who used the SIMPLEC scheme to solve the incompressible Navier–Stokes equations and provided tabulated data on the reattachment length. As has been already observed, the bifurcation was only predicted by the non-CB and MCB scheme for the AC method while all schemes bifurcated for the FSAC-PP method. Figure 5a shows that the AC method predicted the onset of bifurcation prematurely. The FSAC-PP method in Figure 5b, on the other hand, shows that the bifurcation was predicted slightly after the critical Reynolds number. The FSAC-PP method, however, does follow the experimental and numerical data more closely, especially for higher Reynolds numbers. The reattachment length in the AC method becomes weakly invariant to the Reynolds numbers at Re = 80 while the numerical data suggest that the difference between the upper and lower reattachment point should continue to grow. The behaviour is the same for the MCB and non-CB scheme. The hyperbolic nature and thus finite information propagation speed in the AC method could explain this phenomena where the distance between the reattachment point and the upstream convex corner points becomes too large so that disturbances are not propagated fast enough for them to have an effect. In the FSAC-PP method, on the other hand, the disturbances are propagated instantaneously and so even at larger distances, or higher values of DX, the disturbances are still felt upstream. It is interesting to note here that all RS-based approaches in the FSAC-PP method do initially predict the onset of bifurcation later than the non-CB or MCB scheme. It could be argued that the inherent numerical dissipation to the RS delays bifurcation. At higher Reynolds number and sufficiently far away from the critical one, all schemes eventually converge towards the same solution.

Table 3. L_0 and L_1 error norm of the velocity profiles for the AC method using different combinations of CB schemes and the Rusanov Riemann solver at Re = 34.6 and Re = 80.

Re		No Rusanov RS			Rusanov RS	
		No CB	SCB	MCB	No CB	MCB
34.6	$L_0(u)$ [%]	5.05	4.38	4.94	4.65	4.65
	$L_1(u)$ [%]	2.97	2.18	2.89	2.36	2.36
80	$L_0(u)$ [%]	6.43	41.31	7.14	41.19	41.19
	$L_1(u)$ [%]	3.75	25.10	4.19	24.87	24.87

Table 4. L_0 and L_1 error norm of the velocity profiles for the FSAC-PP method using different combinations of CB schemes and the Rusanov Riemann solver at Re = 34.6 and Re = 80.

Re		No Rusanov RS			Rusanov RS	
		No CB	SCB	MCB	No CB	MCB
34.6	$L_0(u)$ [%]	4.10	4.98	4.19	4.05	4.03
	$L_1(u)$ [%]	1.96	2.57	2.03	1.93	1.91
80	$L_0(u)$ [%]	4.25	5.62	4.74	5.18	5.13
	$L_1(u)$ [%]	1.99	3.15	2.33	2.51	2.47

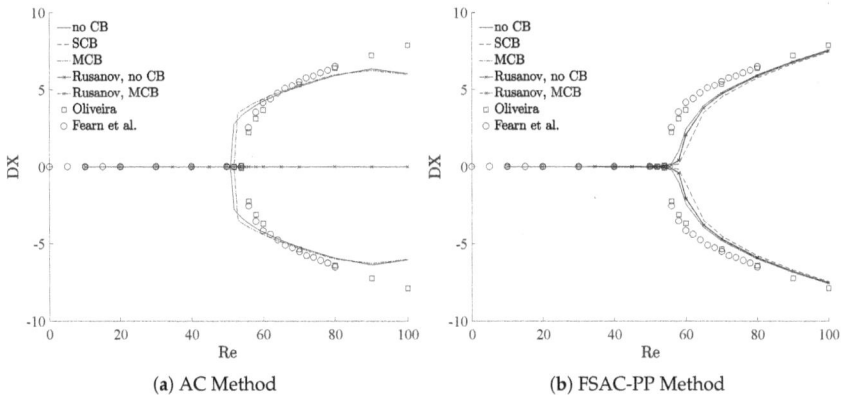

(a) AC Method   (b) FSAC-PP Method

Figure 5. Bifurcation diagram for the AC (**a**) and FSAC-PP (**b**) method using different combinations of CB schemes and the Rusanov Riemann solver.

The results in Figure 5 is summarised in Tables 5 and 6 for the AC and FSAC-PP method, respectively. Here we give both the upper and lower reattachment point and compare against the tabulated data of Oliveira [38]. We further give the number of iterations it took to get converged results up to our convergence criterion of ε. We can confirm from the data given that the non-CB and MCB scheme may indeed match the reference data more closely, especially at high Reynolds numbers. The difference are, however, minute in the case of the FSAC-PP method.

The computational cost is shown in Figure 6, which shows an interesting trend. The AC method is know to have slow convergence properties at low Reynolds numbers, see for example [22,23]. We see the same trend in Figure 6a, where the computational time required grows exponentially as the Reynolds number approaches zero. The non-CB and MCB scheme require substantially more CPU time than the SCB and RS-based approaches, however, the latter did not predict the bifurcation at all. The reason here is that the flow will always develop as a symmetrical flow, even for Reynolds number above Re_{crit}. After the residuals have dropped to about $\varepsilon = 10^{-10}$, the numerical fluctuations become small enough so that the physical fluctuations can promote a different and more stable equilibrium position. At this stage, the reattachment point at the upper and lower wall start to interact with the upstream pressure at the convex corner points and slight physical fluctuations—which obtain their energy through the non-linear term—promote the change to a non-symmetrical flow pattern, which is also discussed by Oliveira [38]. Therefore, results obtained for the bifurcated state may in general require more computational time. At the critical Reynolds number, both the AC and FSAC-PP method peak in terms of the CPU time. It is here that the physical fluctuations become just important enough for the flow to register the change to a non-symmetrical state. At higher reynolds numbers, their effects may be felt stronger and earlier during the calculation which may force the bifurcation to occur earlier and thus with increasing Reynolds numbers, the computational time decreases again. When only considering the non-CB and MCB scheme of the AC method, which eventually bifurcate, it can be seen that the same simulation using the FSAC-PP method requires 2–3 times less CPU time sufficiently far away from the critical Reynolds number. This is consistent with findings of previous works in which it was highlighted that the FSAC-PP method generally performs faster than the AC method while retaining a high level of accuracy [22,23,28,30]. At Reynolds number close to zero, the computational savings are even more significant. This was one of the reasons the FSAC-PP method was developed in the first place, to overcome the stiff nature of the AC and Pressure Projection method for low Reynolds number flows, for example in microfluidic applications, see [23] for a detailed discussion. At the onset of bifurcation, however, the AC method seems to be slightly more cost efficient than the FSAC-PP method. This can be explained by the elliptic behaviour of the Poisson solver, where any pressure change is propagated through the domain instantaneously. These pressure waves require longer to be damped while the hyperbolic behaviour of the AC method induces less pressure waves and hence a converged solution is found quicker.

Table 5. Prediction of the reattachment length X_{r1} and X_{r2} and number of iterations required for convergence for different Reynolds numbers using the AC method for various combinations of CB schemes and the Rusanov Riemann solver.

Re		No Rusanov RS			Rusanov RS		Oliveira [38]
		No CB	SCB	MCB	No CB	MCB	
10	iteration	3,631,653	690,039	4,532,716	696,382	696,476	-
	X_{r1}	2.655	1.130	2.634	1.117	1.117	1.211
	X_{r2}	2.655	1.130	2.634	1.117	1.117	1.211
20	iteration	2,113,766	163,702	1,547,448	166,414	166,411	-
	X_{r1}	3.387	1.968	3.364	1.936	1.936	2.111
	X_{r2}	3.387	1.968	3.364	1.936	1.936	2.111
30	iteration	1,221,133	107,248	785,612	106,580	106,591	-
	X_{r1}	4.213	2.879	4.194	2.828	2.828	3.080
	X_{r2}	4.213	2.879	4.194	2.828	2.828	3.080
40	iteration	791,709	100,955	478,125	100,464	100,460	-
	X_{r1}	5.085	3.824	5.070	3.753	3.753	4.075
	X_{r2}	5.085	3.824	5.070	3.753	3.753	4.075
50	iteration	568,705	91,784	330,816	91,432	91,429	-
	X_{r1}	5.990	4.785	5.985	4.697	4.697	5.080
	X_{r2}	5.990	4.785	5.985	4.697	4.697	5.081
52	iteration	858,132	91,683	318,747	90,572	90,572	-
	X_{r1}	4.330	4.978	6.169	4.886	4.886	5.279
	X_{r2}	7.089	4.978	6.169	4.886	4.886	5.285
54	iteration	630,303	91,897	476,622	90,803	90,803	-
	X_{r1}	4.088	5.172	3.838	5.076	5.076	5.445
	X_{r2}	7.369	5.172	7.452	5.076	5.076	5.523
56	iteration	553,873	92,342	430,348	91,269	91,273	-
	X_{r1}	3.939	5.366	3.749	5.266	5.266	4.440
	X_{r2}	7.604	5.366	7.663	5.266	5.266	6.678
58	iteration	533,040	97,869	403,268	96,932	96,928	-
	X_{r1}	3.839	5.561	3.683	5.457	5.457	4.107
	X_{r2}	7.814	5.561	7.857	5.457	5.457	7.208
60	iteration	515,231	99,678	383,592	98,563	98,556	-
	X_{r1}	3.770	5.755	3.635	5.648	5.648	3.935
	X_{r2}	8.008	5.755	8.040	5.648	5.648	7.609
70	iteration	455,579	101,927	328,784	100,829	100,832	-
	X_{r1}	3.629	6.732	3.550	6.605	6.605	3.669
	X_{r2}	8.843	6.732	8.844	6.605	6.605	9.019
80	iteration	411,266	104,035	304,704	104,958	104,951	-
	X_{r1}	3.626	7.713	3.562	7.566	7.566	3.658
	X_{r2}	9.553	7.713	9.538	7.566	7.566	10.060
90	iteration	388,146	110,098	292,355	109,346	109,341	-
	X_{r1}	3.668	8.697	3.610	8.529	8.529	3.708
	X_{r2}	10.039	8.697	9.876	8.529	8.529	10.930
100	iteration	365,681	110,430	295,891	109,700	109,701	-
	X_{r1}	3.730	9.683	3.676	9.493	9.493	3.781
	X_{r2}	9.771	9.683	9.681	9.493	9.493	11.660

Table 6. Prediction of the reattachment length X_{r1} and X_{r2} and number of iterations required for convergence for different Reynolds numbers using the FSAC-PP method for various combinations of CB schemes and the Rusanov Riemann solver.

Re		No Rusanov RS			Rusanov RS		Oliveira [38]
		No CB	SCB	MCB	No CB	MCB	
10	iteration	68,164	88,652	77,805	85,234	76,780	-
	X_{r1}	1.218	1.160	1.199	1.211	1.212	1.211
	X_{r2}	1.217	1.161	1.198	1.211	1.211	1.211
20	iteration	63,567	77,596	71,278	76,451	70,299	-
	X_{r1}	2.052	1.960	2.030	2.049	2.050	2.111
	X_{r2}	2.049	1.958	2.028	2.047	2.048	2.111
30	iteration	58,896	68,241	64,767	67,870	63,793	-
	X_{r1}	2.961	2.839	2.936	2.958	2.960	3.080
	X_{r2}	2.954	2.834	2.930	2.953	2.955	3.080
40	iteration	54,291	61,483	58,382	61,441	58,369	-
	X_{r1}	3.902	3.747	3.874	3.897	3.901	4.075
	X_{r2}	3.888	3.758	3.862	3.887	3.891	4.075
50	iteration	72,955	57,272	62,362	66,594	64,421	-
	X_{r1}	4.866	4.705	4.836	4.856	4.861	5.080
	X_{r2}	4.829	4.682	4.807	4.830	4.836	5.081
52	iteration	103,038	68,241	887,76	94,788	91,491	-
	X_{r1}	5.068	4.866	4.994	5.054	5.060	5.279
	X_{r2}	5.013	4.898	5.035	5.017	5.023	5.285
54	iteration	160,834	98,874	134,456	143,614	139,714	-
	X_{r1}	5.190	5.094	5.239	5.200	5.207	5.445
	X_{r2}	5.279	5.048	5.176	5.257	5.262	5.523
56	iteration	323,916	158,368	243,044	260,922	257,718	-
	X_{r1}	5.333	5.221	5.342	5.368	5.374	4.440
	X_{r2}	5.522	5.299	5.461	5.475	5.480	6.678
58	iteration	878,294	334,541	768,416	865,548	901,939	-
	X_{r1}	5.084	5.362	5.377	5.415	5.405	4.107
	X_{r2}	6.095	5.534	5.802	5.805	5.825	7.208
60	iteration	355,946	888,435	481,042	514,272	486,624	-
	X_{r1}	4.427	5.051	4.641	4.655	4.628	3.935
	X_{r2}	6.861	6.149	6.688	6.700	6.726	7.609
70	iteration	114,816	126,413	119,971	127,121	125,149	-
	X_{r1}	3.734	3.782	3.779	3.785	3.782	3.669
	X_{r2}	8.531	8.278	8.499	8.475	8.485	9.019
80	iteration	91,835	94,097	92,946	99,882	98,683	-
	X_{r1}	3.653	3.650	3.679	3.681	3.679	3.658
	X_{r2}	9.622	9.416	9.618	9.572	9.579	10.060
90	iteration	96,853	95,949	97,232	114,014	105,340	-
	X_{r1}	3.676	3.656	3.696	3.703	3.702	3.708
	X_{r2}	10.525	10.346	10.539	10.480	10.486	10.930
100	iteration	106,408	109,105	119,534	108,770	102,731	-
	X_{r1}	3.749	3.705	3.762	3.754	3.753	3.781
	X_{r2}	11.308	11.147	11.336	11.263	11.269	11.660

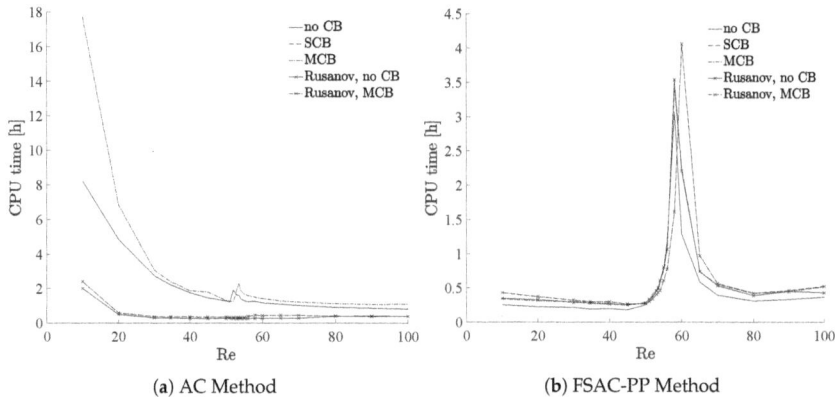

(a) AC Method   (b) FSAC-PP Method

Figure 6. Total number of iterations required for the AC (**a**) and FSAC-PP (**b**) method for different Reynolnds numbers.

6. Conclusions

In this work, we presented results to predict the highly non-linear behaviour produced by the Navier–Stokes equations at low Reynolds numbers inside a channel with a sudden expansion. In particular, we investigated the sub- and super-critical range of Reynolds numbers where the flow bifurcates from a symmetric to a non-symmetric state. We investigated the performance of a non-characteristic-based (CB) scheme, single- and multi-directional characteristics-based scheme (SCB/MCB), as well as the Rusanov Riemann solver (RS) and combinations of these schemes and tested all approaches with the Artificial Compressibility (AC) and Fractional-Step, Artificial Compressibility with Pressure-Projection (FSAC-PP) method.

We found that only the non-CB and the MCB scheme were capable of predicting the bifurcation using the AC method where the RS-based approaches showed too much numerical dissipation to correctly predict the flow patterns. A significant difference between the SCB and MCB scheme could be observed in the AC method, where we showed that the multi-directional nature of the MCB scheme is required to predict the bifurcation at all. The added Rusanov RS showed too much numerical dissipation for the current approach to predict the bifurcation and similar results were obtained using the hybrid Rusanov and MCB scheme. In the FSAC-PP method, however, all schemes correctly predicted the symmetry breaking and overall showed better agreement with reference data in terms of the reattachment length. The incompressible flow method itself could overcome the inherent numerical dissipation of the Rusanov RS, as well as the SCB scheme which uses Godunov's RS by default, showing that the mathematical classification of the method's partial differential equations play a dominant role. The fully hyperbolic behaviour of the AC method was not always capable of capturing the bifurcation phenomena correctly while the mixed hyperbolic-elliptic equations of the FSAC-PP method ensured always a physically correct solution. This comes at a slightly increased computational cost near the bifurcation, however, sufficiently far away from the critical Reynolds number, the FSAC-PP method required 2–3 times less CPU resources compared to the AC method.

Aerospace applications are presented with flow features that are highly non-linear, as discussed in Section 1. The bifurcation shares many similarities to the onset to turbulence and predicting both phenomena correctly is mandatory to gain highly accurate computations in which stall, strong crossflow gradients and lift slope hysteresis are of interest. We have demonstrated here that the success of predicting non-linear flow features is not just scheme dependent, but the incompressible method used for the computation also plays a dominant role. The MCB scheme by itself showed that its multi-directional nature was capable of predicting non-linear phenomena correctly, regardless of the incompressible flow method used, while the SCB scheme was only capable of predicting correctly the flow phenomena when the FSAC-PP method was used. Although only laminar flows were investigated in this study, at higher Reynolds numbers, as is characteristic of aerospace applications, the flow becomes turbulent. In these flow regimes, the numerical dissipation produced by different schemes and RS becomes important. The Rusanov RS provides a sufficient amount of inherent numerical dissipation to tackle high Reynolds number turbulent flows. In future work, we will investigate the proposed framework for high Reynolds number flows, however, the framework present in the current study is directly applicable to such flows for aerospace applications.

Acknowledgments: The present research work was financially supported by the Centre for Computational Engineering Sciences at Cranfield University under the project code EEB6001R. The authors would like to acknowledge the IT support and the use of the High Performance Computing (HPC) facilities at Cranfield University, UK. We would also like to acknowledge the constructive comments of the reviewers of the Aerospace Journal.

Author Contributions: Tom-Robin Teschner, László Könözsy and Karl W. Jenkins contributed equally to this article.

Conflicts of Interest: The authors declare no conflict of interest.

Abbreviations

The following abbreviations are used in this manuscript:

AC	Artificial Compressibility
CB	Characteristics-based
FSAC-PP	Fractional-Step, Artificial Compressibility with Pressure-Projection
FS-PP	Fractional-Step, Pressure-Projection
LES	Large-Eddy Simulation
MCB	multi-directional characteristics-based
RANS	Reynolds-averaged Navier–Stokes
RS	Riemann solver
SCB	single-directional characteristics-based

References

1. Mittal, S.; Saxena, P. Prediction of Hysteresis Associated with the Static Stall of an Airfoil. *AIAA J.* **2000**, *38*, 933–935, doi:10.2514/2.1051.
2. Truong, K. Modeling Aerodynamics, Including Dynamic Stall, for Comprehensive Analysis of Helicopter Rotors. *Aerospace* **2016**, *4*, 1–24, doi:10.3390/aerospace4020021.
3. Traub, L. Semi-Empirical Prediction of Airfoil Hysteresis. *Aerospace* **2016**, *3*, 1–8, doi:10.3390/aerospace3020009.
4. Panaras, A. Turbulence Modeling of Flows with Extensive Crossflow Separation. *Aerospace* **2015**, *2*, 461–481, doi:10.3390/aerospace2030461.
5. Drikakis, D.; Rider, W. *High-Resolution Methods for Incompressible and Low-Speed Flows*; Springer: Heidelberg, Germany, 2005; ISBN 978-3-540-22136-4.
6. Toro, E.F. *Riemann Solvers and Numerical Methods for Fluid Dynamics*; Springer: Heidelberg, Germany, 2009; ISBN 978-3-540-25202-3.
7. Rusanov, V.V. Characteristics of the General Equations of Gas Dynamics. *Noi Math. Math. Fiz.* **1963**, *3*, 508–527, doi:10.1016/0041-5553(63)90294-5.
8. Chorin A.J. A Numerical Method for Solving Incompressible Viscous Flow Problems. *J. Comput. Phys.* **1967**, *2*, 12–26, doi:10.1006/jcph.1997.5716.
9. Drikakis, D.; Govatsos, P.A.; Papantonis, P.E. A Characteristic-based Method for Incompressible Flows. *Int. J. Numer. Methods Fluids* **1994**, *19*, 667–685, doi:10.1002/fld.1650190803.
10. Zienkiewicz, O.C.; Codina, R. A General Algorithm for Compressible and Incompressible Flow—Part I. The Split, Characteristic-Based Scheme. *Int. J. Numer. Methods Fluids* **1995**, *20*, 869–885, doi:10.1002/fld.1650200812.

11. Zienkiewicz, O.C.; Morgan, K.; Satya Sai, B.V.K.; Codina, R.; Vasquez, M. A General Algorithm for Compressible and Incompressible Flow—Part II. Tests on the Explicit Form. *Int. J. Numer. Methods Fluids* **1995**, *20*, 887–913, doi:10.1002/fld.1650200813.

12. Neofytou, P. Revision of the Characteristics-based Scheme for Incompressible Flows. *J. Comput. Phys.* **2007**, *222*, 475–484, doi:10.1016/j.jcp.2006.10.009.

13. Su, X.; Zhao, Y.; Huang, X. On the Characteristics-based ACM for Incompressible Flows. *J. Comput. Phys.* **2007**, *227*, 1–11, doi:10.1016/j.jcp.2007.08.009.

14. Razavi, S.E.; Zamzamian, K.; Farzadi, A. Genuinely Multidimensional Characteristic-based Scheme for Incompressible Flows. *Int. J. Numer. Methods Fluids* **2008**, *57*, 929–949, doi:10.1002/fld.1662.

15. Zamzamian, K.; Razavi, S.E. Multidimensional Upwinding for Incompressible Flows based on Characteristics. *J. Comput. Phys.* **2008**, *227*, 8699–8713, doi:10.1016/j.jcp.2008.06.018.

16. Hashemi, M.Y.; Zamzamian, K. A Multidimensional Characteristic-Based Method for Making Incompressible Flow Calculations on Unstructured Grids. *J. Comput. Appl. Math.* **2014**, *259*, 752–759, doi:10.1016/j.cam.2013.06.008.

17. Hashemi, M.Y.; Zamzamian, K. Efficient and Non-Reflecting Far-Field Boundary Conditions for Incompressible Flow Calculations. *Appl. Math. Comput.* **2014**, *230*, 248–258, doi:10.1016/j.amc.2013.12.089.

18. Zamzamian, K.; Hashemi, M.Y. Multidimensional Characteristic-Based Solid Boundary Condition for Incompressible Flow Calculations. *Appl. Math. Model.* **2015**, *39*, 7032–7044, doi:10.1016/j.apm.2015.02.043.

19. Fathollahi, R.; Zamzamian, K. An Improvement for Multidimensional Characteristic-based Scheme by Using Different Selected Waves. *Int. J. Numer. Methods Fluids* **2014**, *76*, 722–736, doi:10.1002/fld.3958.

20. Razavi, S.E.; Adibi, T. A Novel Multidimensional Characteristic Modeling of Incompressible Convective Heat Transfer. *J. Appl. Fluid Mech.* **2016**, *9*, 1135–1146, doi:10.18869/acadpub.jafm.68.228.24295.

21. Razavi, S.E.; Hanifi, M. A Multi-Dimensional Virtual Characteristic Scheme for Laminar and Turbulent Incompressible Flows. *J. Appl. Fluid Mech.* **2016**, *9*, 1579–1590, doi:10.18869/acadpub.jafm.68.235.25149.

22. Könözsy, L. Multiphysics CFD Modelling of Incompressible Flows at Low and Moderate Reynolds Numbers. Ph.D. Thesis, Cranfield University, Cranfield, UK, 2012.

23. Könözsy, L.; Drikakis, D. A Unified Fractional-Step, Artificial Compressibility and Pressure-Projection Formulation for Solving the Incompressible Navier–Stokes Equations. *Commun. Comput. Phys.* **2014**, *16*, 1135–1180, doi:10.4208/cicp.240713.080514a.

24. Chorin, A.J. Numerical Solution of the Navier–Stokes Equations. *Math. Comput.* 22, 745–762, doi:10.1090/S0025-5718-1968-0242392-2.

25. Temam, R. Sur l'approximation de la Solution des Equations de Navier–Stokes par la Methode des pas Fractionnaires. *Arch. Ration. Mech. Anal.* **1969**, *32*, 377–385.

26. Könözsy, L.; Drikakis, D. A Coupled High-Resolution Fractional-Step Artificial Compressibility and Pressure-Projection Formulation for Solving Incompressible Multi-Species Variable Density Flow Problem at Low Reynolds Numbers. In Proceedings of the European Congress on Computational Methods in Applied Sciences and Engineering, Vienna, Austria, 10–14 September 2012.

27. Könözsy, L.; Drikakis, D.; Ashcroft, M.; Dixon, A.; Perrson, J. Experimental and Numerical Investigation for Trapping and Positioning Cryogenic Propellants. In Proceedings of the 8th European Symposium on Aerothermodynamics for Space Vehicles, Lisbon, Portugal, 3–5 March 2015.

28. Teschner, T.-R.; Könözsy, L.; Jenkins, K.W. Numerical Investigation of an Incompressible Flow over a Backward Facing Step Using a Unified Fractional-Step, Artificial Compressibility and Pressure-Projection (FSAC-PP) Method. In Proceedings of the MultiScience—XXX. microCAD International Multidisciplinary Scientific Conference, Miskolc, Hungary, 21–22 April 2016.

29. Tsoutsanis, P.; Kokkinakis, I.; Ioannis, W.; Könözsy, L.; Drikakis, D.; Williams, R.J.R.; Youngs, D.L. Comparison of Structured- and Unstructured-Grid, Compressible and Incompressible Methods Using the Vortex Pairing Problem. *Comput. Methods Appl. Mech. Eng.* **2015**, *293*, 207–231, doi:10.1016/j.cma.2015.04.010.

30. Teschner, T.-R.; Könözsy, L.; Jenkins, K.W. On Godunov-Type Multi-Directional Characteristic-based Schemes for Hyperbolic Incompressible Flow Solvers. In Proceedings of the IV ECCOMAS Young Investigator Conference, Milan, Italy, 13–15 September 2017.

31. Smith, K.; Teschner, T.-R.; Könözsy, L. On Approximate Riemann Solvers within the Concept of the Unified Fractional-Step, Artificial Compressibility and Pressure Projection (FSAC-PP) Method. In Proceedings of the MultiScience—XXX. microCAD International Multidisciplinary Scientific Conference, Miskolc, Hungary, 21–22 April 2016.

32. Zucrow, M.J.; Hoffman, J.D. *Gas Dynamics, Vol 1*; John Wiley & Sons, Inc.: Hoboken, NJ, USA, 1976; ISBN 978-0471984405.

33. Zucrow, M.J.; Hoffman, J.D. *Gas Dynamics, Vol 2: Multidimensional Flow*; John Wiley & Sons, Inc.: Hoboken, NJ, USA, 1977; ISBN 978-0471018063.

34. Delaney, R.A. A Second-Order Method of Characteristics for Two-Dimensional Unsteady Flow with Application to Turbomachinery Cascades. Ph.D. Thesis, Iowa State University, Ames, IA, USA, 1974.

35. Rusanov, V.V. The Calculation of the Interaction of Non-Stationary Shock Waves and Obstacles. *USSR Comput. Math. Math. Phys.* 1961, *1*, 304–320.

36. Davis, S. Simplified Second-Order Godunov-Type Methods. *SIAM J. Sci. Stat. Comput.* **1988**, *9*, 445–473, doi:10.1137/0909030.

37. Roache, P.J. Perspective: A Method for Uniform Reporting of Grid Refinement Studies. *J. Fluids Eng.* **1994**, *116*, 405–413, doi:10.1115/1.2910291.

38. Oliveira, P.J. Asymmetric Flows of Viscoelastic Fluids in Symmetric Planar Expansion Geometries. *J. Non-Newton. Fluid* **2003**, *114*, 33–63, doi:10.1016/S0377-0257(03)00117-4.

39. Fearn, R.M.; Mullin, T.; Cliffe, K.A. Nonlinear Flow Phenomena in a Symmetric Sudden Expansion. *J. Fluid Mech.* **1990**, *211*, 595–608, doi:10.1017/S0022112090001707.

40. Teschner, T.-R.; Könözsy, L.; Jenkins, K.W. A Three-Stage Algorithm for Solving Incompressible Flow Problems. In Proceedings of the MultiScience—XXXI. microCAD International Multidisciplinary Scientific Conference, Miskolc, Hungary, 20–21 April 2017.

41. Drikakis, D. Bifurcation Phenomena in Incompressible Sudden Expansion Flows. *Phys. Fluids* **1997**, *9*, 76–87, doi:10.1063/1.869174.

Project Report

Aircraft Geometry and Meshing with Common Language Schema CPACS for Variable-Fidelity MDO Applications

Mengmeng Zhang [1,*], **Aidan Jungo** [2], **Alessandro Augusto Gastaldi** [1] and **Tomas Melin** [1]

1   Airinnova AB, 18248 Stockholm, Sweden; aa.gastaldi@protonmail.ch (A.A.G.); tomas.melin@airinnova.se (T.M.)
2   CFS Engineering, 1015 Lausanne, Switzerland; aidan.jungo@cfse.ch
*   Correspondence: mengmeng.zhang@airinnova.se

Received: 1 March 2018; Accepted: 19 April 2018; Published: 24 April 2018

Abstract: This paper discusses multi-fidelity aircraft geometry modeling and meshing with the common language schema CPACS. The CPACS interfaces are described, and examples of variable fidelity aerodynamic analysis results applied to the reference aircraft are presented. Finally, we discuss three control surface deflection models for Euler computation.

Keywords: geometry; meshing; aerodynamics; CPACS; MDO; VLM; Euler; CFD; variable fidelity

---

1. Introduction

The design of aircraft is inherently a multi-disciplinary undertaking, during which data and information must be exchanged between multiple teams of engineers, each with expertise in a specific field. Managing the transmission, possibly translation and storage of data between collaborating groups is complex and error-prone. The adoption of a standardized, data-centric scheme for storage of all data improves consistency and reduces the risk of misconceptions and conflicts. In order to achieve this effectively, an initial effort must be made to develop suitable interfaces between the analysis modules and the data archive.

Furthermore, each phase of the design process poses different requirements on the fidelity and resolution of the design and analysis tools. For stability and control analysis, as well as for flight simulation, look-up tables for aerodynamic forces, moments and derivatives need to be generated. Different flight analysis tools require different tables/input formats. For example, the flight analyzer and simulator PHALANX [1–4] developed by Delft University of Technology requires a set of three-dimensional tables of force and moment coefficients with the effect of each control channel acting individually. Multi-fidelity aerodynamic modeling aims to cover the flight state parameter space of the entire flight envelope with an optimal distribution of computational resources. This again requires a standardized, data-centric scheme to host the data, which can be used for variable fidelities.

The label Li, where $i = 0, 1, 2, 3$, is used to classify the fidelity level of a computational model and its software implementation:

L0:   handbook methods, based on statistics and/or empirical design rules;

L1:   based on simplified physics, can model and capture a limited amount of effects. For example, the linearized-equation models, the Vortex-Lattice Method (VLM) or the panel method in aerodynamics;

L2:   based on accurate physics representations. For example, the non-linear analysis, Euler-based CFD;

L3:   represents the highest end simulations, usually used to capture detailed local effects, but do not allow wide exploration of the design space due to computational cost. Additionally, the modeling may require extensive ad hoc manual intervention. For example, the highest fidelity methods, RANS-based CFD.

To construct a reasonable variable fidelity CFD analysis system, one should consider the variable fidelity of the geometrical representations corresponding to the CFD tools. The level of detail in the geometry gathered from a CAD system needs to match the CFD model fidelity. The chosen high fidelity model must be as accurate as possible and can reflect all considered complex flow characteristic; the chosen low-fidelity model must reflect the basic flow characteristics and be as effective as possible. In the conceptual design stage, the usual practice, for example, in the RDS [5], the AAA [6] and the VSP [7] software systems, is to use a purpose-specific CAD that is simpler than the commercial systems, and fewer parameters need to be used for the configuration layout at this stage in the design cycle [8]. However, for some innovative configurations, different ranges of flight conditions or more detailed analyses, the simplified CAD is not sufficient for a higher fidelity CFD analysis; thus, an enriched geometry definition with more parameters is needed. The Common Parametric Aircraft Configuration Schema (CPACS) [9,10], defining the aircraft configuration parametric information in a hierarchical way, gives the opportunity to incorporate different fidelity CFD tools with one single CPACS file. For different fidelity tools to be used, the corresponding geometry information can be imported/retrieved from the common CPACS file to match the model fidelity.

SUAVE, Stanford University Aerospace Vehicle Environment [11–13], which is also a multi-fidelity design framework developed at Stanford University, stores the aircraft geometry information using an inherent defined data class, which can be easily modified. The aerodynamic solutions can be generated from simple models within SUAVE or easily imported from external sources like CFD or wind tunnel results. The aircraft analysis in SUAVE is calculated with a so-called "fidelity zero" VLM to predict lift and drag, with a number of corrections such as the compressibility drag correction, parasite drag correction, etc. [12], to adapt the VLM prediction to a wider range (transonic and supersonic flow regions). It incorporates the "multi-fidelity" aerodynamics through the provided response surface by combining the different fidelity data. However, currently, SUAVE is still working on connecting higher fidelity models directly to it; the response surfaces are only available to incorporate higher fidelity lift and drag data from the external sources [12]. At this point, one cannot guarantee that the geometry information used for different fidelity tools is consistent during data exchanging, transferring and translating. Moreover, the prediction is only limited to lift and drag, so that it might not be easy for engineers to look into the physical details for a better design, for example, the pressure isobars and distributions, the laminar flows, transitions and the shock forming, etc. Thus, a dataset that can store complete and consistent information for different fidelity tools to solve the physical flows is desired. The CPACS-based multi-fidelity aerodynamic tools show a great consistency due to the one data-centric schema, and the automation of the progressive process can thus be implemented and realized with minimum data loss.

With all the computed aerodynamic data at hand, an important task is to construct surrogate models that integrate all analysis results computed by tools of different fidelity. Such data fusion applications are enabled by standardization of the data—format, syntax and semantics—of the aerodynamic simulation tools. The work in [14] describes the workflow of an automatic data fusion process for CPACS [9,10]. The application was developed in the EU research project Aircraft 3rd Generation MDO for Innovative Collaboration of Heterogeneous Teams of Experts (AGILE) [15], where every module (the aerodynamic module, the sampling module, the surrogate modeling module) communicates by CPACS files.

This paper will address other aspects of the work in [14], namely how the different fidelity tools in the aerodynamic module communicate and how a look-up table of the aero-dataset can be obtained automatically from the tools (L1 and L2 in this paper). Section 2 describes the CPACS file definition in more detail, especially the geometry definitions, which are important for CFD mesh generators. Section 3 details the CPACS interfaces for variable fidelity analysis. Section 4 gives an overview of the CFD flow solvers used in the work. Section 5 presents the applications to the test case using variable fidelity tools. Section 6 discusses different modeling approaches for control surfaces used for Euler simulations, and finally, Section 7 summarizes the conclusions of the work.

1.1. Background

AGILE is an EU-funded Horizon 2020 project coordinated by the Institute of Air Transportation Systems of the German Aerospace Center (DLR). Its objective is to implement a third generation multidisciplinary optimization process through efficient collaboration by international multi-site design teams. The 19 partners bring different knowledge and competences regarding aircraft design and optimization. As mentioned above, such a collaboration is enabled by the adoption of a common data storage format. To this end, AGILE relies on the XML format CPACS (Common Parametric Aircraft Configuration Schema) [9,10] in development at DLR since 2005.

The RCE (Remote Component Environment) integration environment and workflow manager [16] controls and executes the sequence of analysis modules and manages the data transport and translation, as well as logging the process. RCE makes it easy to set up an MDO workflow also with modules running on remote hosts. That is handled by the BRICS (Building blocks for mastering network Restrictions involved in Inter-organisational Collaborative engineering Solutions) [17] system, which supports remote execution and data transport. The request can be with "engineer in the loop" for a remote expert to run the calculation or for an automated workflow to be run without user intervention. The input is generally a CPACS file containing all the information required. The new data generated are added to the CPACS file and sent back to the requester. More details about the AGILE collaborative approach can be found in [18,19].

The variable fidelity aerodynamic tools read a CPACS file, analyze the corresponding information extracted from the file, run the calculation and store the new data (e.g., aero-data tables) back to the CPACS file. CPACS supports a very flexible user-defined node feature (cpacs.toolspecific) to handle parameters for the computational models, which are relevant only for a specific tool [20].

1.2. Aerodynamic Model Description

The test case is the reference aircraft used in AGILE, a regional jet-liner, which was analyzed and simulated using the AGILE MDO system, without experimental data. This virtual aircraft is similar to an Airbus 320 or a Boeing 737. The reference aircraft is defined in CPACS [9] format, shown in Figure 1. Its aspect ratio is 9.5, and the detailed information of the airfoils along the aircraft span is shown in Figure 2. Figure 2a shows the plots of the airfoil along the three stations of the semi-span ($b/2$), with the root at 0%, the kink at 40% and the tip at 100%. Figure 2b shows the maximum thickness and cambers per chord along the semi-span, as well as the corresponding locations of the local chords. It should be noted that the design exercises are carried out as if in an early design stage, so for instance, no engine is modeled. The configuration was also used in previous studies, to benchmark the conceptual design software CEASIOM [20] and to validate the AGILE data fusion tool [14] for building multi-fidelity aero-datasets. Some of the results shown in this paper are consistent with the previous simulations in [14,20], that the configuration is unchanged and the same CPACS file for geometry definition is used to assure a consistent and continuous investigation of the tools and methods.

Figure 1. The reference aircraft, rendered by the CPACS visualization tool TIGLViewer [21].

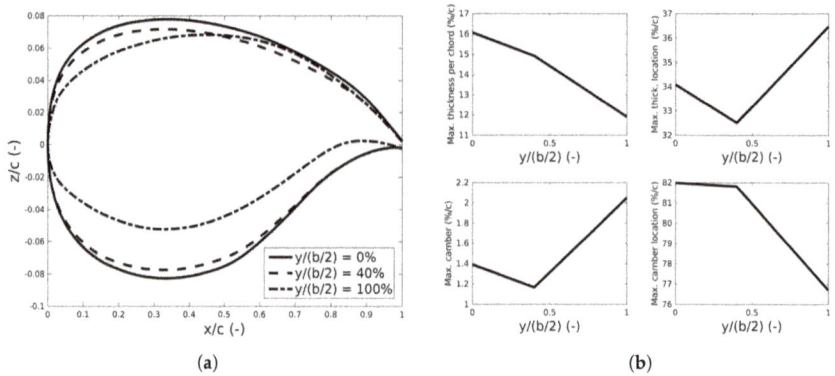

Figure 2. The airfoil details of the reference aircraft. (**a**) The airfoils of the reference aircraft; (**b**) the thickness and camber information of the airfoils.

2. CPACS File Description

2.1. The CPACS Hierarchical Data Definition Structure

Thanks to its hierarchical structure, CPACS is capable of hosting the entire aircraft geometry, as well as additional design information relating to flight missions, airports, propulsion systems and aerodynamic datasets; see Figure 3.

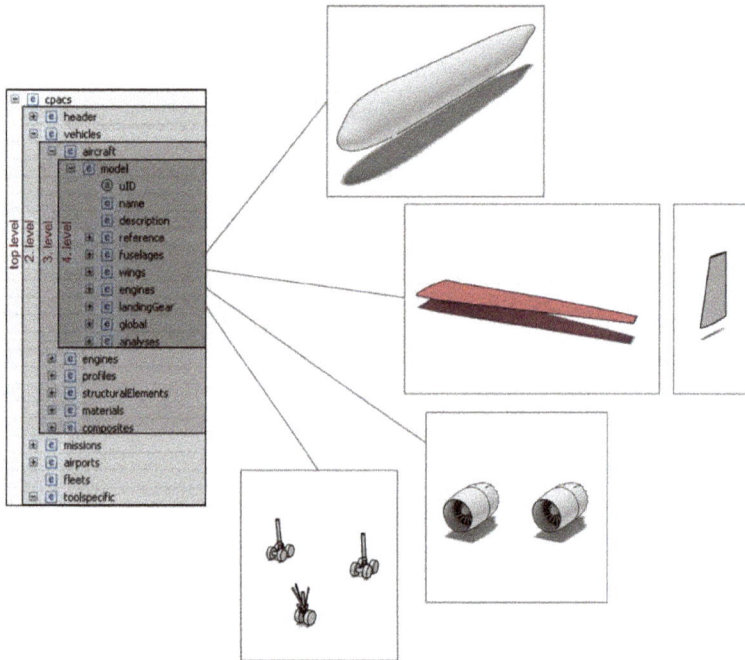

Figure 3. The CPACS hierarchical structure (image from the CPACS website [9]).

The hierarchical data structure is used to define the order of "construction" of each aircraft component (fuselage, wing, etc.). Figure 4 shows, for example, the construction of a wing geometry from the CPACS data. Other parts of the aircraft are treated similarly. For wings, the construction begins with ordered lists of points, which define airfoils. A library of airfoils identified by `profileUIDs` can be stored in the CPACS geometry definition and from these, a list of wing `elements`. An `element` is defined by its `profileUID` and a transformation: the scaling along coordinate directions, a 3D rotation and a translation. Two such elements define a `section`, the `positioning` of which is effected by a `length`, `sweep angle` and a `dihedral angle`. The sections are assembled to form a wing, to which it is possible once more to apply a transformation. Symmetries can be used to create instances of wings already defined, A single CPACS file holds a set of named lifting surfaces defined in this way. It must be noted that a given wing geometry allows multiple distinct CPACS definitions.

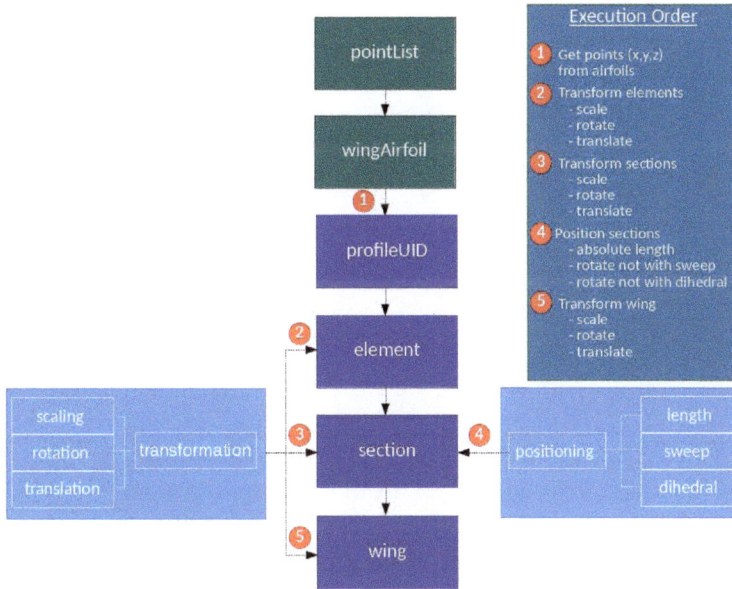

Figure 4. Adapted from CPACS documentation [9]; schema for the construction of an aircraft wing from its XML file definition.

2.2. The CPACS Control Surface Definition

On each wing, several types of control surfaces may be defined: leading edge devices, spoilers and Trailing Edge Devices (TEDs). A detailed explanation of their definition is given here only for TEDs. It is analogous for the other devices. In order to define a TED, a `componentSegment` first needs to be created. Each `componentSegment` is defined from two, not necessarily contiguous, wing elements. Each wing must have at least one `componentSegment` to define the wing structure, fuel tanks, control surfaces, etc. Each corner of the outer shape of the control surface is defined by its relative position in the span- and chord-wise directions of the `componentSegment`, as shown in Figure 5, requiring eight values to be specified. For TEDs, corner points that are not explicitly defined lie on the trailing edge of the wing. In addition to these points, the `hingeLine` must also be defined, by the relative position of its inner and outer points in the span- and chord-wise directions, as well as their "vertical" position from 0 = lower wing surface to 1 = upper wing surface.

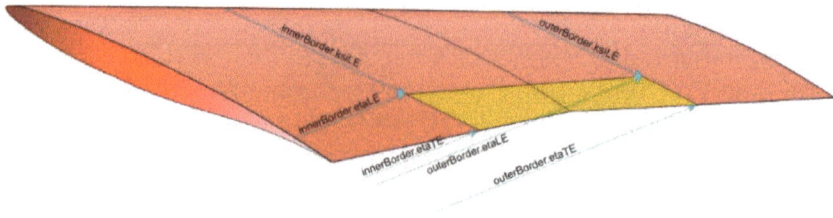

Figure 5. CPACS trailing edge device definition for node outerShape [9,22].

A control surface movement, called a step, is defined by an angle of rotation around the hinge line and a translation of the rotated surface, to allow the definition of flap movements. A deployment is then defined by interpolation in the table of steps.

The deflected control surfaces are modeled without gaps in this work. Thus, the coordinates of a control surface quadrilateral (see a TED example in Figure 5) are imported from the outerShape node affiliated with each defined control surface. The information is then interpreted by the different models in the CFD tools.

3. Geometry and CPACS Interfaces for Variable Fidelity Tools

Our focus is on multi-fidelity aerodynamic analysis, as a sub-task of MDO. The CPACS database acts as a single source of information in our multi-site collaborative environment.

Aircraft geometry modeling and mesh generation tools for L1 and L2, respectively, Tornado [23] and Sumo/TetGen [24,25], will be discussed next. Tornado is a VLM implementation for assessing aero-forces and moments on rigid lifting surfaces. Sumo/TetGen is an automatic volume mesh generator for CFD. It is fully automatic for the generation of isotropic tetrahedral grids for Euler solvers. Its Pentagrow [8] module provides semi-automatic mesh generation for RANS.

3.1. CPACS-Tornado Interface

Tornado [23], originally written in MATLAB, computes the aerodynamic coefficients and their first order derivatives for lifting surfaces at low speeds. The lifting surfaces are modeled as cambered lamina. The horseshoe vortices can be defined with seven segments to model the geometry of trailing edge movable surfaces. Leading edge movable surfaces can be similarly modeled, but seldom are, since such devices are for high-lift, high-alpha, augmentation, which VLM cannot reliably predict. The steady wake can be chosen fixed relative to the wing or to follow the free stream. Effects of compressibility at high Mach numbers (<0.75) are included through the Prandtl–Glauert correction [26]. The induced drag can be calculated by the Kutta–Joukowski law (default) or Trefftz-plane integration [27]. In the latest version, some additional features are included:

- Aircraft configuration visualization including fuselage representation and control surface identifications;
- Fast MEX-compiled version of core-functions for matrix computations;
- All-moving surfaces and overlapped movable surfaces.

Tornado can import/export CPACS files via a separate wrapper also written in MATLAB. The wrapper reads the geometric information, as well as the paneling and flight conditions from CPACS, translates them into the Tornado native data structures and writes the computed results back to CPACS. Figure 6 shows the visualization of the configuration and panel distributions for the reference aircraft.

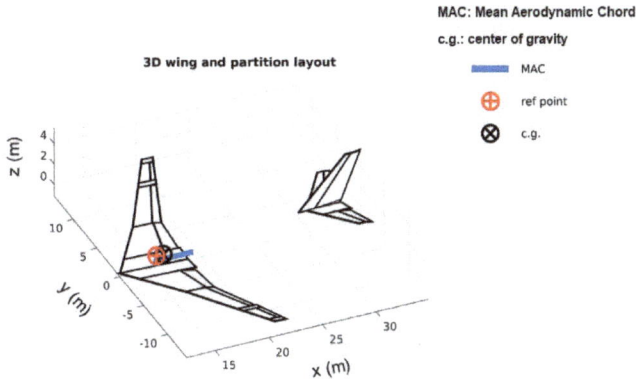

Figure 6. Tornado partition layout with control surfaces for the reference aircraft, imported from CPACS.

3.1.1. PyTornado: A VLM Solver with Native CPACS Compatibility

The Tornado internal geometry definition differs from the hierarchical geometry definition of CPACS. This is the leading motivation for the development of a VLM solver with native CPACS compatibility. The outcome is the PyTornado implementation written in Python and C++.

It inherits the essential analysis capabilities of the mature MATLAB Tornado code. Nonetheless, it can be considered as an independent program for VLM aerodynamics, with its own definition of input and output systems.

PyTornado is structured as two parts:

- A Python wrapper, dedicated to high-level tasks such as communication with CPACS, pre- and post-processing for VLM, as well as visualization of the model and generated results; see Figure 7a,b,
- The actual VLM solver, re-structured and re-written in C++ from the MATLAB Tornado VLM solver with performance in mind.

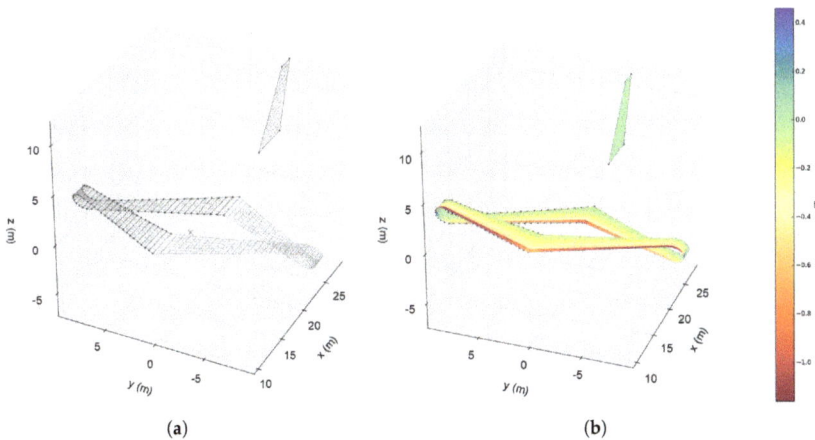

(a) (b)

Figure 7. Panel distributions and the C_p visualization in PyTornado for a box-wing aircraft, imported from CPACS. (a) Panel layout; (b) the C_p simulation for U = 100 m/s, $\alpha = 5°$.

The user or an external program can use PyTornado as a computational service through the wrapper, which controls the execution steps.

PyTornado is a lightweight, fast and flexible VLM code. Its native compatibility with CPACS and the choice of Python as a main programming language make it a promising candidate for effective integration in larger analysis and optimization frameworks. Future development will be aimed at extending the Python wrapper functionality and completing the C++ solver with further analysis features. The core functions of the solver are exposed to the wrapper through the built-in C++ API of Python and NumPy (The fundamental package for scientific computing with Python, http://www.numpy.org/). Thus, data transfer between the components occurs effectively in-memory. This design leverages the performance of C++ where required and the flexibility of Python for its high-level features and interface. Seamless integration of PyTornado with CPACS is enabled by its internal geometry definition, which closely corresponds to the hierarchical structure of the file format. Thus, operations on engineering/design parameters in CPACS can be translated to geometric data for VLM analysis without additional user effort.

Figure 8 shows the computational workflows. PyTornado, in its present state of completeness, is already an improvement over the MATLAB implementation both in performance and flexibility. The seamless integration of CPACS is a significant merit, leaving it open for extension and coupling with other tools, e.g., for structural analysis. PyTornado is currently still under development for further validations, with more features to be imported from CPACS files.

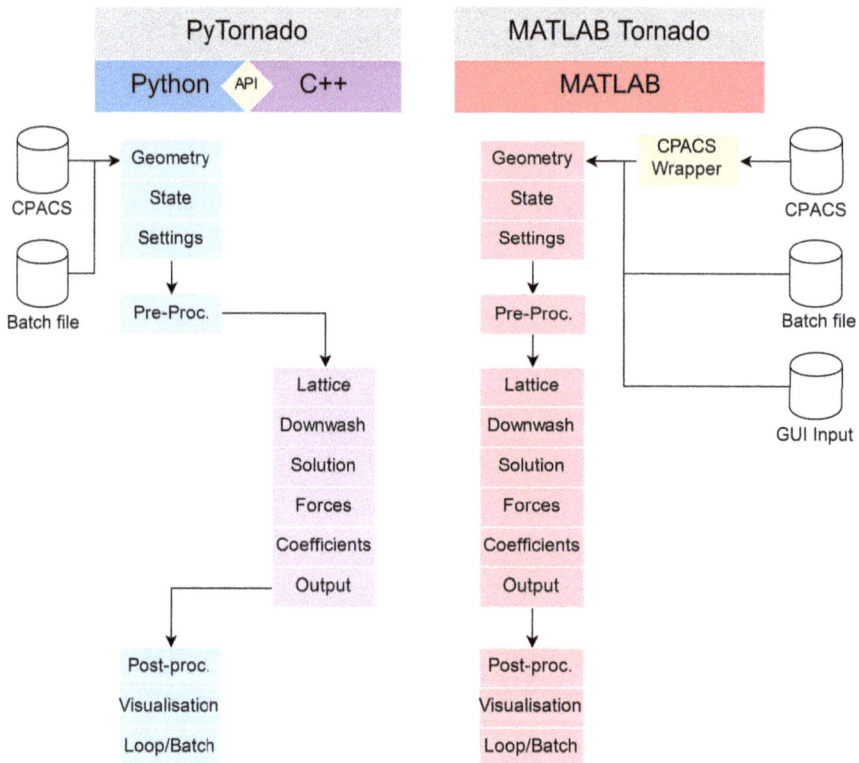

Figure 8. Tornado program workflow: user interface in Python and the core functions in C++.

3.2. CPACS-Sumo Interface

The Euler (L2) and RANS (L3) flow models need a computational mesh adapted to the solver fidelity. This section introduces the mesh generation tool Sumo for L2 analysis and its interface with CPACS. Sumo can also generate RANS (L3) meshes with Pentagrow [8]; note that the RANS simulations are only used for validation in this paper (see Section 5.1).

3.2.1. Sumo: A Gateway from CPACS to Higher-Fidelity Aerodynamics

Sumo [24] is a graphic tool for rapid modeling of aircraft geometries and automatic unstructured surface mesh generation. It is not a full-fledged CAD system, but rather an easy-to-use sketchpad, highly specialized towards aircraft configurations in order to streamline the workflow. Isotropic tetrahedral volume meshes for Euler computation can be generated from the surface mesh, by the tetrahedral mesh generator TetGen [25].

Pentahedral boundary layer elements for RANS solvers can also be (semi-)automatically generated by the Pentagrow [8] module in Sumo after the surface mesh is generated, before creation of the volume mesh by TetGen. Pentagrow sets up the prismatic element layers on the configuration surface from a configuration file with a list of user-defined parameters such as the first cell height, the total number of layers, the growth rate, etc. The volume mesh can be exported in various formats including CGNS (the CFD General Notation System), TetGen's plain ASCII format and native formats for the CFD solvers Edge [28] and SU2 [29–31]. Mesh examples are shown in Figure 9a,b.

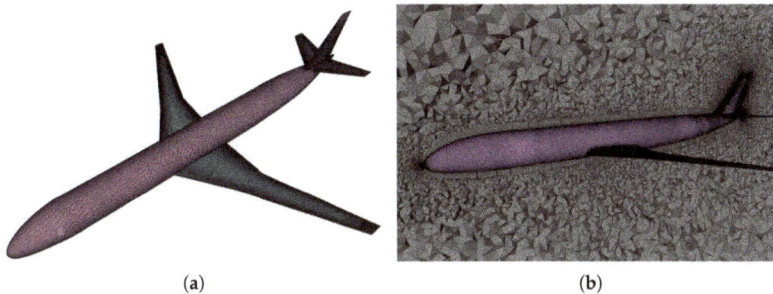

(a)   (b)

Figure 9. The surface and volume meshes of the reference aircraft generated by Sumo with TetGen and Pentagrow. (a) Sumo surface mesh; (b) Sumo-Pentagrow RANS mesh.

3.2.2. The Interface CPACS2SUMO

The aircraft configuration defined in a CPACS XML file is converted into a Sumo [24] native .smx file by the CPACS2SUMO Python converter without manual intervention. This conversion is relatively straightforward since both formats define aircraft in a similar way. Fuselage and wings are created from a gathering of sections placed in a certain order. Each section is defined by a 2D profile written as a list of points. Then, these profiles can be scaled, rotated and translated to form the desired shape. Figure 10 shows how Sumo represents a wing as a stack of airfoils. The 3D wing surface is lofted from the sections by Bézier or B-spline surfaces.

The CPACS format allows quite a general definition of cross-sections. For instance, a CPACS cross-section may be placed in the global coordinate system via reference to any other section (see Figure 4), whereas Sumo uses the order of the sections as they appear in the file.

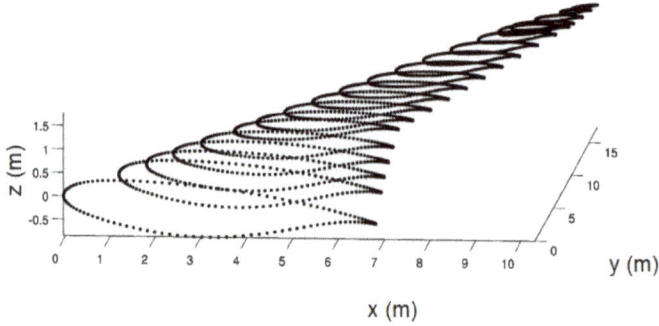

Figure 10. Sumo creates a wing from a stack of airfoils.

In Sumo (Figure 11), transformation by scaling, rotation and translation of an entity is executed at one level and with limitation. For example, a Sumo fuselage profile is assumed perpendicular to the *x*-axis. Furthermore, CPACS and Sumo formats use different definitions of 3D rotation angles.

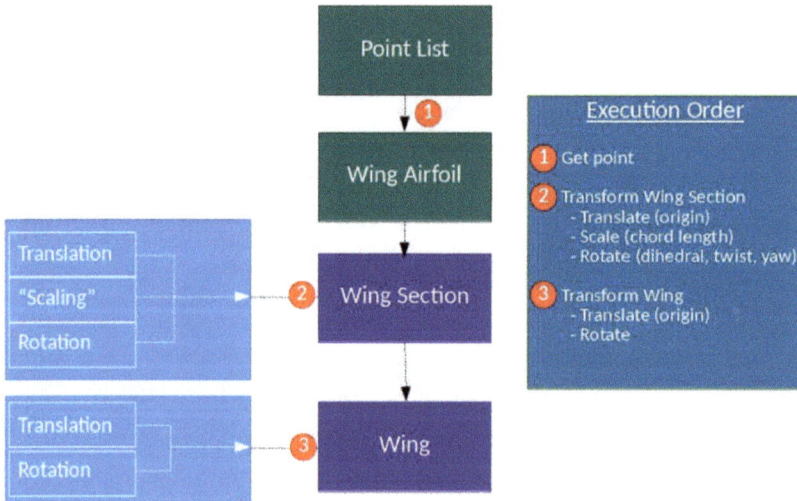

Figure 11. Schema to construct an aircraft from its Sumo XML file format.

4. Flow Solvers

The CFD (fidelity level L2–3) codes SU2 [29,31] and Edge [28] are used for Euler and RANS flow modeling.

Edge is the Swedish national CFD code for external steady and unsteady compressible flows. Developed by the Swedish Defense Research Agency (FOI), it uses unstructured grids with arbitrary elements and an edge-based formulation with a node-centered finite-volume technique to solve the governing equations. Edge supports a number of turbulence models, as well as LES and DES simulations.

The SU2 [29] software suite from Stanford University is an open-source, integrated analysis and design tool for complex, multi-disciplinary problems on unstructured computational grids. The built-in optimizer is a Sequential Least Squares Programming (SLSQP) algorithm [32] from the SciPy Python scientific library. The gradient is calculated by continuous adjoint equations of the flow equations [29,31]. SU2 is in continued development. Most examples pertain to inviscid flow, but also, RANS flow models with the Spalart–Allmaras and the Menter's Shear Stress Transport (SST)$k - \omega$ turbulence models can be treated.

Figure 12 shows the comparison of the aerodynamic coefficients computed by Euler equations in SU2 and Edge for two different meshes of the reference aircraft. "Mesh-4p" has 4.0 million cells, and "mesh-2p" has 1.9 million cells. The meshes have the same meshing parameter settings as described in [20]. Both have refined wing leading and trailing edges, and "mesh-4p" has settings for even smaller minimum dimensions of the cells. According to the mesh study in [20], the predictions from both solvers converge as the mesh resolution increases, and "mesh-4p" was selected in [20] for all the simulations carried out by SU2 by considering both computational accuracy and efficiency. In this paper, more simulations are made for both "mesh-2p" and "mesh-4p" using both Edge and SU2 for different flight conditions. For "mesh-2p", Edge and SU2 give fairly close predictions for the aerodynamic coefficients C_L, C_D and C_m at Mach = 0.78 and 0.9. In the current study, we will use "mesh-4p" for the Euler solutions by SU2 in Section 5 and "mesh-2p" for the calculations on control surface modeling in Section 6, where SU2 and Edge are used to compare the geometry modeling approaches.

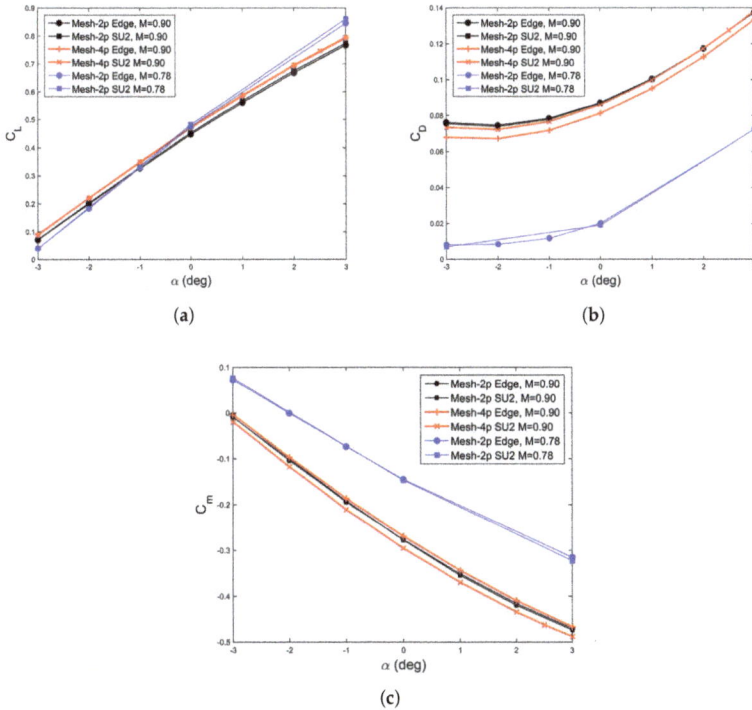

Figure 12. Euler simulations for the reference aircraft for C_L, C_D and C_m computed by Edge and SU2 for meshes "mesh-2p" and "mesh-4p", at a flight altitude of 10,000 m, Mach 0.78 and Mach 0.9 respectively. (**a**) Lift coefficient C_L for Mach number of 0.78 and 0.9; (**b**) drag coefficient C_D for Mach numbers of 0.78 and 0.9; (**c**) pitching moment coefficient C_m for Mach numbers of 0.78 and 0.9.

5. Applications

5.1. Aerodynamic Results Comparison

The aerodynamic coefficients obtained with L1 and L2 fidelity tools are compared in Figure 13. A Mach number of 0.6 was used to avoid transonic effects (at low angles of attack) that are not well predicted by L1 (Tornado). The flight condition used for this comparison is an altitude of 5000 m and a side slip angle $\beta = 0$ deg. The L3 (RANS) simulations for a fully-turbulent flow [33] using the Spalart–Allmaras turbulence model are also shown at the same flight condition, as the highest fidelity data for verification. The RANS mesh is generated by Pentagrow using the same Sumo surface mesh for the L2 Euler computations, which has 792,900 triangles on the surface. The RANS mesh has 8.2 million cells. The first layer height is 3.8×10^{-6} ($y^+ = 1$); the growth rate is 1.2; the number of layers is 40, with the corresponding Reynolds number 32.4 million of the reference chord 3.7317 m. The airspeed is 192 m/s, which corresponds to Mach = 0.6 and altitude = 5000 m. Figure 14 shows the computed y^+ diagram over the reference aircraft at $\alpha = 1°$ with airspeed 192 m/s and Reynolds number 32.4 million.

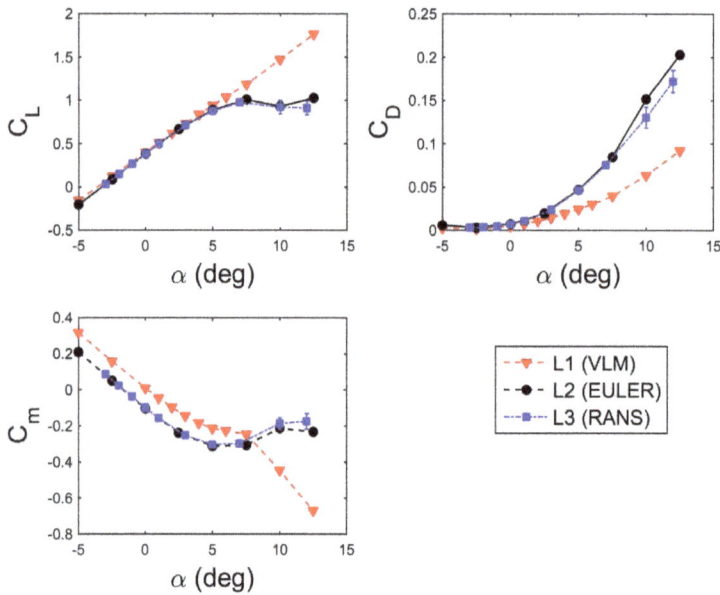

Figure 13. L1, L2 and L3 simulations for the test case aircraft, α sweep at Mach = 0.6, altitude = 5000 m and $\beta = 0°$.

Note that the L2 and L3 simulations agree quite well, so that we can safely assume that for this configuration under the corresponding flight conditions, L2 simulations are "as best as" the L3 simulations. In this paper, we only discuss L1 and L2 tools and their simulations. The numerical flow for $\alpha \geq 10°$ is highly unsteady and is not entirely converged, the aerodynamic forces calculated by the L2 and L3 tools are the mean values of the iterations in the search of a steady flow.

The lift coefficient C_L is well predicted by both L1 and L2 tools between angles of attack of $-5°$ and $+5°$. Above this range, "computational" stall occurs at α of approximately 8°, which is clearly visible in the L2 results.

The drag polar shows that the minimum drag coefficient is obtained for an angle of attack of about $-2.5°$ where C_L vanishes. The minimum C_D is very small with both Tornado (L1) and SU2 Euler (L2) because they do not include skin friction drag in their physical model and there is no wave drag.

The L2 prediction of high C_D for high angles of attack is due to wave drag. The pitching moment coefficient C_m on the left corner shows that the aircraft is longitudinally stable ($\partial C_m/\partial \alpha < 0$) for angles of attack from $-5°$ to $+5°$. The breaks in the curves after $+5°$ are different. For L2, it is due to "computational" stall of the horizontal stabilizer. As a reminder, this aircraft is only in the first phase of its design, so it has not been optimized in terms of stability.

The good agreement for some ranges obtained by different fidelity aerodynamic tools supports the idea of building a surrogate model trained by an automatic sampling approach that takes advantage of each method according to their fidelity levels and limitations. For example, it is useless to spend computational time with Euler calculations in the linear aerodynamic region where Tornado can give reliable results. This computational time is better spent on higher Mach number or angles of attack where the cheapest tools fail. An application of the "variable fidelity" technique is to fuse the data from different fidelity levels of tools by kriging and co-kriging [34]; see also [14].

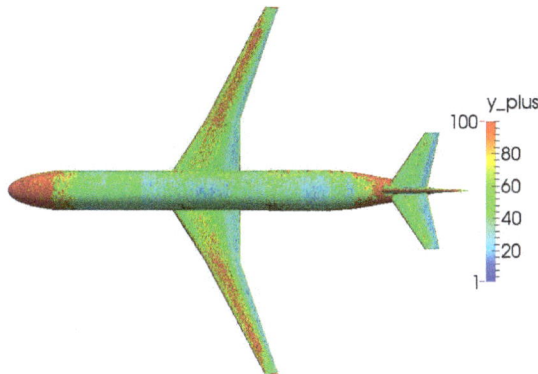

Figure 14. The y^+ diagram over the reference aircraft at $\alpha = 1°$ with airspeed 192 m/s and Reynolds number 32.4 million.

5.2. Multi-Fidelity Aerodynamics for Data Fusion

A surrogate model with automatic sampling fuses the data obtained by the different aerodynamic tools. This is useful for constructing a look-up aero-table for quality analysis and flight simulation. This section shows which are the final surrogate models of the static aerodynamic coefficients for horizontal flight, and more results are also shown in [14]. The aircraft handling qualities are also predicted and analyzed; see the details in [14]. The multi-fidelity of the aerodynamic tools used to generate the various data for constructing the surrogate models is executed via BRICS remotely at different sites by importing/exporting the common CPACS file through the interfaces described in this paper.

Figure 15 shows the fused C_L, C_D and C_m aero-coefficient results of the reference aircraft model from the L1 Tornado) and L2 SU2 Euler tools with the elevator deflection $\delta = 0°$ over the flight envelope. • represents the Tornado samples, and × represents the SU2 Euler samples. Figure 15a,c,e shows the response surfaces from the surrogate models, as well as the sampled data over the flight envelope in the three-dimensional space. Figure 15b,d,f represents the variation with α and δ_e for Mach numbers 0.5 (black) and 0.78 (blue) from the response surfaces and their corresponding sampled data.

Figure 15a,b shows the surrogate models (response surfaces) for C_L produced by co-kriging [34]. The non-linear behavior at higher angles of attack is captured as the L2 samples indicate.

Figure 15c,d shows the prediction for C_D. The surrogate model predicts higher drag than the L1 samples, since they do not predict wave drag. The surrogate model picks up the compressibility phenomena from the L2 samples.

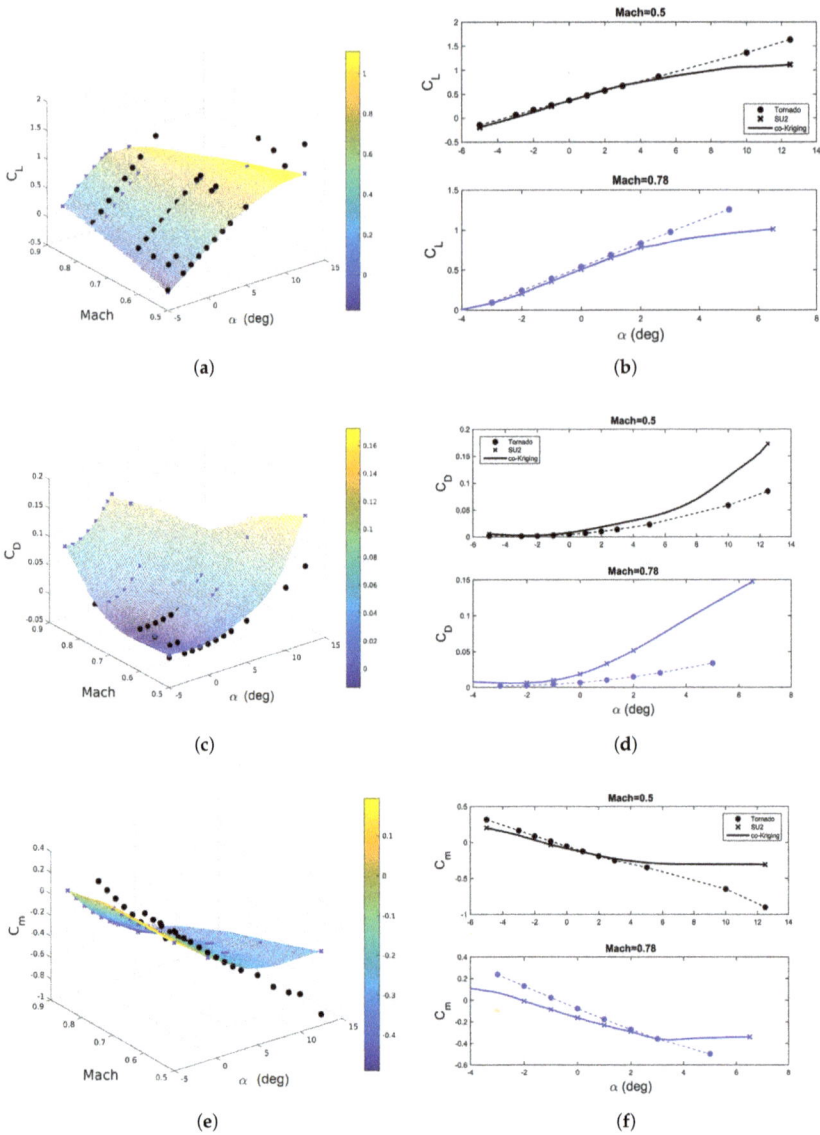

Figure 15. Surrogate model results of the reference aircraft for C_L, C_D and C_m, with no elevator deflection. Notations: dot: Tornado (L1) samples; cross: SU2 (L2) Euler samples; line: the response surfaces. (**a,c,e**) The response surfaces and sampled data over the flight envelope. (**b,d,f**) The cuts for Mach number 0.5 (black) and 0.78 (blue) from the response surfaces and their corresponding sampled data; the results are also shown in [14]. (**a**) Lift coefficient C_L surface and the sampled data; (**b**) fused lift coefficient C_L for Mach numbers of 0.5 and 0.78; (**c**) drag coefficient C_D surface and the sampled data; (**d**) fused drag coefficient C_D for Mach numbers of 0.5 and 0.78; (**e**) pitching moment coefficient C_m surface and the sampled data; (**f**) fused pitching moment coefficient C_m for Mach 0.5 and 0.78.

The surrogate model for C_m is shown in Figure 15e,f. Note again that the surrogate model predicts the non-linear trends at high AoA, as expected. The coarse L2 samples correct the response surfaces significantly.

The computation time of the surrogate model is ≈0.05 s on a desktop computer with four CPUs. The reliability of the surrogate as indicated by the root mean square error $max(RMSE) = 0.048 < 5\%$.

5.3. Aero-Data for Low Speed by the L1 Tool

Tornado computes all static and quasi-static aerodynamic coefficients, including the effects of trailing edge device deflections. In the following paragraphs, the sizing of the rudder and horizontal trim and handling qualities are discussed based on the Tornado calculations.

5.3.1. Sizing the Fin and Rudder for the One-Engine-Out Case

An aircraft must have an established minimum control speed V_{MC}, legally defined in, e.g., [35], as the lowest calibrated airspeed at which the aircraft is controllable. It may not be larger than 1.13-times the reference stall speed. Aircraft with engines set far from the centerline will experience large yawing moments if an engine fails. The sizing of the vertical tail, and its rudder, is usually determined by a one-engine-out case. Flying with side-slip and rudder deflection, at a certain airspeed, the fin and rudder produce just enough yawing moment to counteract the asymmetric thrust. This speed is essentially the minimum control speed, although certification requires a few more parameters.

In this section, an exercise of sizing the rudder of the aircraft is carried out by Tornado, using a simplified method as described by Torenbeek [36]. Side-slip and roll response were neglected, and the tail volume [37] was held constant.

As an initial estimate, the minimum control speed for different control surface sizes was computed until the requirement of 1.13 of the stall speed was achieved. The selected rudder deflection was 25 degrees to allow five degrees of maneuver margin. Figure 16 shows the predicted minimum control velocity.

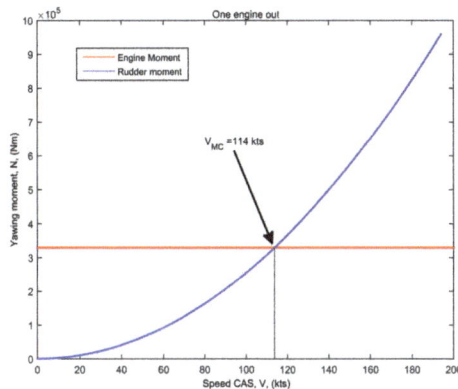

Figure 16. Computed minimum control speed V_{MC}, from Tornado solutions.

5.3.2. Handling Qualities

The horizontal sea level trim at low speed can be estimated from the aero-coefficients and the mass distribution whose reference values are available from CPACS. The straight and level flight trim results at sea level, calculated by Tornado in [20] are shown in Figure 17.

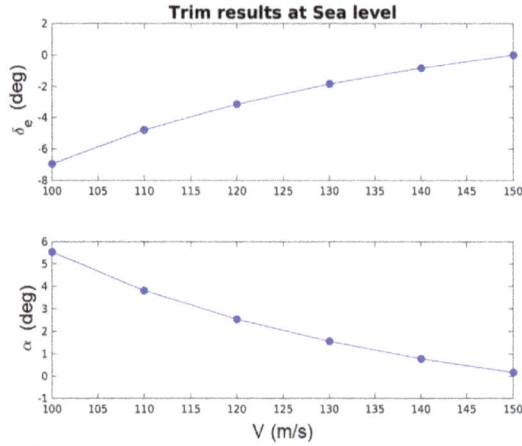

Figure 17. Trimmed angle of attack and elevator deflection at sea level, from Tornado solutions.

The classical modes of motion indicate the linear stability of the aircraft, i.e., its responses to (infinitesimally) small disturbances. Flight simulation allows the full range of stability of the aircraft to be assessed. The time history in Figure 18 shows how the attitude angle θ, the angle of attack α and the pitch rate q oscillate when excited by a step-function-type elevator movement. The PHALANX [1–4,38], a flight simulation tool from Delft University of Technology, produced the time histories.

The time domain simulation starts as trimmed straight and level flight at sea level conditions with True Air Speed 130 m/s. After 1 s, the pilot executes a 2-3-1-1 maneuver in pitch, namely, stick 2 s nose down, 3 s nose up, 1 s nose down and 1 s nose up.

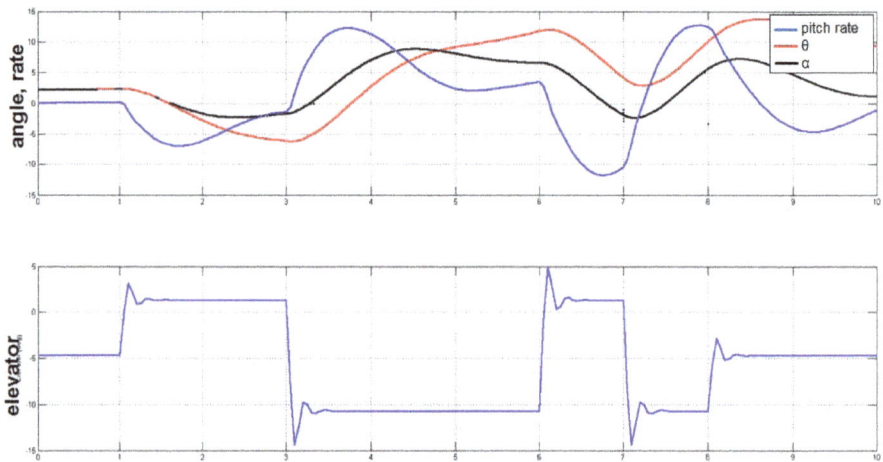

Figure 18. Flight time domain simulation from PHALANX, trimmed flight with True Air Speed (TAS) = 130 m/s at sea level.

6. Euler Computation for Various Control Surface Models

Analysis for control surface deflection is an important capability of L2-3 CFD tools. In this section, elevator deflection by mesh deformation, morphing trailing edge shape and transpiration boundary conditions are described. Euler results by different flow solvers are compared. The surfaces are deflected by deforming the surfaces, and no gaps are modeled.

6.1. Modeling Movable Surfaces

SU2 uses a mesh deformation defined by a set of control points for a Free-Form Deformation (FFD) box [39], thus only the clean configuration mesh is generated from scratch. The SU2_DEF built-in mesh deformation function is then used to deform the mesh around the elevator locations on the horizontal tail (see Figure 19a). An FFD box is defined at the elevator locations. The FFD box of control points can be rotated around the hinge line. Owing to the affine invariance of the map from control points to the mesh, the surface mesh follows. According to the authors' experience, with a deflection of less than eight degrees, the deformed mesh is still well formed enough to function for Euler simulation isotropic grids.

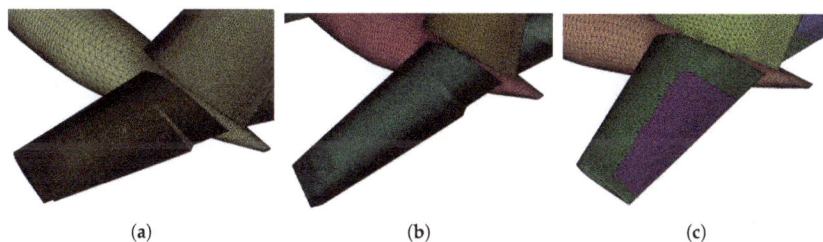

(a)	(b)	(c)

Figure 19. The mesh generated by Sumo with different modeling technologies to be computed for elevator deflection at 6°. (**a**) Surface mesh with mesh deformed by FFD on the elevator for $\delta_e = 6$ degrees; (**b**) surface mesh with morphed elevator modeled by Sumo for $\delta_e = 6$ degrees, which includes a linear type 5% transition zone of the morphing elevator length; (**c**) tailplane and the elevators marked in the surface mesh in Sumo.

It is also possible to morph the shape, represented by control point technology, and re-generate the mesh for each deflection. Auto-morphing scripts (re-)generating the surface grids with deflected control surface(s) in Sumo are described in [40].

The Trailing Edge (TE) morphing starts by extracting the camber curve and thickness distribution from the airfoil data [40]. The auto-morphing scripts create new sections (as required) at control surface edges. The new section camber curves are morphed according to the deflection angle, to which the thickness is added. Figure 20a illustrates the camber controlled by quadratic Bézier curves for morphed leading and trailing edges with the thickness distribution preserved. The quadratic Bézier control points are chosen to give (at least) G1-continuity of the deformed camber curves. Issues related to crossing of camber curve normals close to control surface edges must be addressed. The camber curves for morphed leading and trailing edges are based on the Class–Shape function Transformation (CST) approach [41]. It enables easy geometry manipulation by user-defined parameters such as the hinge line, rotation angle and the transition zones; see Figure 20b.

(a)

(b)

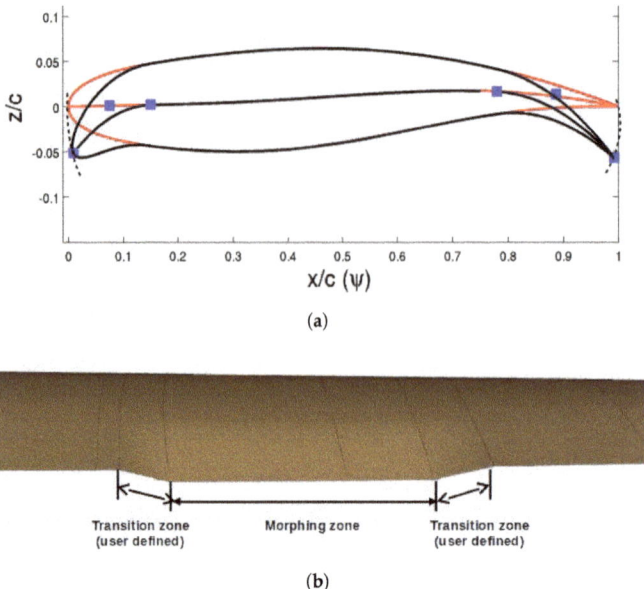

Figure 20. The demonstration of the automatically gapless movable surfaces morphing technology by Sumo. (**a**) The geometric parameters describing the morphing airfoil [40], morphing leading edge and trailing edge and a fixed central area using a variable camber; (**b**) illustration of the morphing trailing edge with the morphing zone and transition zones modeled in Sumo.

Figure 19b shows the Sumo surface mesh for a deflection angle of six degrees. Both mesh deformation by FFD (Figure 19a) and morphing of the camber lines produce gapless meshes according to the control surface deflections. The morphing technology in addition supports a user-defined transition zone (length and type) to obtain a smoother surface and avoid bad tetrahedral cells. The smooth transition feature makes the trailing edge morphing technology possible for a RANS simulation; however, the mesh deformation by FFD tends to produce high aspect ratio cells at the deformed junctions, so the mesh may work well only for coarser Euler simulations.

Edge calculates the aerodynamics of control surface deflections by transpiration boundary conditions. The grid does not move; only the normals used in the no-flow boundary condition are deflected; see Figure 19c.

6.2. Results Comparison

This section shows the results of the control surface modeling approaches with different CFD solvers. Figure 21 and Table 1 show the comparisons for C_L, C_m and their derivatives for elevator deflection $\pm 6°$, at Mach 0.78, altitude 10,000 m, $\alpha = 0°$. We expect close agreement for morphing elevators computed by SU2 and Edge, by mesh deformation, as well as morphing. The solutions for mesh deformation by SU2 FFD and transpiration boundary condition by Edge are quite comparable. The morphing control surface model predicts slightly higher slopes for both C_L and C_m and a smoother flow on the horizontal tail (see Figure 22), probably due to the transition part of the morphed shape. Note that the transition zone as defined in this case gives a slightly larger deflected area.

Note that the abbreviations used in Table 1, Figures 21 and 22 are:

Mesh-Def(orm): Mesh deformation using FFD;
Morph. (cs): Morphing the control surfaces by Sumo;
Transp. b.c.: transpiration boundary conditions in Edge.

Table 1. Result for different modeling of elevator deflections using different solvers, at Mach 0.78, flight altitude 10,000 m, $\alpha = 0°$.

Model Type	Solver	$C_{L,\delta}$ (deg)	$C_{m,\delta}$ (deg)
Mesh-deform	SU2	0.0092	−0.0399
Morph. cs	SU2	0.0130	−0.0565
Morph. cs	Edge	0.0117	−0.0557
Transp. b.c.	Edge	0.0095	−0.0411

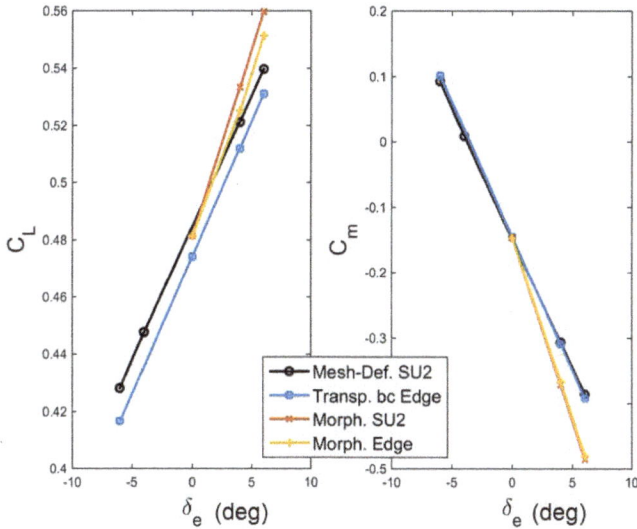

Figure 21. Euler solutions for elevator deflection $\delta_e = 6°$ at Mach 0.78, flight altitude 10,000 m, $\alpha = 0°$, by three control surface modeling methods, for SU2 and Edge.

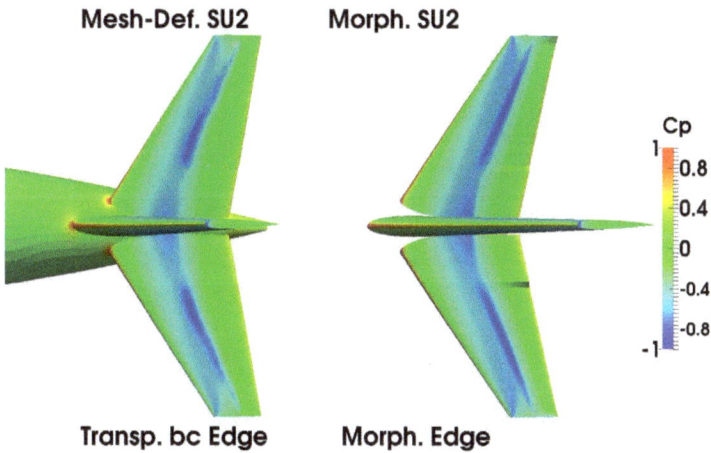

Figure 22. C_p contours of Euler solutions for the elevator deflection $\delta_e = 6°$ at Mach 0.78, flight altitude 10,000 m, $\alpha = 0°$, by three control surface modeling methods, SU2 and Edge.

7. Conclusions

Seamless connection by the CPACS data schema enables integration and implementation of variable fidelity tools into the aircraft design multi-disciplinary analysis and optimization workflow. This paper presents L1 and L2 fidelity aerodynamic design tools using auto-meshing. Applications to the defined test case aircraft includes aerodynamic analysis, data fusion of variable fidelity data and handling quality analysis by flight simulation. A comprehensive discussion is given of different models for control surface deflections for Euler computation.

Acknowledgments: The research presented in this paper has been performed in the framework of the AGILE project (Aircraft 3rd Generation MDO for Innovative Collaboration of Heterogeneous Teams of Experts) and has received funding from the European Union Horizon 2020 Programme (H2020-EU.3.4) under Grant Agreement No. 636202. The Swiss participation in the AGILE project was supported by the Swiss State Secretariat for Education, Research and Innovation (SERI) under Contract Number 15.0162. The authors are grateful to the partners of the AGILE consortium for their contributions and feedback; Mark Voskuijl, Delft University of Technology, for the flight simulation analysis from PHALANX; Jesper Oppelstrup, Royal Institute of Technology (KTH), Sweden, for the valuable comments and suggestions.

Author Contributions: Mengmeng Zhang developed the CPACS-Tornado interface and made the corresponding calculations, carried out the Euler calculations by SU2 and Edge for solver comparisons and for the control surface model analyses, analyzed the data and wrote the paper. Aidan Jungo developed the CPACS2SUMO interface, carried out the Euler calculations by SU2 and wrote the corresponding part of the paper and the CPACS description. Alessandro Augusto Gastaldi developed the PyTornado, wrote the corresponding part of the paper and helped write the paper. Tomas Melin carried out the Tornado calculations for sizing the fin and rudder and wrote the corresponding part of the paper.

Conflicts of Interest: The authors declare no conflict of interest. The sponsors had no role in the design of the study; in the collection, analyses or interpretation of data; in the writing of the manuscript; nor in the decision to publish the results.

Abbreviations

AGILE	Aircraft 3rd Generation MDO for Innovative Collaboration of Heterogeneous Teams of Experts
API	Application Programming Interface
CAD	Computer Aided Design
CFD	Computational Fluid Dynamics
CPACS	The Common Parametric Aircraft Configuration Schema
CST	Class-Shape function Transformation
DES	Detached Eddy Simulation
FFD	Free-Form Deformation
GUI	Graphic User Interface
LES	Large-Eddy Simulation
MAC	Mean Aerodynamic Chord
MDA	Multidisciplinary Analysis
MDO	Multidisciplinary Design and Optimization
RANS	Reynolds-Averaged Navier–Stokes equations
TAS	True Air Speed
TE(D)	Trailing Edge (Device)
UI	User Interface
VLM	Vortex Lattice Method
XML	Extensible Markup Language

Symbols

α or AoA	Angle of Attack (deg)
δ_e	Elevator deflection angle (deg)
θ	Attitude angle (deg)
q	Pitch rate (deg/s)
C_p	Pressure Coefficient (-)
C_L	Lift coefficient (-)
C_D	Drag coefficient (-)
C_m	Pitching moment coefficient (-)

Aerospace **2018**, *5*, 47

References

1. Voskuijl, M.; de Klerk, J.; van Ginneken, D. *Flight Mechanics Modelling of the Prandtl Plane for Conceptual and Preliminary Design*; Springer: London, UK, 2012; Volume 66.
2. Voskuijl, M.; La Rocca, G.; Dircken, F. Controllability of blended wing body aircraft. In Proceedings of the 26th Congress of International Council of the Aeronautical Sciences, Anchorage, AK, USA, 14–19 September 2008.
3. Fengnian, T.; Voskuijl, M. Automated Generation of Multiphysics Simulation Models to Support Multidisciplinary Design Optimization. *Adv. Eng. Inform.* **2015**, *29*, 1110–1125.
4. Foeken, M.J.; Voskuijl, M. Knowledge-Based Simulation Model Generation for Control Law Design Applied to a Quadrotor UAV. *Math. Comput. Model. Dyn. Syst.* **2010**, *16*, 241–256. [CrossRef]
5. Raymer, D. Conceptual design modeling the RDS-professional aircraft design software. In Proceedings of the AIAA Aerospace Sciences Meeting, Orlando, FL, USA, 4–7 January 2011.
6. Anemaat, W.A.; Kaushik, B. Geometry design assistant for airplane preliminary design. In Proceedings of the 49th AIAA Aerospace Sciences Meeting including the New Horizons Forum and Aerospace Exposition, Orlando, FL, USA, 4–7 January 2011.
7. Gloudemans, J.R.; Davis, P.C.; Gelausen, P.A. A rapid geometry modeler for conceptual aircraft. In Proceedings of the 34th AIAA Aerospace Sciences Meeting and Exhibit, Reno, NV, USA, 15–18 January 1996.
8. Tomac, M. Towards Automated CFD for Engineering Methods in Aircraft Design. Ph.D. Thesis, Royal Institute of Technology, KTH, Stockholm, Sweden, 2014; ISSN 1651-7660.
9. CPACS—A Common Language for Aircraft Design. Available online: http://www.cpacs.de/ (accessed on 7 February 2018).
10. Böhnke, D.; Nagel, B.; Zhang, M.; Rizzi, A. Towards a collaborative and integrated set of open tools for aircraft design. In Proceedings of the 51st AIAA Aerospace Sciences Meeting including the New Horizons Forum and Aerospace Exposition, Grapevine, TX, USA, 7–10 January 2013; pp. 7–10.
11. SUAVE—An Aerospace Vehicle Environment for Designing Future Aircraft. Available online: http://suave.stanford.edu/ (accessed on 20 March 2018).
12. Lukaczyk, T.; Wendorff, A.; Botero, E.; MacDonald, T.; Momose, T.; Variyar, A.; Vegh, J.M.; Colonno, M.; Economon, T.; Alonso, J.J.; et al. SUAVE: An Open-Source Environment for Multi-Fidelity Conceptual Vehicle Design. In Proceedings of the 16th AIAA Multidisciplinary Analysis and Optimization Conference, Dallas, TX, USA, 22–26 June 2015.
13. MacDonald, T.; Clarke, M.; Botero, E.M.; Vegh, J.M.; Alonso, J.J. SUAVE: An Open-Source Environment Enabling Multi-fidelity Vehicle Optimization. In Proceedings of the 18th AIAA Multidisciplinary Analysis and Optimization Conference, Denver, CO, USA, 5–9 June 2017.
14. Zhang, M.; Bartoli, N.; Jungo, A.; Lammen, W.; Baalbergen, E. Data Fusion and Aerodynamic Surrogate Modeling for Handling Qualities Analysis. *Prog. Aerosp. Sci. Spec. Issue* **2018**, submitted.
15. AGILE EU Project Portal. Available online: http://www.agile-project.eu (accessed on 15 January 2018).
16. Seider, D.; Fischer, P.; Litz, M.; Schreiber, A.; Gerndt, A. Open Source Software Framework for Applications in Aeronautics and Space. In Proceedings of the IEEE Aerospace Conference, Big Sky, MT, USA, 3–10 March 2012.
17. Baalbergen, E.; Kos, J.; Louriou, C.; Campguilhem, C.; Barron, J. Streamlining cross-organisation product design in aeronautics. *Proc. Inst. Mech. Eng. Part G J. Aerosp. Eng.* **2016**, [CrossRef]
18. Moerland, E.; Ciampa, P.D.; Zur, S.; Baalbergen, E.; D'Ippolito, R.; Lombardi, R. Collaborative Architecture supporting the next generation of MDO within the AGILE Paradigm. *Prog. Aerosp. Sci. Spec. Issue* **2018**, submitted.
19. Van Gent, I.; Aigner, B.; Beijer, B.; Jepsen, J.; Rocca, G.L. Knowledge architecture supporting the next generation of MDO in the AGILE paradigm. *Prog. Aerosp. Sci. Spec. Issue* **2018**, submitted.
20. Jungo, A.; Vos, J.; Zhang, M.; Rizzi, A. Benchmarkig New CEASIOM with CPACS Adoption for Aerodynamic Analysis and Flight Simulation. *Aircr. Eng. Aerosp. Technol.* **2017**, *90*, -11-2016-0204. [CrossRef]
21. TIGL—Geometry Library to Process Aircraft Geometries in Pre-Design. Available online: https://github.com/DLR-SC/tigl (accessed on 30 November 2017).
22. CPACS Documentation. Available online: https://github.com/DLR-LY/CPACS/tree/develop/documentation (accessed on 20 April 2018).

23. Melin, T. Using Internet Interactions in Developing Vortex Lattice Software for Conceptual Design. Ph.D. Thesis, Department of Aeronautics, Royal Institute of Technology, KTH, Stockholm, Sweden, 2003.

24. Tomac, M.; Eller, D. From Geometry to CFD Grids: An Automated Approach for Conceptual Design. *Prog. Aerosp. Sci.* **2011**, *47*, 589–596, [CrossRef]

25. Si, H. *TetGen: A Quality Tetrahedral Mesh Generator and 3D Delaunay Triangulator*; Technical Report, User's Manual; Technical Report No. 13; Numerical Mathematics and Scientific Computing, Weierstrass Institute for Applied Analysis and Stochastics (WIAS): Berlin, Germany, 2013.

26. Anderson, J.D. *Modern Compressible Flow with Historical Perspective*, 3rd ed.; McGraw-Hill: New York, NY, USA, 2004.

27. Katz, J.; Plotkin, A. *Low-Speed Aerodynamics: From Wing Theory to Panel Methods*; McGraw-Hill, Inc.: New York, NY, USA, 1991.

28. Eliasson, P. Edge, a Navier-Stokes Solver for Unstructured Grids. In *Finite Volumes for Complex Applications III*; Elsevier: Amsterdam, The Netherlands, 2002; pp. 527–534.

29. Palacios, F.; Colonno, M.R.; Aranake, A.C.; Campos, A.; Copeland, S.R.; Economon, T.D.; Lonkar, A.K.; Lukaczyk, T.W.; Taylor, T.W.R.; Alonso, J. Stanford University Unstructured (SU2): An open-source integrated computational environment for multi-Physics simulation and design. In Proceedings of the 51st AIAA Aerospace Sciences Meeting including the New Horizons Forum and Aerospace Exposition, Grapevine, TX, USA, 7–10 January 2013; AIAA 2013-0287.

30. Economon, T.D.; Palacios, F.; Copeland, S.R.; Lukaczyk, T.W.; Alonso, J.J. SU2: An Open-Source Suite for Multiphysics Simulation and Design. *AIAA J.* **2015**, *54*, 828–846. [CrossRef]

31. Palacios, F.; Economon, T.D.; Wendorff, A.D.; Alonso, J. Large-Scale Aircraft Design Using SU2. In Proceedings of the 53rd AIAA Aerospace Sciences Meeting, AIAA 2015-1946, Kissimmee, FL, USA, 5–9 January 2015.

32. Griva, I.; Nash, S.G.; Sofer, A. *Linear and Nonlinear Optimization*, 2nd ed.; Society for Industrial Applied Mathematics: Philadelphia, PA, USA, 2009.

33. Yousefi, K.; Razeghi, A. Determination of the Critical Reynolds Number for Flow over Symmetric NACA Airfoils. In Proceedings of the 2018 AIAA Aerospace Sciences Meeting, AIAA 2018–0818, Kissimmee, FL, USA, 8–12 January 2018.

34. Forrester, A.; Keane, A. *Engineering Design via Surrogate Modelling: A Practical Guide*; John Wiley & Sons: Hoboken, NJ, USA, 2008.

35. Federal Aircraft Administration. *CFR 25.149*; Technical Report, Retrieved 18 January 2018; Federal Aircraft Administration: Washington, DC, USA, 2018.

36. Torenbeek, E. *Synthesis of Subsonic Airplane Design*; Delft University Press: Delft, The Netherlands, 1982.

37. Etkin, B.; Reid, L.D. *Dynamics of Flight: Stability and Control*; John Wiley & Sons: Hoboken, NJ, USA, 1996.

38. Pfeiffer, T.; Nagel, B.; Böhnke, D.; Voskuijl, M.; Rizzi, A. *Implementation of a Heterogeneous, Variable-Fidelity Framework for Flight Mechanics Analysis in Preliminary Aircraft Design*; Deutscher Luft-und Raumfahrtkongress: Bremen, Germany, 2011.

39. Sederberg, T.W.; Parry, S.R. Free-Form Deformation of Solid Geometric Models. In Proceedings of the SIGGRAPH '86 13th Annual Conference on Computer Graphics and Interactive Techniques, Atlanta, GA, USA, 1–5 August 1986; Volume 20, pp. 151–160.

40. Zhang, M. Contributions to Variable Fidelity MDO Framework for Collaborative and Integrated Aircraft Design. Ph.D. Thesis, Royal Institute of Technology KTH, Stockholm, Sweden, 2015.

41. Kulfan, B. Universal Parametric Geometry Representation Method. *J. Aircr.* **2008**, *45*, 142–158. [CrossRef]

Article

Experimental Study and Neural Network Modeling of Aerodynamic Characteristics of Canard Aircraft at High Angles of Attack

Dmitry Ignatyev * and Alexander Khrabrov

Central Aerohydrodynamic Institute, 140180 Zhukovsky, Moscow Region, Russia; khrabrov@tsagi.ru
* Correspondence: d.ignatyev@mail.ru

Received: 29 December 2017; Accepted: 28 February 2018; Published: 2 March 2018

Abstract: Flow over an aircraft at high angles of attack is characterized by a combination of separated and vortical flows that interact with each other and with the airframe. As a result, there is a set of phenomena negatively affecting the aircraft's performance, stability and control, namely, degradation of lifting force, nonlinear variation of pitching moment, positive damping, etc. Wind tunnel study of aerodynamic characteristics of a prospective transonic aircraft, which is in a canard configuration, is discussed in the paper. A three-stage experimental campaign was undertaken. In the first stage, a steady aerodynamic experiment was conducted. The influence of a reduced oscillation frequency and angle of attack on unsteady aerodynamic characteristics was studied in the second stage. In the third stage, forced large-amplitude oscillation tests were carried out for the detailed investigation of the unsteady aerodynamics in the extended flight envelope. The experimental results demonstrate the strongly nonlinear behavior of the aerodynamic characteristics because of canard vortex effects on the wing. The obtained data are used to design and test mathematical models of unsteady aerodynamics via different popular approaches, namely the Neural Network (NN) technique and the phenomenological state space modeling technique. Different NN architectures, namely feed-forward and recurrent, are considered and compared. Thorough analysis of the performance of the models revealed that the Recurrent Neural Network (RNN) is a universal approximation tool for modeling of dynamic processes with high generalization abilities.

Keywords: wind tunnel; neural networks; modeling; unsteady aerodynamic characteristics; high angles of attack

---

1. Introduction

Modern transport airplanes use angles of attack close to stall during take-off and landing. Different trigger factors such as possible pilot mistakes, equipment faults and atmospheric turbulence and their combinations can cause loss of control, stall and spin. Different statistical surveys reported (see, for example, [1]) that loss of control in flight (LOC-I) was the major cause of fatal transport aviation accidents. Thus, many intensive studies are aimed at modeling of aerodynamics in the extended flight envelope in order to support investigations of aircraft dynamics, control system design [2] and to provide realistic pilot training using ground-based simulators in upset conditions [3,4].

Flow over an aircraft at high angles of attack is complicated by the dynamics of flow separation and reattachment, the development and breakdown of vortical flow, and their interaction with dynamics of the aircraft. This causes significant nonlinearities of unsteady aerodynamic characteristics—for example, non-uniqueness of stability derivatives and hysteresis of aerodynamic characteristics.

Growth in computing capacity and the development of numerical techniques has recently led to significant progress in finding solutions for Navier–Stokes equations coupled with the dynamics equations governing the aircraft motion, facilitating flight dynamics studies [3,5–10]. However, at present the problems of fluid mechanics and flight dynamics cannot be solved simultaneously in certain flight mechanical applications—for example, in semi-realistic simulation of the aircraft flight using ground-based flight simulators or control system design [3,11]. Solving such flight dynamics problems demands Reduced-Order Models (ROM) of unsteady aerodynamics describing nonlinear phenomena observed in extended range of flight parameters. Experimental data obtained from wind tunnel tests of Computational Fluid Dynamics (CFD) data are commonly used for the development of such models.

In flight dynamics problems, the aerodynamic forces and moments are usually represented in the form of look-up tables [12,13]. For example, for small disturbed motion about a trim incidence α_0, the longitudinal coefficients are represented in the following form:

$$
\begin{aligned}
C_N &= C_N(\alpha_0) + C_{N_\alpha}(\alpha - \alpha_0) + C_{N_q}q\bar{c}/2V + C_{N_{\dot{\alpha}}}\dot{\alpha}\bar{c}/2V, \\
C_m &= C_m(\alpha_0) + C_{m_\alpha}(\alpha - \alpha_0) + C_{m_q}q\bar{c}/2V + C_{m_{\dot{\alpha}}}\dot{\alpha}\bar{c}/2V.
\end{aligned}
\tag{1}
$$

This representation can be successfully applied only for small angles of attack, namely in the range where the aerodynamic derivatives exist and are unique, and can be represented as linear dependences on the kinematic parameters. Application of this technique for high angles of attack can lead to significant errors.

The general technique for modeling unsteady aerodynamic characteristics is based on a nonlinear indicial function representation [14]. To develop an aerodynamic model based on the nonlinear indicial functions, a large amount of unsteady aerodynamics data should be used. Nevertheless, it requires a set of serious simplifications when applied to a real problem, so that a final mathematical model is formulated in a simple form of first-order linear differential equations [15,16].

The state-space-based phenomenological approach [17] takes into account delays of flow structure development. The authors of [17] proposed a first-order delay differential equation for an additional internal state variable x, which accounts for the unsteady effects associated with separated and vortex flow. The variable x may, for example, represent the location of flow separation or that of vortex breakdown. The form of the differential equation governing x is:

$$
\tau_1\frac{dx}{dt} + x = x_0\left(\alpha - \tau_2\frac{d\alpha}{dt}\right),
\tag{2}
$$

where α is the angle of attack, x_0 describes the steady dependency of x on α, and τ_1 and τ_2 are characteristic times of the flow structure development. This approach was shown to be effective in accurate prediction of the unsteady aerodynamic effects, including unsteady flow over an airfoil with separation [18], a delta wing with vortex breakdown [19], and a maneuvering fighter aircraft [17]. Furthermore, this approach was improved in order to take into account more complicated flow effects. Following this technique, aerodynamic loads can be divided into linear non-delaying and nonlinear delaying components [20,21]. Ordinary differential equations are responsible for the internal dynamics of the nonlinear components of aerodynamic characteristics. The characteristic time constants can be identified using the dynamic wind tunnel [20,21] or CFD [22] test results. Such an approach enables us to describe quite precisely the nonlinear behavior of unsteady aerodynamic characteristics at high angles of attack, namely, the dependence of aerodynamic derivatives on frequency and amplitude of oscillations and the aerodynamic hysteresis. Nevertheless, application of the state-space-based phenomenological approach in an arbitrary case is complicated because of non-formalized and expert-based procedure of the model structure design and identification of the nonlinear components of unsteady aerodynamic characteristics.

Surrogate modeling approaches, which use mathematical approximations of the true responses of the system, are a cost-effective tool for unsteady aerodynamics. The most popular surrogate modeling techniques are artificial neural networks [23–27], Radial Basis Function (RBF) interpolation [9,10], and kriging [28]. Neural Networks (NN) have been recently shown to be a formal and effective tool for modeling nonlinear unsteady aerodynamics regardless of the aircraft configurations. The main reason for such a successful application of NN is the universal approximation properties [29], which enable the NN to be used for an arbitrary aircraft without significant simplifying assumptions. NN was found to be capable of reproducing histories of unsteady aerodynamic loads on the suction side of pitching airfoils in real time [23,24]. Faller et al. [23] utilized experimental data to train a RNN for predicting the pressure coefficient readings along three spanwise positions on the upper surface of the wing. They concluded that RNNs could be applied for time-dependent problems. Reisenthel used a RNN to generate the response function for a nonlinear indicial model in [25].

Several nonlinear models of unsteady aerodynamics are considered in this paper. The mathematical models are developed and tested using the experimental data, obtained for the pitch moment coefficient of the generic Transonic CRuiser (TCR), which was a canard configuration. TCR aircraft was studied in the SimSAC project of 6th European Framework Program. Significant experimental and numerical investigations aimed at understanding the flow over such complex configuration and aerodynamic loads acting on the TCR model were carried out previously [5,6,21,26,27,30,31]. In the present paper, results of the intensive experimental campaign, which was undertaken in order to investigate the main static and dynamic properties of the pitch moment coefficient for TCR, are considered. The campaign included steady and dynamic experiments. The behavior of the steady aerodynamic characteristics versus angle of attack is obtained during the steady tests. The influence of the reduced oscillation frequency and the angle of attack on the unsteady aerodynamic characteristics was studied using small-amplitude forced oscillations. Finally, the forced large-amplitude oscillation tests were carried out for detailed investigation of the unsteady aerodynamics at high-angle-of-attack departures. The details of the conducted experiments are also given in the paper. The experimental data presented in the current paper extend the previously published results [27,30].

Present study is also focused on comparison of ROM for unsteady aerodynamics. Two NN architectures suitable for the reduced-order modeling of unsteady aerodynamic characteristics in the extended angle-of-attack range are considered, namely, a Feed-Forward Neural Network (FFNN) and a Recurrent Neural Network. One of the paper contributions is application of the phenomenological approach used in [17,20,21,27] in order to take into account nonlinear effects due to the complex canard-wing vortex flow in the nonlinear pitch moment coefficient model. This model is used as a benchmark model and compared with the results obtained for the NNs. As concerns the phenomenological state-space model of unsteady aerodynamics, an ordinary differential equation is utilized to describe the effects associated with delay of the vortical flow formation. The paper also deals with comparison of two regularization techniques for NN training that improve the NN performance. Both techniques use the Bayesian rule but one of the techniques implies that the experimental data are heteroscedastic. The results of the experimental data simulation using both the state-space and NN models are presented and compared.

2. Experiments

The prospective civil transport aircraft called TransCruiser (TCR) was designed to operate at transonic speeds. The conceptual design of TCR was implemented by SAAB (Sweden) within the SimSAC project of The Sixth European Framework Program. The aircraft is a configuration with a high-sweep wing with leading edge extension (LEX) and a high-sweep canard surface. The canard is an all moving surface and a close-coupled type. The main geometrical parameters of the tested TCR model were as follows: reference area $S = 0.3056$ m^2, wing span $b_a = 1.12$ m, mean aerodynamic chord $\bar{c} = 0.2943$ m. The general view of the TCR model is given in Figure 1a, and a scheme is shown in Figure 1b, where the model conventional center of gravity is marked. The experiments were conducted in the TsAGI T-103 wind tunnel with the flow velocity $V = 40$ m/s with corresponding Reynolds number Re $= 0.78 \times 10^6$.

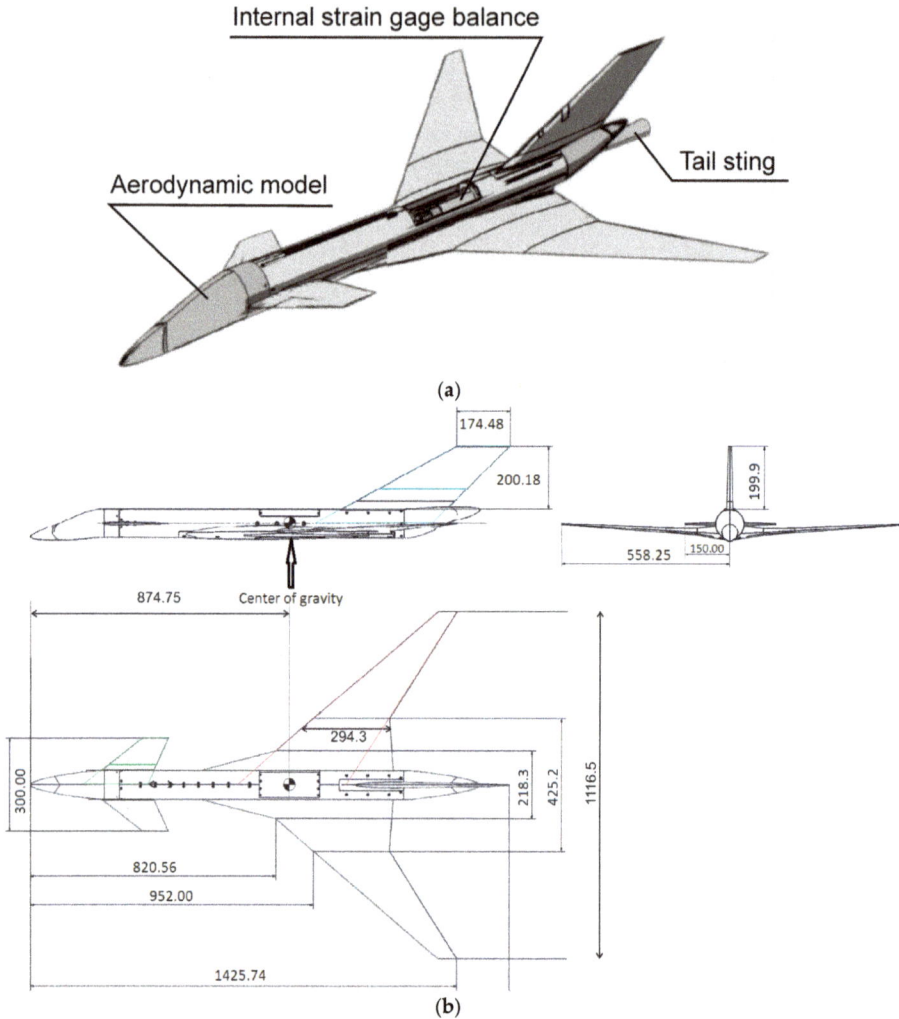

Figure 1. TCR aircraft model (**a**) 3D view of the model mounted on the supporting device; (**b**) 3 views of the model: side (up-left), front (up-right) and top (bottom).

The wind tunnel experimental campaign was carried out in three stages. The tests were performed with the model installed on the tail sting, with the bank angle being equal to 90° (Figure 2). At the first stage, the static aerodynamic characteristics in a wide range of angles of attack were studied. The incidence angle was varied from −10° to 40° with the step of 2°. The static experiments were performed for various configurations of the model, namely, with and without canard. For the full configuration the canard deflection angle φ_c varied from −30° to +10° with a step of 5°. Rotation of the wind tunnel turn table provided variation of the angle of attack. In steady experiments for each angle of attack data sampling time was 2 s and the sampling rate was 100 Hz.

A five-component internal strain gauge balance was used for measurements of forces and moment acting on the aircraft model (a drag force was not measured). Reference point of the balance coincided with the model conventional center of gravity.

In the second stage, the stability derivatives were determined through the small-amplitude forced oscillations. During this experiment a harmonic motion in pitch with a fixed center of gravity is implemented:

$$\alpha = \alpha_0 + A_\alpha \sin(2\pi f t + \vartheta_0),$$
$$\dot{\alpha} = q = 2\pi f A_\alpha \cos(2\pi f t + \vartheta_0). \tag{3}$$

The oscillation amplitude was $A_\alpha = 3°$, frequencies f were 0.5, 1.0, and 1.5 Hz (corresponding reduced frequencies $k = 2\pi f \bar{c}/2V$ were 0.012, 0.023, and 0.035) with the mean angles of attack α_0 varying from $-10°$ to $40°$. For small amplitude forced oscillation experiments the data were sampled 128 times per period of oscillation, each oscillation was repeated 8 times. No special adjustments of the sampling rates depending on the oscillation frequency were carried. These experiments were performed on the forced angular oscillations dynamic rig used in the TsAGI T-103 wind tunnel. The rig is shown in Figure 2. The rig kinematical scheme is shown on the left side, and the TCR model installed on the rig during the wind tunnel tests is shown on the right side. The mean angle of attack α_0 was specified with rotation of the wind tunnel turn table, and variation of the angle of attack α was obtained via oscillation of the supporting sting.

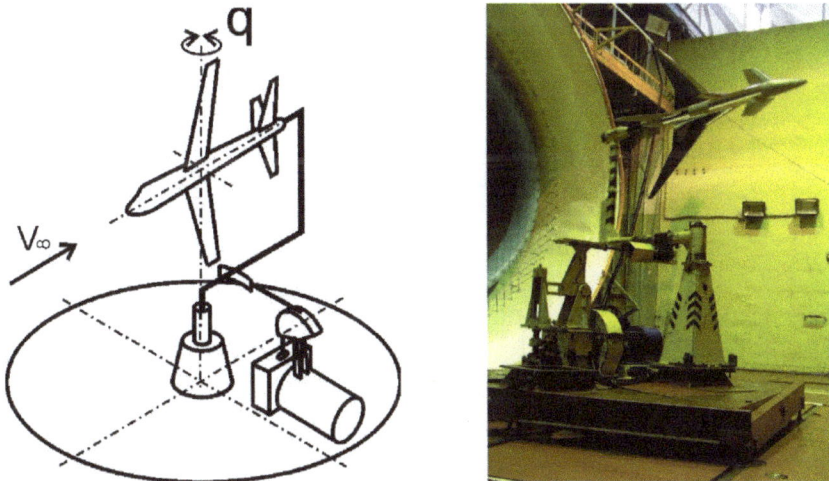

Figure 2. Small-amplitude angular oscillations dynamic rig used in the TsAGI T-103 wind tunnel. The rig configuration is shown on the **left**, the TCR model inside the test section of the wind tunnel is shown on the **right**.

In the third stage, nonlinear unsteady aerodynamic coefficients at high angles of attack were investigated through the large-amplitude forced oscillations in pitch. A view of the TCR model inside the wind tunnel during this stage is shown in Figure 3. These experiments were intended to obtain the additional experimental data for the more comprehensive models in the extended flight envelope.

Figure 3. TCR model at the large amplitude oscillation rig at wind tunnel test section.

The scheme of the rig is demonstrated in Figure 4. Kinematics is also given with the Equation (3). One can see from the figure that during these experiments the model bank angle was 0°. The mean angle of attack (3) was specified with an inclination of the sting support, and variation of the angle of attack was provided with oscillation of the sting with respect to its mean position. Oscillation amplitudes were 10° and 20°, frequencies were 0.5, 1, and 1.5 Hz (corresponding reduced frequencies k = 0.012, 0.023, and 0.035), and the mean angles of attack were 8° and 18°.

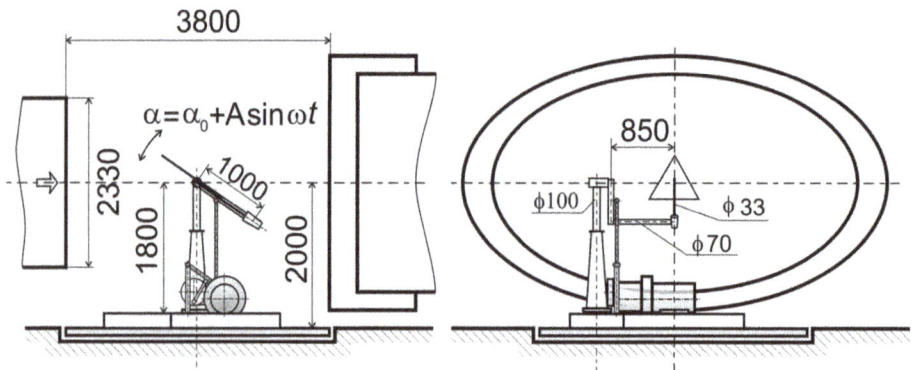

Figure 4. Large-amplitude angular oscillations dynamic rig used in the TsAGI T-103 wind tunnel.

Data sampling rate was similar to the small-amplitude test rate, namely, 128 times per period, which is sufficient enough to capture abrupt variations of the measured aerodynamic characteristics during both types of the experiments. Each oscillation was repeated 16 times.

2.1. Static Aerodynamics Characteristics

The influence of the canard and canard deflection angle φ_c on the coefficients of normal force and pitching moment in steady conditions is shown in Figure 5. The analysis of the experiments shows that influence of the canard on the normal force coefficient C_N is not so significant up to angle of attack α = 10°. At the angles of attack α > 10° the normal force is higher for the canard configurations. The detailed analysis in [23] reveals that, at small angles of attack, the wing in the presence of the canard has less slightly lift than a wing-only configurations; this is mainly due to canard downwash effects on the wing. However, the total lift remains the same because of the additional lift generated on the canard. At higher angles of attack, the wing behind the canard produces more lift than a wing-only geometry [23].

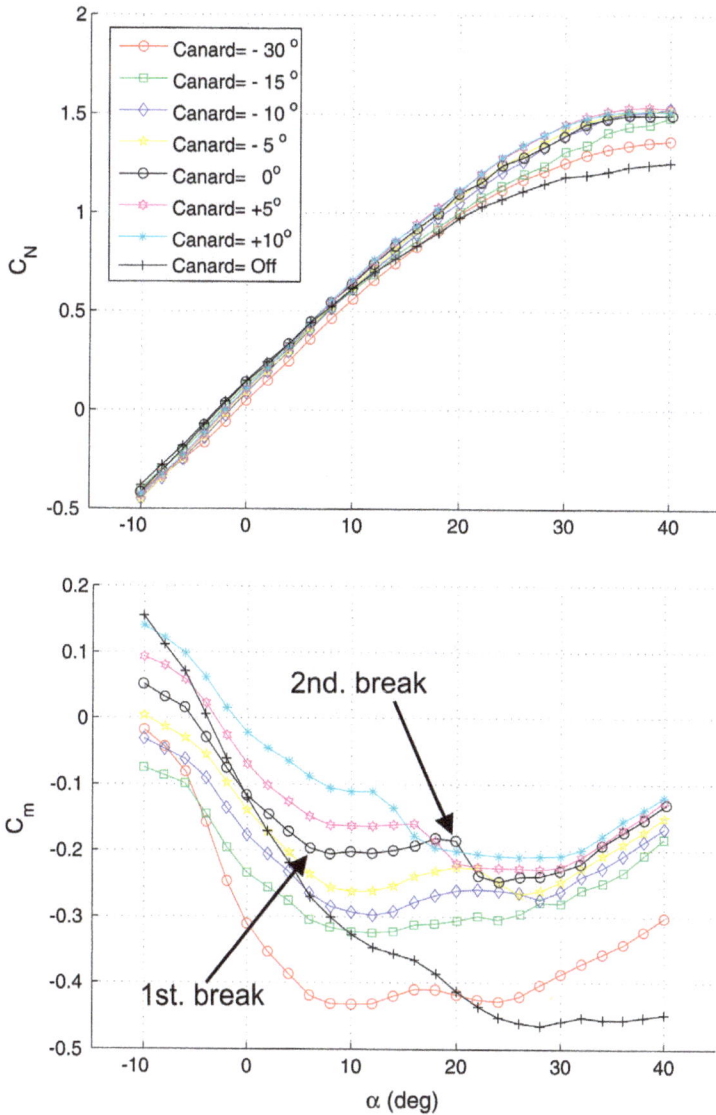

Figure 5. Influence of the canard and canard deflection angle φ_c on the TCR aerodynamic characteristics.

The canard significantly contributes to the total pitching moment coefficient of the TCR model bringing a destabilizing effect. For the case of zero canard deflection, the pitching moment evolution with the angle of attack presents a negative slope (nose down when α increases) up to $\alpha = 6°$, then a first break, after which the slope sign changes, due to the continuously increasing lift of the canard, upstream the reference point (nose up). Then a second break takes place, with a loss of efficiency at about $\alpha = 20°$. The locations of these two breaks depend on the canard deflection angle.

Ghoreyshi et al. [23] reported some flow features of TCR at different angles of attack and at low subsonic speeds. Both the LEX, wing, and canard have rounded leading edges and are swept back at and more than 50°, that causes a complex vortex formation over these surfaces at moderate to high angles of attack. At about $\alpha = 12°$ a canard vortex and an inboard (LEX) and outboard wing vortex are present. The wing in the presence of the canard shows smaller inboard vortices than the canardless configuration; this is due to canard downwash effects that reduce the local angle of attack behind the canard span. On the other hand, the wing outboard vortex is slightly bigger in the presence of the canard. The canard vortex becomes larger with increasing angle of attack. At about $\alpha = 18°$ the wing vortices merge. At about $\alpha = 20°$, the inboard and outboard vortices interact and merge. At $\alpha = 24°$ angle of attack, the canard vortex lifts up from the surface as well. At higher angles, the canard in the TCR aircraft has favorable effects on the wing aerodynamic performance.

2.2. Small-Amplitude Forced Oscillation Characteristics

The small amplitude oscillations are dedicated to determine the aerodynamic derivatives in Equation (1). The experimentally measured aerodynamic derivatives $C_{N_q} + C_{N_{\dot{\alpha}}}$ and $C_{m_q} + C_{m_{\dot{\alpha}}}$ are shown in Figure 6.

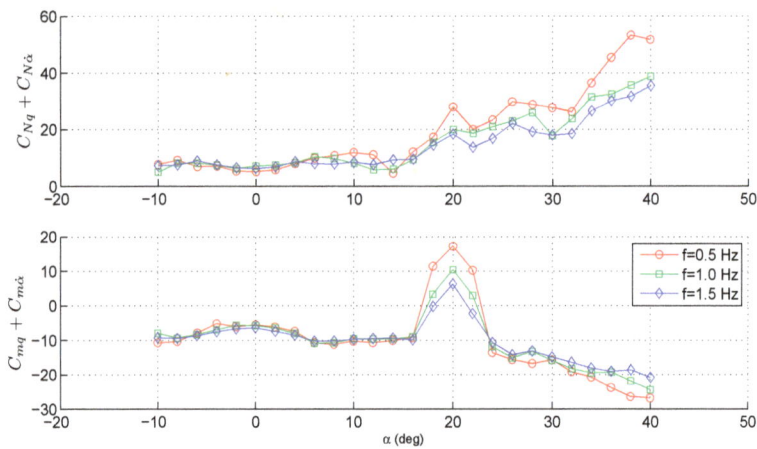

Figure 6. Influence of the oscillation frequency on the unsteady derivatives $C_{N_q} + C_{N_{\dot{\alpha}}}$ and $C_{m_q} + C_{m_{\dot{\alpha}}}$.

These dependencies were obtained at various frequencies of the aircraft model oscillations inside the wind tunnel. It is seen that the influence of the oscillation frequency on the aerodynamic derivatives is small, excluding the incidence region in the vicinity of $\alpha = 20°$. In this region, a dependency of dynamic derivatives values versus the frequency of oscillations is observed. A comparison of the unsteady derivatives obtained for the canard and canardless configurations is given in Figure 7. The influence of the canard on the normal force derivative is significant for angles of attack larger than $\alpha = 32°$; for pitch damping derivative it is also relatively small, except for the region of incidences near $\alpha = 20°$ (Figure 7), where a positive damping for the TCR model is observed.

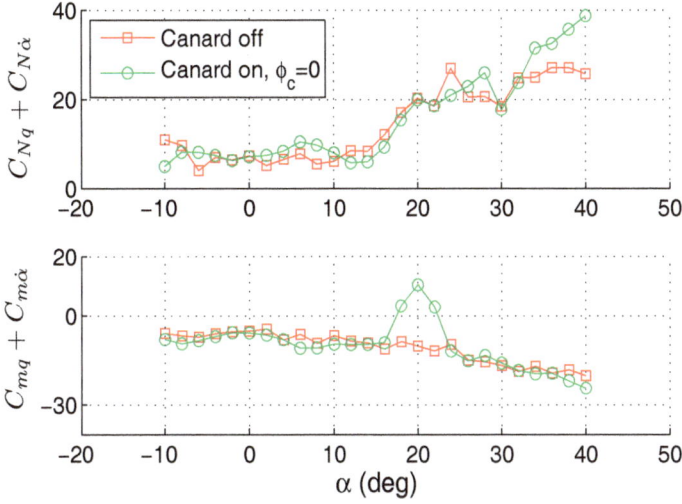

Figure 7. Influence of canard on unsteady aerodynamic derivatives ($k = 0.023$).

The influence of the canard deflection angle φ_c was also investigated: the positive canard deflection moved the positive damping region to lower incidences, with the amplification of the phenomenon as compared to the case of $\varphi_c = 0°$ (Figure 8). The negative canard deflection moved this region to higher angles of attack, with the positive damping being weakened. For the canard deflection angle $\varphi_c = -30°$, the positive damping moved to $\alpha \approx -5°$. For the normal force derivative, such a considerable effect is not observed.

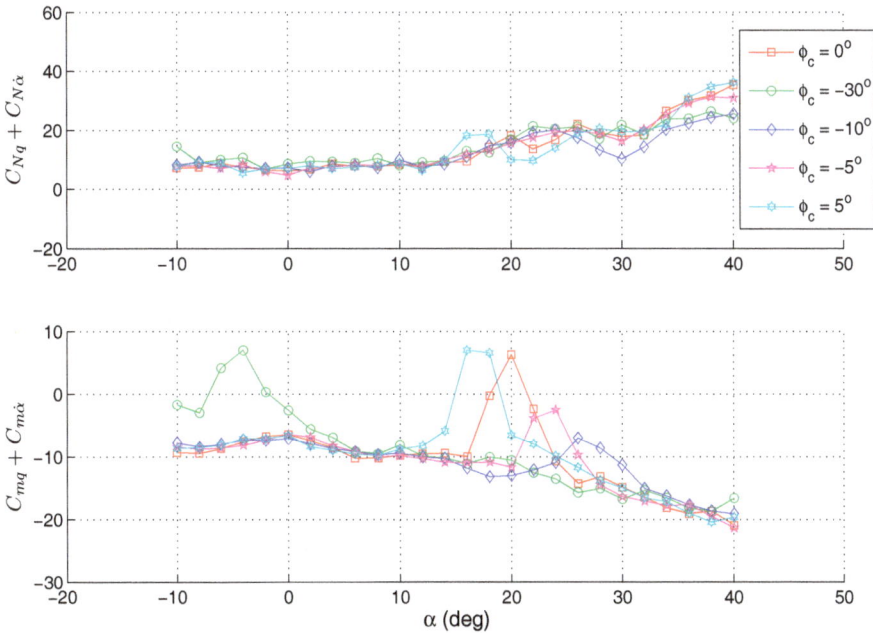

Figure 8. Influence of canard deflection angle on pitch damping derivative ($k = 0.023$).

2.3. Large Amplitude Oscillations Characteristics

In order to investigate the vortex dynamics effect on unsteady aerodynamic characteristics at high angles of attack under high oscillation rates the large amplitude oscillations were carried out. As far as pitch oscillations were concerned, the canard-off TCR configuration revealed the classical linear dynamic effects without any strong nonlinearities. The addition of the canard leaded to severe unsteady effects, not only for angles of attack in the region of $\alpha = 20°$ but also for lower angles of attack (Figure 9).

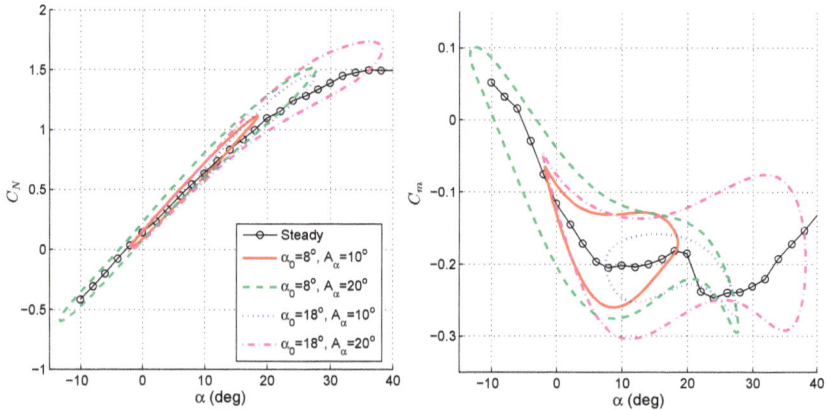

Figure 9. Pitching moment and normal force coefficients evolutions for two sets of large amplitude pitch oscillations—TCR with canard, $\varphi_c = 0°$, $k = 0.035$.

3. Models of Unsteady Aerodynamic Characteristics

3.1. Aerodynamic Derivative Modeling

In order to quantify whether the linear model based on the look-up tables of aerodynamic derivatives is applicable the large amplitude oscillations results are simulated. Since the linear mathematical model in Form of Equation (1) is not valid for the large deviations from the trim incidence α_0 the mathematical model is written in the following form:

$$
\begin{aligned}
C_N(t) &= C_N^{st}(\alpha(t)) + (C_{N_q} + C_{N_{\dot{\alpha}}})\dot{\alpha}(t)\bar{c}/2V \\
C_m(t) &= C_m^{st}(\alpha(t)) + (C_{m_q} + C_{m_{\dot{\alpha}}})\dot{\alpha}(t)\bar{c}/2V
\end{aligned}
\tag{4}
$$

where $C_N^{st}(\alpha(t))$, $C_m^{st}(\alpha(t))$, $C_{N_q} + C_{N_{\dot{\alpha}}}$ and $C_{m_q} + C_{m_{\dot{\alpha}}}$ are derived from the look-up tables of characteristics through the linear approximation. While modeling large-amplitude oscillations, the complexes $C_{N_q} + C_{N_{\dot{\alpha}}}$ and $C_{m_q} + C_{m_{\dot{\alpha}}}$ determined for the same oscillation frequency are used. The results of simulation large amplitude pitch oscillations are shown in Figure 10. One can see that the large amplitude oscillation results for the normal force coefficient can be described with a good precision using the look-up table approach. However, while the simulation of the pitching moment coefficient evolutions fits sufficiently well with the experimental data practically in the overall range of angle-of-attack range, there is a region of the incidences in the vicinity of $\alpha = 20°$, for which this approach is failed to predict the experimental results. These modeling results are in good agreement with the small-amplitude test data given in Figure 6. Particularly, one can see that canard introduce the nonlinear behavior mostly for the pitching moment derivative $C_{m_q} + C_{m_{\dot{\alpha}}}$, while it effect on the normal force derivative $C_{N_q} + C_{N_{\dot{\alpha}}}$ is not so vivid. The canard influence is observed in the vicinity of $\alpha = 20°$, where the nonlinear dependency of the pitch moment derivatives on pitch rate is observed in the experiment (Figure 6), and the linear model failed to describe the experimental results.

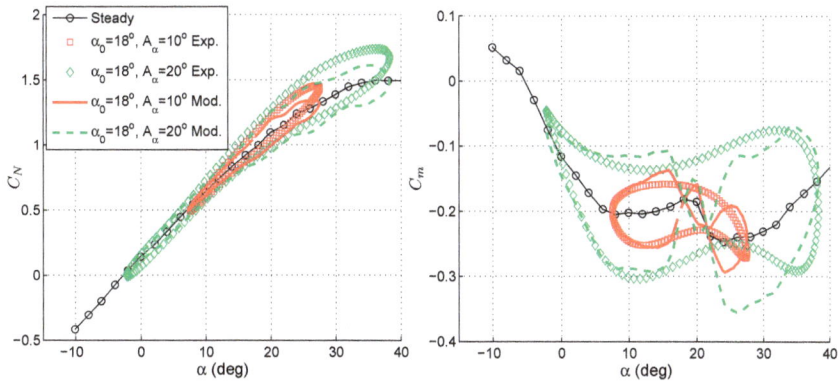

Figure 10. Linear simulations based on the measured aerodynamic derivatives compared to large amplitude oscillation measurements.

Thus, nonlinear approaches should be applied in order to design comprehensive models of the pitch moment coefficient. Below, the NN approach will be used for modeling of unsteady aerodynamics under such conditions. In order to evaluate performance of the NN models they are compared with the state-space modeling approach [17].

3.2. State-Space Model

The state-space model of the pitch moment coefficient was developed through an analysis of the obtained experimental data. The pitch moment coefficient of the TCR model $C_m(\alpha)$ is considered as a sum of the pitching moment of the canardless configuration $C_{m0}(\alpha)$ and the corresponding contribution from the canard $\Delta C_m(\alpha)$ as

$$C_m(\alpha) = C_{m0}(\alpha) + \Delta C_m(\alpha). \tag{5}$$

The canard contribution under the static condition is divided into a term ΔC_{m1}, which is linear in angle of attack, and a nonlinear term ΔC_{m2} which is caused by the canard influence as follows:

$$\Delta C_m(\alpha) = \Delta C_{m1}(\alpha) + \Delta C_{m2}(\alpha) = \Delta C_{m_\alpha}\alpha + \Delta C_m^{nonlin}(\alpha). \tag{6}$$

This representation of the pitch moment coefficient is demonstrated in Figure 11. In the present study an unknown constant of the mathematical model ΔC_{m_α} and nonlinear function $\Delta C_m^{nonlin}(\alpha)$ were determined using the static test results.

In order to describe the internal dynamics due to the vortex structure development the following dynamic equation is applied:

$$\tau_1 \frac{d\Delta C_m^{dyn}}{dt} + \Delta C_m^{dyn} = \Delta C_m^{nonlin}\left(\alpha - \tau_2 q\frac{\bar{c}}{2V}\right), \tag{7}$$

where ΔC_m^{dyn} is the dynamic value of the canard influence due to development of vortex structure, and C_m^{nonlin} is its steady-state value.

This equation is a first-order filter with the time constant τ_1, which is in the left side of this equation. Additionally, incidence delay $\tau_2 q\frac{\bar{c}}{2V}$ is introduced in the function argument in the right-hand side of the equation. For small values of time delays it follows from Equation (7) that $\Delta C_m^{dyn}(t) = \Delta C_m^{nonlin}(\alpha)$, which enables the steady dependences to be satisfied identically.

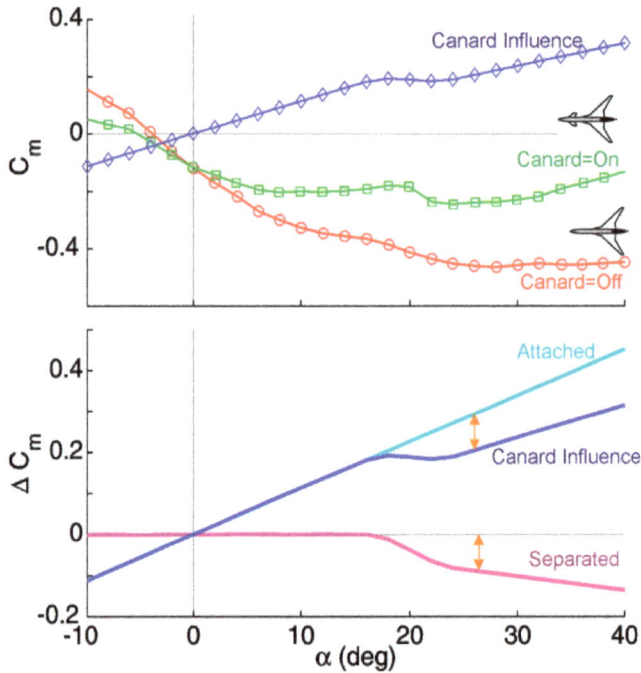

Figure 11. The pitch moment coefficient representation.

The resulting pitch moment coefficient model is the following:

$$C_m = C_{m0}(\alpha) + \left(C_{m_q} + C_{m_{\dot{\alpha}}}\right)_0 \cdot q\frac{\bar{c}}{2V} + \Delta C_{m_\alpha}\alpha + \Delta C_m^{dyn}. \tag{8}$$

The equation combines the linear terms derived from the look-up tables for the canardless and canard TCR configurations (the first, the second and the third terms) and the nonlinear term responsible for the canard influence (the forth terms). Thus, the complete model for the pitch moment coefficient contains Equations (7) and (8) with unknown constants τ_1 and τ_2. The constants C_{m0}, ΔC_{m_α} and the nonlinear function ΔC_m^{nonlin} were determined by means of steady tests of the canardless and the canard configurations. The damping derivative $\left(C_{m_q} + C_{m_{\dot{\alpha}}}\right)$ is a function of the angle of attack and can be determined using the experimental results of the small-amplitude forced oscillations of the canardless TCR configuration.

For identification of the unknown parameters τ_1 and τ_2 the experimental results of small-amplitude pitch oscillations of the canard configuration of TCR model at various frequencies were used. The solution of Equation (7) for the small-amplitude harmonic oscillations in pitch $\alpha(t) = \alpha_0 + A_\alpha \sin kt$ can be linearized. After the substitution of the results into relationship (8) it leads to the following expressions for the aerodynamic derivatives:

$$\begin{aligned}
C_{m_\alpha} &= C_{m0_\alpha} + \Delta C_{m_\alpha} + \frac{d\Delta C_m^{nonlin}}{d\alpha}\frac{1 - \tau_1\tau_2 k^2}{1 + \tau_1^2 k^2} \\
C_{m_q} + C_{m_{\dot{\alpha}}} &= \left(C_{m_q} + C_{m_{\dot{\alpha}}}\right)_0 - \frac{d\Delta C_m^{nonlin}}{d\alpha}\frac{\tau_1 + \tau_2}{1 + \tau_1^2 k^2}.
\end{aligned} \tag{9}$$

It is seen that the aerodynamic derivatives depend on the oscillation frequency in the range of the angles of attack where the nonzero derivative $\frac{dC_m^{nonlin}}{d\alpha}$ exists. The dependencies of aerodynamic derivatives C_{m_α} and $C_{m_q} + C_{m_{\dot{\alpha}}}$ versus oscillation frequency are determined by the characteristic times

τ_1 and τ_2. These values are supposed to be functions of the angle of attack and determined as the smooth cubic spline interpolations in the range of $\alpha = 10 \div 30°$ with the spline maximum in the center of the range (see Figure 9). It is considered that $\tau_2(\alpha) = 0$ beyond this range. The same assumption for τ_1 leads to the degeneracy of differential Equation (4); therefore, $\Delta\tau_1 = 2$ is added to the spline function τ_1. This small addition does not influence significantly the filter characteristics in the left side of Equation (4), but enables the coefficient in front of derivative to be positive. For identification of these constants the following penalty function is introduced:

$$\Phi(\tau_1, \tau_2) = \sum_{i=1}^{n}\sum_{j=1}^{m}[C_{m_\alpha\,test}(\alpha_i, \overline{\omega}_j) - C_{m_\alpha\,sim}(\alpha_i, k_j)]^2 + \sum_{i=1}^{n}\sum_{j=1}^{m}[C^*_{m_q\,test}(\alpha_i, k_j) - C^*_{m_q\,sim}(\alpha_i, k_j)]^2. \quad (10)$$

This function represents the sum of squared differences between the simulation results and the experimental results for the dynamic derivatives, determined in the entire investigated range of angles of attack α_i for three values of the reduced oscillation frequency k_j. For short, the designation $C^*_{m_q} = C_{m_q} + C_{m_{\dot\alpha}}$ is introduced in the expression. To determine the values of τ_1 and τ_2 the function $\Phi(\tau_1, \tau_2)$ should be minimized. It is seen in Figure 12 that this function has a flat minimum, which can be found using the conventional minimization techniques. The resulting functions $\tau_1(\alpha)$ ($\tau_{1\,max} \approx 32.7$) and $\tau_2(\alpha)$ ($\tau_{2\,max} \approx 3.9$) are shown in Figure 12.

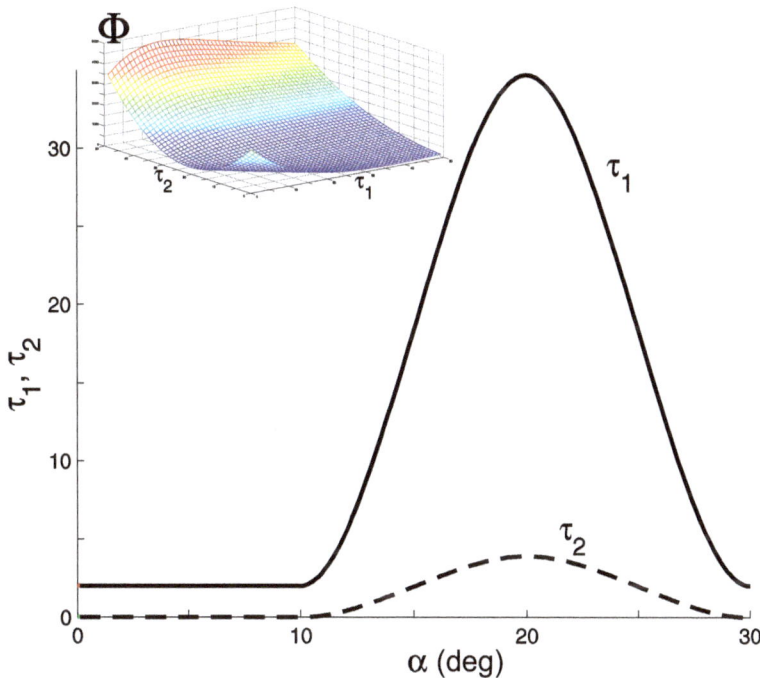

Figure 12. The results of identification of $\tau_1(\alpha)$ and $\tau_2(\alpha)$. Penalty function $\Phi(\tau_1, \tau_2)$ is shown in the upper left part of the figure.

The aerodynamic derivatives C_{m_α} and $C_{m_q} + C_{m_{\dot\alpha}}$ versus angles of attack, simulated with the proposed mathematical model, are shown in Figure 13 with lines. The simulation results for various oscillation frequencies are demonstrated by lines of different types. The corresponding experimental results are shown with different markers. The developed state-space model describes adequately the results observed in the dynamic experiment in the entire ranges of the angles of attack and oscillation frequencies. It is important that the model describes the positive damping zone within the range of the angles of attack of $\alpha = 15$–$25°$ and the dependencies of the derivatives versus oscillation frequency, which are observed in the experiment.

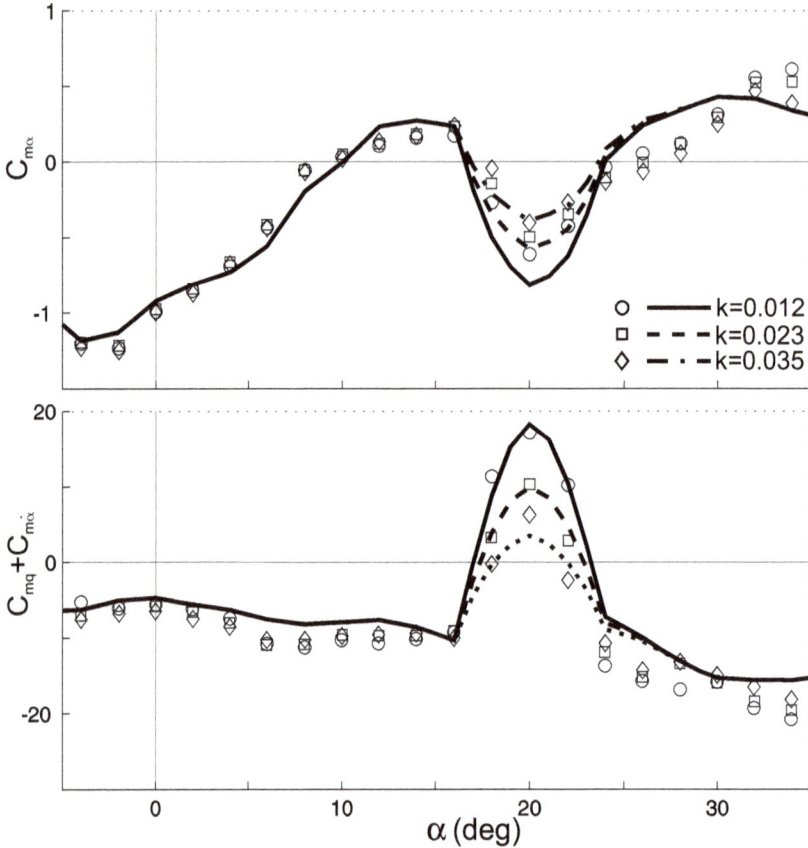

Figure 13. Aerodynamic derivatives obtained for different oscillation frequencies: experiments (markers) and state-space model simulation (lines).

The state-space model (Equations (7) and (8)) developed only with the data from the forced small-amplitude oscillation tests was also applied to simulate the forced large-amplitude oscillations. The results obtained for three test cases are shown in Figure 14. The results of simulation (solid lines) and the unsteady experiment data (markers) for the dynamic values of $C_m(t)$ are compared in the upper plots. The measured static values of $C_m(\alpha)$ are shown with dashed lines. The evaluations of dynamic components caused by the canard vortex flow formation $\Delta C_m^{nonlin}(t)$ (solid lines) are shown in the bottom plots. The static components of the vortex flow influence $\Delta C_m^{nonlin}(\alpha)$ are shown with dashed lines in the same plots. The bottom graphs demonstrate the contribution of the differential equation with delay (Equation (7)) to the general mathematical model (Equation (8)).

The modeling results of the dynamic effects at the mean angle of attack $\alpha_0 = 18°$ of the pitch oscillations with large amplitude $A_\alpha = 10°$ and small reduced frequency $k = 0.012$ are shown in Figure 14a. The positive damping in the sense of the linear mathematical model (1) is observed at angles of attack in the vicinity of $\alpha_0 = 18°$. The additional kink in the dynamic loop demonstrates this fact. While the oscillation amplitude increasing, the positive damping practically vanishes in both the experiment and the simulation (see Figure 14b). Further oscillation frequency growth leads to a significant expansion of the hysteresis loop. This effect in the experimental and simulation results is shown in Figure 14c.

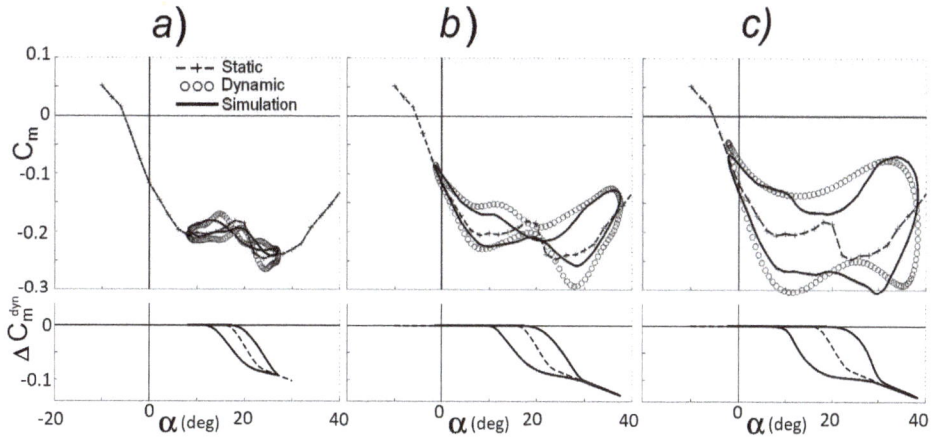

Figure 14. Simulation results for the forced pitch oscillations with large amplitude: (a) $\alpha_0 = 18°$, $A_\alpha = 10°$, $k = 0.012$; (b) $\alpha_0 = 18°$, $A_\alpha = 20°$, $k = 0.012$; (c) $\alpha_0 = 18°$, $A_\alpha = 20°$, $k = 0.012$.

The NN techniques described and applied below are compared with the state-space approach.

4. Neural Network Modeling

FFNN and RNN are considered in the paper. The NN model of unsteady pitch moment coefficient of TCR using RNN was developed in this paper and compared with the model obtained using FFNN [19].

4.1. NN Architectures

The FFNN, which scheme is given in Figure 15a, can be considered as a directed graph with neurons placed in it nodes. The neurons of the first layer do not implement nonlinear mapping but distribute input signals between neurons of the first hidden layer. Neuron of the hidden layer is an elementary calculating unit. A set of signals $S_j, j = 1 \ldots n$ from the input layer are fed into the neuron of the hidden layer. Coefficients w_{ik} correspond to the signal transmit connections and are the weight factor while summing the input signals. Neuron bias b_k is added to the weighted sum of the input signals, and the resulting sum is mapped through nonlinear activation function f_k. Mapped signal ϕ_k goes forward to the next-layer neurons, which implement the same operations and transmit the signal further. The signal from the last layer is output from the NN.

RNN can be represented as FFNN with feedback connections. NARX (Nonlinear AutoRegressive model with eXogenous variables) architecture [32], which is given in Figure 15b, is used in the present study. For modeling variable y at time t, the state vector $\mathbf{x}(t)$ and a series of its former values $\mathbf{x}(t-1), \mathbf{x}(t-2) \ldots \mathbf{x}(t - D_{in})$ are fed into the NN. The values of the modeling variable $y(t-1), y(t-2) \ldots y(t - D_{out})$ calculated by the NN earlier are also added to input signal. The resulting NN model can be presented in the following form:

$$y(t) = M(\mathbf{x}(t), \mathbf{x}(t-1), \ldots, \mathbf{x}(t - D_{in}), y(t-1), \ldots, y(t - D_{out})), \tag{11}$$

where M is the function of NN mapping.

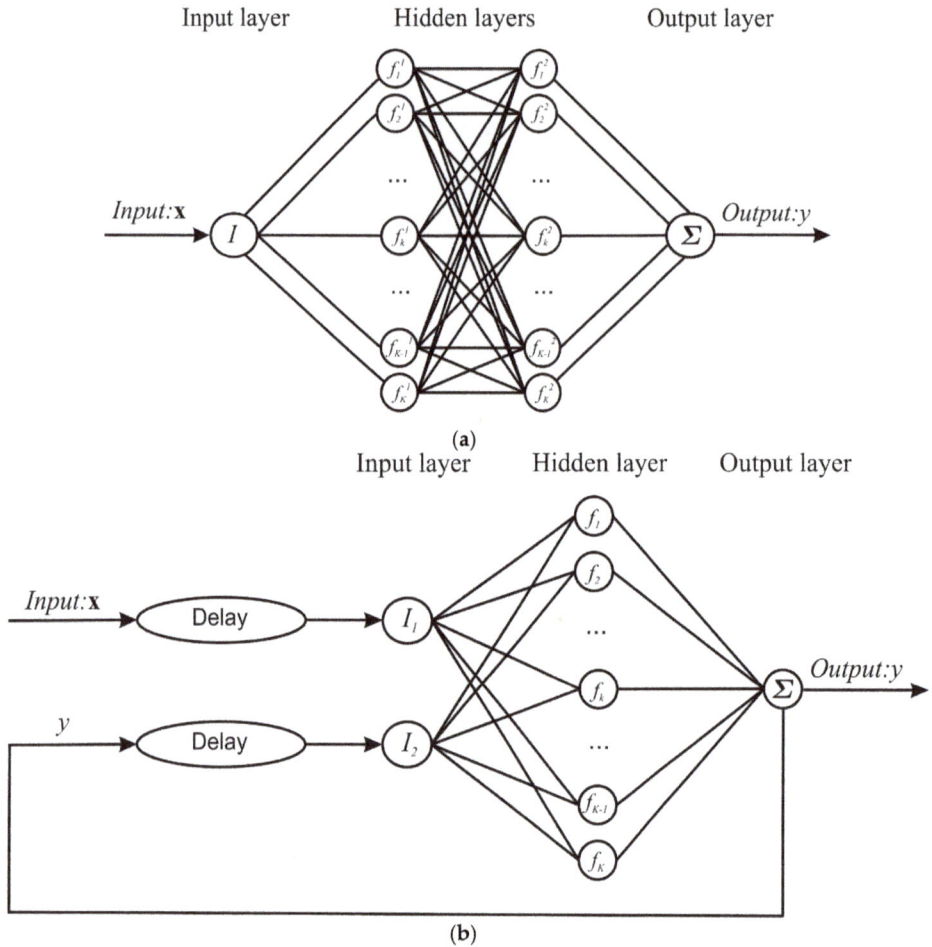

Figure 15. NN architectures: (**a**) FFNN architecture; (**b**) NARX architecture.

4.2. Regularization Techniques during NN Training

The problem considered in the present study is to develop a NN model using a restricted set of experimental data and to apply the model for flight dynamics applications, implying arbitrarily aircraft maneuvers; thus, generalization ability of the model is crucial. Regularization is one of the popular techniques preventing overfitting of a regression model and improving its generalization ability [33]. A short introduction to regularization techniques is given below.

Connection weights w_{ik} and biases b_k are adjusted during NN training when the examples of training set are presented through minimization of the difference between NN operation results y_i and target value a_i for each example ($i = 1 \ldots N$) from the training set

$$E_D = \frac{1}{2} \sum_{i=1}^{N} (y_i - a_i)^2. \tag{12}$$

One of the important problems of NN training is called overfitting. The error in the training set is driven to a very small value, but when new data are presented to the network the error is large. The network has memorized the training examples, but it has not learned to generalize to new situations. A part of the whole initial data can be used for model testing. The test set error should be made as small as possible and must not be significantly higher than the training set error. When this condition is valid, the NN is considered to have good generalization performance.

Regularization is one of the techniques for improving generalization. According to this technique, a term penalizing the NN for weight increase is added in the objective function besides the error measure E_D (12). The sum of squares of the weights can be used for this purpose:

$$E_W = \frac{1}{2} \sum_{j=1}^{K} w_j^2, \tag{13}$$

where K is a number of neural network weights; the objective function takes the form

$$F = \eta E_D + \rho E_W, \tag{14}$$

where η and ρ are objective function parameters. To develop the mathematical model with high generalization ability, Bayes' rule was proposed to define the objective function parameters [34]. Algorithm of Gauss–Newton approximation to Bayesian regularization (GNBR) for training NN was further implemented in [33].

GNBR algorithm is the effective tool to improve NN generalization, but it supposes the model error to be the same on different subsets of initial data. The unsteady aerodynamic models for flight dynamics problems are developed using different dynamic experiments, in the various ranges of kinematic parameters, and with different accuracies. To obtain more precise models the data could be considered as heteroscedastic. The GNBR algorithm was modified for the case of heteroscedastic data and Bayesian Regularization to NN training on Heteroscedastic Data (BRHD) was proposed [35]. Below the proposed algorithm is briefly discussed.

4.3. Bayesian Regularization to NN Training on Heteroscedastic Data (BRHD)

Let us suppose that experimental data to be approximated are obtained in n types of different experiments $(\mathbf{x_1}, \mathbf{a_1}), (\mathbf{x_2}, \mathbf{a_2}), \ldots, (\mathbf{x_n}, \mathbf{a_n})$, where $\mathbf{x}_i = \left(x_{i_1} \ldots x_{i_{N_i}} \right)$ is the vector of values of the controlled phenomenon parameter, obtained in i-th type of experiment, $\mathbf{a}_i = \left(a_{i_1} \ldots a_{i_{N_i}} \right)$ is the vector of values of the observed variable obtained in i-th type of experiment, $D_i = \left\{ x_{i_{m_i}} a_{i_{m_i}} \right\}$, $m_i = 1 \ldots N_i$ is the dataset obtained at the same type of experiment.

The problem is to identify the NN function y that describes the obtained experimental data $D_i = \left\{ x_{i_{m_i}} a_{i_{m_i}} \right\}$, $m_i = 1 \ldots N_i$:

$$
\begin{aligned}
a_{1_{m_1}} &= y\left(x_{1_{m_1}} \right) + v_{1_{m_1}}, & m_1 &= 1 \ldots N_1, \\
a_{2_{m_2}} &= y\left(x_{2_{m_2}} \right) + v_{2_{m_2}}, & m_2 &= 1 \ldots N_2, \\
&\quad \cdots \\
a_{n_{m_{n-1}}} &= y\left(x_{n_{m_{n-1}}} \right) + v_{n_{m_{n-1}}}, & m_{n-1} &= 1 \ldots N_{n-1} \\
a_{n_{m_n}} &= y\left(x_{n_{m_n}} \right) + v_{n_{m_n}}, & m_n &= 1 \ldots N_n
\end{aligned}
\tag{15}
$$

The errors in each experiment $v_{i_{m_i}}$, $m_i = 1 \ldots N_i$ are supposed to be independent and normal with zero statistical expectation but with different standard deviations σ_i. Using Bayes' rule, the following objective function can be obtained:

$$F = \frac{1}{2} \eta \mathbf{w}^T \mathbf{w} + \frac{1}{2} \mathbf{e}^T \mathbf{R} \mathbf{e}, \tag{16}$$

where $\mathbf{w} = (w_1\ w_2\ \dots\ w_K)^\mathrm{T}$ is the vector of weights, $\mathbf{e} = (e_1\ \dots\ e_N)^\mathrm{T}$ is the vector of errors, $e_j = y(x_j) - a_j$ is the error of approximation of j-th data pair, $\mathbf{R}$ is the matrix $N \times N$; the objective function parameters ρ_i are placed on the main diagonal of matrix $\mathbf{R}$, the other elements of this matrix are equal to zero:

$$
\mathbf{R} = \begin{pmatrix}
\rho_1 & 0 & \cdots & & & & & 0 \\
0 & \rho_1 & 0 & \cdots & & & & 0 \\
& & \cdots & & & & & \\
0 & \cdots & 0 & \rho_i & 0 & \cdots & & 0 \\
0 & & \cdots & 0 & \rho_i & 0 & \cdots & 0 \\
& & & & \cdots & & & \\
0 & & & & & 0 & \rho_n & 0 \\
0 & & & & & & 0 & \rho_n
\end{pmatrix}.
\tag{17}
$$

Note that the objective function in form (16), which is used instead of (14), contains the weighted sum of errors on each subset $\mathbf{e}^T\mathbf{Re}$, with weights ρ_i, corresponding to each subset. Following Bayes' rule, the expressions for the objective function parameter η (16) is obtained [35]:

$$
\eta \approx \frac{K - \eta\,\mathrm{Sp}(\mathbf{H}^{-1})}{\mathbf{w}^T\mathbf{w}},
\tag{18}
$$

where K is the total number of parameters in the network, $\mathbf{H} = \nabla^2 F$ is the Hessian matrix of the objective function, and Sp is the matrix trace.

The following expressions can be obtained for ρ_i:

$$
\rho_i = \frac{N_i}{\mathbf{e}^T \frac{d\mathbf{R}}{d\rho_i}\mathbf{e} + \mathrm{Sp}\left(\frac{d\mathbf{H}}{d\rho_i}\mathbf{H}^{-1}\right)},
\tag{19}
$$

where N_i is the number of patterns of the i-th training subset.

Within this approach, the parameters of the objective function corresponding to the data subset are adjusted subject to their approximation errors.

The algorithm for implementation of the described training technique was developed [35]. To obtain values of the objective function parameters it is required to calculate Hessian matrix in the minimum point of objective function F. The Gauss–Newton method is applied to approximate Hessian matrix with modified Levenberg–Marquardt optimization algorithm used to locate the minimum point:

$$
\mathbf{w}_i = \mathbf{w}_{i-1} - \left(\mathbf{J}^T\mathbf{RJ} + (\alpha + \mu)\mathbf{E}\right)^{-1}\left(\mathbf{J}^T\mathbf{Re} + \alpha\mathbf{w}_{i-1}\right).
\tag{20}
$$

Let us consider the Levenberg–Marquardt algorithm in more detail. When the scalar μ is zero, this is just Newton's method, using the approximate Hessian matrix. When μ is large, this becomes gradient descent with a small step size. Newton's method is faster and more accurate near an error minimum, so the aim is to shift toward Newton's method as quickly as possible. Thus, μ is decreased after each successful step (reduction in performance function) and is increased only when a tentative step would increase the performance function. In this way, the performance function is always reduced at each iteration of the algorithm [36].

The modification proposed in the present paper improves the Levenberg–Marquardt algorithm convergence in the case of heteroscedastic data in the vicinity of the minimum point.

4.4. Modeling

The RNN model of unsteady pitch moment coefficient, which has a NARX configuration, is compared with the FFNN model presented in [26]. The RNN has one hidden layer. RNN containing from five to 20 neurons in the hidden layer were tested and 12 neurons were selected because this number provides better generalization. Hidden layer neurons have a sigmoid activation function:

$$f_k(x) = \frac{1}{1 + e^{-x}}. \tag{21}$$

Experimental data, which are used to train the NN, consisted of the oscillation cases corresponding to different amplitudes and frequencies of oscillation. Evolutions of the pitch moment coefficient and kinematic parameters during each oscillating case are discretized in time into 128 steps both for small- and large-amplitude tests. Small-amplitude oscillation cases total 78; large-amplitude oscillation cases total 12. The training patterns are composed of the target data, which are the records of pitch moment coefficient $C_m(i)$, $C_m(i-1)$ at steps i, $i-1$, together with the input vector. In the present study the input vector included the angle of attack $\alpha(i)$ and pitch rate $q(i)$ at the i-th step, and the motion parameters $\alpha(i-1), \alpha(i-2), q(i-1), q(i-2)$ at previous steps $i-1$, $i-2$. Usage of angle of attack and pitch rate as the main input parameters is motivated by the statement of the problem. Namely, the developed NN model should be used for flight dynamics problems, and, hence, we should use only parameters available during a real flight. Influence of the Mach and Reynolds numbers are not considered in the present experimental study and, hence, are not included in the NN model as the input parameters.

To compare only the NN configurations, the regularization technique (GNBR) is selected to be the same as for FFNN in [26].

To train the RNN, a special configuration can be used. Because the true output is available during the training of the network, it is possible to create a feed-forward architecture, in which the true output is used instead of feeding back the estimated output. This has two advantages. The first is that the input to the feed-forward network is more accurate. The second is that the resulting network has a purely feed-forward architecture, and static back propagation can be used for training [32]. The stopping criterion for training was exceeding a threshold value (10^{20}) by the Levenberg–Marquardt algorithm parameter μ, which corresponded to reaching a minimum of the objective function (16).

Thirty-six out of 78 small amplitude test cases and eight out of 12 large amplitude test cases were randomly selected for training; the rest of the data were used for testing.

At the modeling stage, predicting the pitch moment coefficient $C_m(i)$ RNN uses results computed at the previous time step $C_m(i-1)$, along with the current and two previous steps of input signal. Hereby, the model is a nonlinear regression on seven parameters. As is shown in [26], a six-dimensional state vector is enough to specify the harmonic oscillation process.

In the first step, we simulated the aerodynamic derivatives of pitch moment coefficient (1). They were obtained with RNN as follows. First, the forced small-amplitude oscillations of the aircraft model were simulated. Then, the coefficients of the model (1) $C_{m\alpha}$, $C_{mq} + C_{m\dot{\alpha}}$ were identified from the simulated data using the linear regression method. RNN simulation of the pitch moment derivatives, compared with the small-amplitude experiment, is given in Figure 16. It can be seen from the figure that the RNN model captures the dependency of the derivatives on oscillation frequency, which is observed in the angle-of-attack range $16° < \alpha < 24°$ and corresponds to the development of the vortical flow above the wing surface. Here the results from both the training and testing subsets are demonstrated together in order to illustrate that the developed model describes all available small-amplitude test results and can be used for prediction of the unsteady aerodynamics phenomena in the overall studied angle-of-attack range. Nevertheless, a detailed study of the model performance on both training and test subsets is given in Section 4.

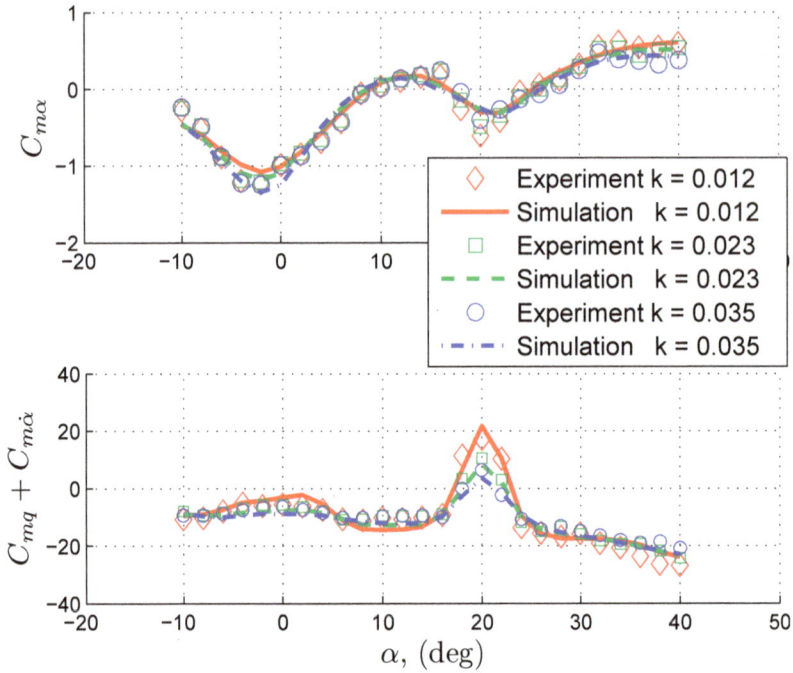

Figure 16. Unsteady aerodynamic derivatives of the pitch moment coefficient, simulated with RNN (lines) and obtained in the experiment (markers).

Hysteresises of the pitch moment coefficient obtained by the RNN simulation of the forced large-amplitude oscillations are given in Figure 17. The experimental results are also plotted in the same figure. These cases are from the testing subset. One can see that there is a good correlation between the experiment and the simulation, and a good generalization is exhibited by RNN.

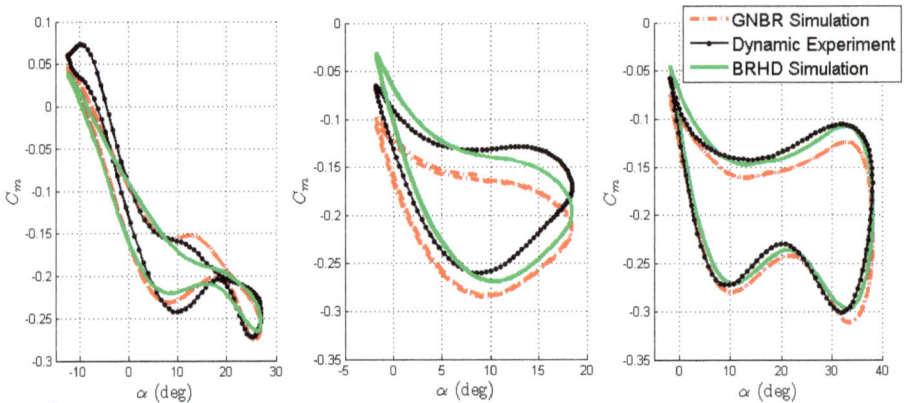

Figure 17. Forced large-amplitude oscillation results for the pitch moment coefficient.

It can be concluded from Figures 16 and 17 that RNN describes the nonlinear behavior of the pitch moment coefficient in the extended range of the angles of attack, which are observed in the experiments.

FFNN, developed in [26], had two hidden layers, with 12 neurons in the first layer and seven in the second. The neuron activation function of the both layers was also chosen sigmoid function. Patterns for the training of the FFNN were composed of the records of pitch moment coefficient C_m, together with the state input $\mathbf{x} = (\alpha(t), q(t), t, \alpha_0, A_\alpha, k)$ determined for the oscillation case. Each oscillating case was discretized in time into 128 steps (similar to the experiment data) and readings from a strain-gauge balance were used for each step. Two thirds of the experimental data were used to train the neural network, and one third of the experimental data were used to test the generalization ability. To simulate the pitch moment coefficient at the time t of an experiment case the whole input vector at this time step should be fed in the NN.

Backpropagation was used to train the NN. Levenberg–Marcquardt algorithm, combined with Bayesian regularization (GNBR), was utilized to minimize the error [26]. The following section gives a comparison of the performance of the discussed NN models.

5. Comparison of the Models

First, let us compare the NN models. The performances of the NN models were tested quantitatively by calculating the errors obtained for the models of pitch moment coefficient C_m and the complex of aerodynamic derivatives $C_{mq} + C_{m\dot{\alpha}}$ separately for the train and test subsets. The error measure is the mean square error divided by the entire range Δy of the measured value y^{test}:

$$err_i = \frac{\sqrt{\frac{1}{N_i-1} \sum_{j=1}^{N_i} \left(y_j^{test} - y_j^{sim}\right)^2}}{\Delta y}. \tag{22}$$

The results are given in Table 1.

Table 1. Errors of unsteady aerodynamics models.

Model	Regularization	Variable	err_i	
			Training Subset, %	Test Subset, %
FFNN	GNBR	$C_{mq} + C_{m\dot{\alpha}}$ (small amplitudes)	3.88	3.96
		C_m (large amplitudes)	15.25	27.42
RNN	GNBR	$C_{mq} + C_{m\dot{\alpha}}$ (small amplitudes)	7.09	8.58
		C_m (large amplitudes)	5.59	8.3
RNN	BRHD	$C_{mq} + C_{m\dot{\alpha}}$ (small amplitudes)	5.65	5.77
		C_m (large amplitudes)	4.53	6.34
State-space model	-	$C_{mq} + C_{m\dot{\alpha}}$ (small amplitudes)	9.11	
		C_m (large amplitudes)	6.87	

Considering the table, one can conclude that the FFNN error for the small amplitude test is approximately the same as for the training and testing subsets. The error for the large amplitude test is very high, approximately four times higher for the training subset and seven times higher for the test subset as compared with the small-amplitude results. Thus, the FFNN provides worse performance for the large amplitude subset. The reason is that the small-amplitude training examples are dominant in the overall training set, namely, 36 out of 44 cases. The FFNN trained better to model the small-amplitude behavior shows poor performance for the large-amplitude subset. This is not satisfactory from the point of view of flight dynamics because a model should guarantee an equal level of model precision in the overall simulation domain.

On the contrary, errors of the RNN determined for small- and large-amplitude subsets are not as high as the FFNN large-amplitude errors and are very close to each other, especially for the testing subsets. This indicates that the RNN is trained to model both small- and large-amplitude results almost equivalently and the shortcomings of the FFNN have been overcome.

Another important remark should be given while comparing the FFNN and RNN architectures. In addition to the aforementioned advantage of RNN, there is another one. It is possible to simulate any consequence of time-dependent states using the RNN thanks to the feedback connection. On the contrary, while using FFNN one should input all parameters of the oscillation cycle, including amplitude and frequency of oscillation, which is not suitable for flight dynamics applications. This comparison reveals that the recurrent configuration is a favorable technique for the simulations of time-dependent unsteady aerodynamic characteristics in flight dynamics problems.

Unsteady aerodynamic characteristics obtained in different types of the experiments are obtained with different errors. Performances of the mathematical models are improved using the BRHD technique that considers the model developed using the experiment data from different types of experiments as heteroskedastic [27].

While comparing the regularization techniques, the recurrent architecture was used because in the previous section this NN configuration was shown to be better for the problem of modeling of aerodynamics in flight dynamics than the feed-forward architecture. The NN was trained using both the GNBR and the BRHD algorithms on the same data. For the BRHD, 8 neurons in the hidden layer were selected, i.e., it is less than for GNBR (12 neurons). The model performances were compared in several ways. First, the testing was done graphically by coplotting C_m values measured in the experiment and predicted by the models. The results of large-amplitude modeling are shown in Figure 17, where it can be concluded that the NN model, trained with the BRHD algorithm, has better coincidence with the experiments.

More thorough analysis of the obtained results was implemented to determine whether the BRHD algorithm helped to improve accuracy of models derived from the different-type experiment data. For this purpose, the NN models of TCR pitch moment coefficient, obtained using the BRHD and GNBR training algorithms were compared. A quantitative comparison of the training techniques was done in the same way as for the configuration comparison through the error calculation (22). The results are also given in Table 1. The comparison reveals the accuracy improvement of the model developed with BRHD. The errors for C_m decreased by 23% and 31% for the train and test subsets, respectively. The errors for $C_{mq} + C_{m\dot{\alpha}}$ decreased by 25% and 49% for the train and test subsets, respectively.

In addition, the scatterograms plotted for the test subsets of $C_{mq} + C_{m\dot{\alpha}}$ and C_m are shown in Figure 18a,b. The BRHD technique yields less scattering.

The analysis presented above shows that the regularization technique (BRHD) improves the model accuracy if two or more subsets obtained in different experiments are used to train the NN.

The error of the state-space model, which is calculated according to Equation (22) in order to evaluate the performance of the NN modeling approach, is given in Table 1. The RNN trained with the conventional GNBR algorithm shows better accuracy as compared to the state-space model for small amplitudes, and for the large-amplitude tests the performance of the RNN (GNBR) is better in the training test and worse in the testing subset. The RNN (BRHD) has better precision for both small- and large-amplitude oscillation data.

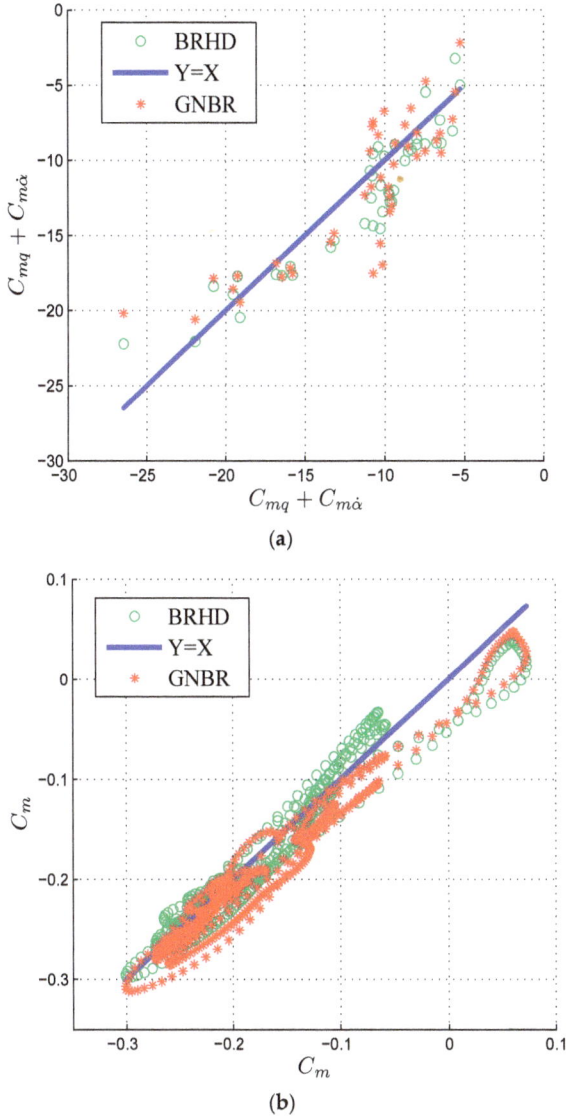

Figure 18. Scattering diagrams: test subsets: (**a**) Small amplitude subtest; (**b**) large amplitude subtest.

6. Conclusions

An experimental investigation of the aerodynamic characteristics of the prospective civil transport aircraft TCR has been carried out in the TsAGI T-103 wind tunnel. The aircraft was a configuration with a high-sweep wing with LEX and the high-sweep canard surface. A three-stage experimental campaign was undertaken. In the first stage, the steady aerodynamic characteristics were under consideration. The influence of the reduced oscillation frequency and the angle of attack on unsteady aerodynamic derivatives was studied in the second stage. In the third stage, forced large-amplitude oscillation tests were carried out for the detailed investigation of the unsteady aerodynamics at high-angle-of-attack departures. The analysis of the experiments revealed that canard had a great impact on the overall aircraft performance. Static experimental results showed that the influence of the canard on the normal force coefficient C_N was not so significant up to the angle of attack $\alpha = 10°$. At angles of attack $\alpha > 10°$ the normal force was higher for the canard configurations. Such behavior at high angles of attack was due to the fact that the wing behind the canard produced more lift than a wing-only geometry because of the canard–wing vortex interaction.

The canard also significantly contributed to the total pitching moment coefficient of the TCR model, making it less stable. In addition, canard-wing vortex interaction phenomena caused the positive damping peak in the stability derivative $C_{mq} + C_{m\dot{\alpha}}$ obtained in a small-amplitude forced oscillation experiment, which was not observed for the wing-only configuration. Changing the canard deflection angle φ_c, one could change the position and amplification of the positive damping peak.

Concerning large-amplitude forced oscillations, the wing-only configuration revealed the classical linear dynamic effects without strong nonlinearities. The addition of the canard led to severe unsteadiness in the form of hysteresises. The delay of complex vortical flow development caused the dependence of the aerodynamic derivatives on the oscillation frequency and the complicated hysteresis loops of the total pitch moment coefficient, observed in the large-amplitude oscillation.

While modeling large-amplitude oscillation results using the look-up tables approach, the large amplitude oscillation results for the normal force coefficient were described with good precision. However, for the pitch moment coefficient the technique failed and several more sophisticated mathematical models obtained via different popular approaches, namely, neural network and the phenomenological state-space modeling technique, were developed.

We compared several approaches for reduced-order modeling that are capable of capturing the observed nonlinear phenomena. In particular, NN of the feed-forward and recurrent architectures were compared with each other and with the state-space model. RNN trained with the BRHD algorithm showed better results in terms of prediction ability. Comparison of the NN models revealed that the recurrent architectures were favorable for modeling of unsteady aerodynamic characteristics in flight dynamics problems. The advantage of the RNN was the feedback connection, which provided prehistory of motion and brought the information required for modeling dynamic processes. In addition, RNN demonstrated better generalization ability, which was an important advantage because the ROM of aerodynamics were designed with a restricted set of kinematic parameters, which could be obtained in the wind tunnel or CFD tests. However, solving of flight dynamic problems, including ground-based simulator studies, supposes simulation of arbitrary aircraft maneuvers.

Acknowledgments: This work has been carried out with funding from the Russian Ministry of Education and Science into the subject "Applying artificial neural networks to promote flight safety," project 14.624.21.0046.

Author Contributions: Dmitry Ignatyev and Alexander Khrabrov conducted the experiments; Dmitry Ignatyev developed NN models; Alexander Khrabrov developed state-space model, Dmitry Ignatyev and Alexander Khrabrov analyzed the data, Dmitry Ignatyev wrote the paper.

Conflicts of Interest: The authors declare no conflict of interest.

Nomenclature

A_α	amplitude of oscillation
y_i	neural network operation results
a_i	target value
b_a	wing span
b_k	neuron bias
S_m	pitch moment coefficient
$\bar{c}$	mean aerodynamic chord
E_D	sum of squared neural network errors
E_W	sum of squared neural network weights
e_j	neural network error

err_i	error measure
F	objective function
f_k	neuron activation function
$\mathbf{H}$	Hessian matrix
$\mathbf{J}$	Jacoby matrix
k	reduced oscillation frequency
M	function of neural network operations
S_j	input signals fed into neuron
t	time
V	airspeed
w_{ik}	weights of the neural network connections
α	angle of attack
α_0	mean angle of attack at the oscillations
η, ρ_i	objective function parameters
φ_c	canard deflection angle
τ_1, τ_2	characteristic times
ϕ_k	signal mapped by the neuron

Subscripts

dyn	dynamic
sep	separated
sim	simulation
st	static
test	testing
T	transpose

Aerodynamic derivatives

$C_{i\alpha}$	$\frac{\partial C_i}{\partial \alpha}$
C_{iq}	$\frac{\partial C_i}{\partial (q\bar{c}/2V)}$
$C_{i\dot{\alpha}}$	$\frac{\partial C_i}{\partial (\dot{\alpha}\bar{c}/2V)}$, where $i = N, m$.

References

1. European Aviation Safety Agency. Annual Safety Review 2011. Available online: https://www.easa.europa.eu/communications/docs/annual-safety-review/2011/EASA-Annual-Safety-Review-2011.pdf (accessed on 2 April 2014).

2. Ignatyev, D.I.; Sidoryuk, M.E.; Kolinko, K.A.; Khrabrov, A.N. Dynamic Rig for Validation of Control Algorithms at High Angles of Attack. *J. Aircr.* **2017**, *54*, 1760–1771. [CrossRef]

3. Abramov, N.B.; Goman, M.G.; Khrabrov, A.N.; Kolesnikov, E.N.; Fucke, L.; Soemarwoto, B.; Smaili, H. Pushing ahead—SUPRA airplane model for upset recovery. In Proceedings of the AIAA Modelling and Simulation Technologies Conference (AIAA 2012-4631), Minneapolis, MN, USA, 13–16 August 2012; American Institute of Aeronautics and Astronautics: Reston, VA, USA, 2012. [CrossRef]

4. Foster, J.V.; Cunningham, K.; Fremaux, C.M.; Shah, G.H.; Stewart, E.C.; Rivers, R.A.; Wilborn, J.E.; Gato, W. Dynamics modelling and simulation of large transport airplanes in upset conditions. In Proceedings of the AIAA Atmospheric Flight Mechanics Conference and Exhibit (AIAA-2005-5933), San Francisco, CA, USA, 15–18 August 2005; American Institute of Aeronautics and Astronautics: Reston, VA, USA, 2005. [CrossRef]

5. Da Ronch, A.; Vallespin, D.; Ghoreyshi, M.; Badcock, K.J. Evaluation of dynamic derivatives using computational fluid dynamics. *AIAA J.* **2012**, *50*, 470–484. [CrossRef]

6. Harrison, S.; Darragh, R.; Hamlington, P.; Ghoreyshi, M.; Lofthouse, A. Canard-wing interference effects on the flight characteristics of a transonic passenger aircraft. In Proceedings of the 34th AIAA Applied Aerodynamics Conference, AIAA AVIATION Forum (Paper 2016-4179), Washington, DC, USA, 13–17 June 2016; American Institute of Aeronautics and Astronautics: Reston, VA, USA, 2016. [CrossRef]

7. Sereez, M.; Abramov, N.; Goman, M. Computational Ground Effect Aerodynamics and Airplane Stability Analysis during Take-Off and Landing. In Proceedings of the 7th European Conference for Aeronautics and Aerospace Sciences (EUCASS), Milan, Italy, 3–6 July 2017. [CrossRef]

8. Schütte, A.; Einarsson, G.; Raichle, A.; Schoning, B.; Mönnich, W.; Forkert, T. Numerical simulation of maneuvering aircraft by aerodynamic, flight mechanics and structural mechanics coupling. *J. Aircr.* **2009**, *46*, 53–64. [CrossRef]

9. Ghoreyshi, M.; Cummings, R.; Da Ronch, A.; Badcock, K. Transonic aerodynamic load modelling of X-31 aircraft pitching motions. *AIAA J.* **2013**, *51*, 2447–2464. [CrossRef]

10. Ghoreyshi, M.; Jirásek, A.; Cummings, R. Computational approximation of nonlinear unsteady aerodynamics using an aerodynamic model hierarchy. *Aerosp. Sci. Technol.* **2013**, *28*, 133–144. [CrossRef]

11. Zaichik, L.; Yashin, Y.; Desyatnik, P.; Smaili, H. Some aspects of upset recovering simulation on hexapod simulators. In Proceedings of the AIAA Modelling and Simulation Technologies Conference, Guidance, Navigation, and Control and Co-Located Conferences (AIAA Paper 2012-4949), Minneapolis, MN, USA, 13–16 August 2012; American Institute of Aeronautics and Astronautics: Reston, VA, USA, 2012. [CrossRef]

12. Bushgens, G.S. (Ed.) *Aerodynamics, Stability and Controllability of Supersonic Aircraft*; Nauka, Fizmatlit: Moscow, Russia, 1998; 816p. (In Russian)

13. Etkin, B. *Dynamics of Atmospheric Flight*; Wiley: New York, NY, USA, 1972; pp. 101–114, 129–154, 219–256, 280–295.

14. Tobak, M.; Schiff, L.B. *On the Formulation of the Aerodynamic Characteristics in Aircraft Dynamic*; NASA TR R-456; National Aeronautic and Space Administration: Washington, DC, USA, 1976.

15. Klein, V.; Noderer, K.D. *Modelling of Aircraft Unsteady Aerodynamic Characteristics. Part 1—Postulated Models*; NASA Technical Memorandum 109120; Langley Research Center: Hampton, VA, USA, 1994.

16. Huang, X.Z.; Lou, H.Y.; Hanf, E.S. Airload Prediction for Delta Wings at High Incidence; ICAS Paper. In Proceedings of the 22nd Congress of the Aeronautical Sciences, Harrogate, UK, 27 August–1 September 2000; pp. 221.1–221.10.

17. Goman, M.G.; Khrabrov, A.N. Space representation of aerodynamic characteristics of an aircraft at high angles attack. *J. Aircr.* **1994**, *31*, 1109–1115. [CrossRef]

18. Jumper, E.J.; Schreck, S.J.; Dimmick, R.L. Lift-curve characteristics for an airfoil pitching at constant rate. *J. Aircr.* **1987**, *24*, 680–687. [CrossRef]

19. Ioselevich, A.S.; Stolyarov, G.I.; Tabachnikov, V.G.; Zhuk, A.N. Experimental Investigation of Delta Wing A = 1.5 Damping in Roll and Pitch at High Angles of Attack. *Proc. TsAGI* **1985**, *2290*, 52–70.

20. Abramov, N.; Goman, M.; Greenwell, D.; Khrabrov, A. Two-step linear regression method for identification of high incidence unsteady aerodynamic model. In Proceedings of the AIAA Atmospheric Flight Mechanics Conference (Paper 2001-4080), Montreal, QC, Canada, 6–9 August 2001. [CrossRef]

21. Vinogradov, Y.A.; Zhuk, A.N.; Kolinko, K.A.; Khrabrov, A.N. Mathematical simulation of dynamic effects of unsteady aerodynamics due to canard flow separation delay. *TsAGI Sci. J.* **2011**, *42*, 655–668. [CrossRef]

22. Luchtenburg, D.M.; Rowley, C.W.; Lohry, M.W.; Martinelli, L.; Stengel, R.F. Unsteady high-angle-of-attack aerodynamic models of a generic jet transport. *AIAA J. Aircr.* **2015**, *52*, 890–895. [CrossRef]

23. Faller, W.E.; Schreck, S.J.; Helin, H.E. Real-time model of three dimensional dynamic reattachment using neural networks. *J. Aircr.* **1995**, *32*, 1177–1182. [CrossRef]

24. Faller, W.E.; Schreck, S.J. Unsteady fluid mechanics applications of neural networks. *J. Aircr.* **1997**, *34*, 48–55. [CrossRef]

25. Reisenthel, P.H. Development of nonlinear indicial model using response functions generated by a neural network. In Proceedings of the 35th Aerospace Sciences Meeting and Exhibit (AIAA 97-0337), Reno, NV, USA, 6–9 January 1997. [CrossRef]

26. Ignatyev, D.I.; Khrabrov, A.N. Application of neural networks in the simulation of dynamic effects of canard aircraft aerodynamics. *TsAGI Sci. J.* **2011**, *42*, 817–828. [CrossRef]

27. Ignatyev, D.I.; Khrabrov, A.N. Neural network modelling of unsteady aerodynamic characteristics at high angles of attack. *Aerosp. Sci. Technol.* **2015**, *41*, 106–115. [CrossRef]

28. Glaz, B.; Liu, L.; Friedmann, P. Reduced-order nonlinear unsteady aerodynamic modelling using a surrogate-based recurrence framework. *AIAA J.* **2010**, *48*, 2418–2429. [CrossRef]

29. Kolmogorov, A.T. *On the Representation of Continuous Functions of Many Variables by Superposition of Continuous Functions of One Variable and Addition*; American Mathematical Society Translations: Providence, RI, USA, 1963; Volume 28, pp. 55–59.

30. Mialon, B.; Khrabrov, A.; Da Ronch, A.; Cavagna, L.; Zhang, M.; Ricci, S. Benchmarking the prediction of dynamic derivatives: Wind tunnel tests, validation, acceleration methods. In Proceedings of the AIAA Atmospheric Flight Mechanic Conference (AIAA Paper 2010-8244), Toronto, ON, Canada, 2–5 August 2010. [CrossRef]

31. Ghoreyshi, M.; Korkis-Kanaan, R.; Jirásek, A.; Cummings, R.; Lofthouse, A. Simulation validation of static and forced motion flow physics of a canard configured TransCruiser. *Aerospace Sci. Technol.* **2016**, *48*, 158–177. [CrossRef]

32. Hagan, M.T.; Demuth, H.B.; Beale, M.H. *Neural Network Design*; PWS Publishing: Boston, MA, USA, 1996.

33. Foresee, F.D.; Hagan, M.T. Gauss-Newton approximation to Bayesian regularization. In Proceedings of the International Joint Conference on Neural Networks, Houston, TX, USA, 12 June 1997; pp. 1930–1935. [CrossRef]

34. MacKay, D.J.C. Bayesian Interpolation. *Neural Comput.* **1992**, *4*, 415–447. [CrossRef]

35. Ignatyev, D.I.; Khrabrov, A.N. Neural network modelling of longitudinal aerodynamic characteristics of aircraft. *Informatsionnye Tekhnologii* **2014**, *3*, 61–69.

36. Hagan, M.T.; Menhaj, M. Training feed-forward networks with the Marquardt algorithm. *IEEE Trans. Neural Netw.* **1994**, *5*, 989–993. [CrossRef] [PubMed]

aerospace

MDPI

Article

AEROM: NASA's Unsteady Aerodynamic and Aeroelastic Reduced-Order Modeling Software

Walter A. Silva

NASA Langley Research Center, Hampton, VA 23681, USA; Walter.A.Silva@nasa.gov

Received: 3 November 2017; Accepted: 6 April 2018; Published: 10 April 2018

Abstract: The origins, development, implementation, and application of AEROM, NASA's patented reduced-order modeling (ROM) software, are presented. Using the NASA FUN3D computational fluid dynamic (CFD) code, full and ROM aeroelastic solutions are computed at several Mach numbers and presented in the form of root locus plots. The use of root locus plots will help reveal the aeroelastic root migrations with increasing dynamic pressure. The method and software have been applied successfully to several configurations including the Lockheed-Martin N+2 supersonic configuration and the Royal Institute of Technology (KTH, Sweden) generic wind-tunnel model, among others. The software has been released to various organizations with applications that include CFD-based aeroelastic analyses and the rapid modeling of high-fidelity dynamic stability derivatives. We present recent results obtained from the application of the method to the AGARD 445.6 wing that reveal several interesting insights.

Keywords: aeroelasticity; reduced-order model; flutter

1. Introduction

In the early days, aeroelasticians typically used linear methods to compute unsteady aerodynamic responses and subsequent aeroelastic analyses [1]. These aeroelastic analyses were usually presented in the form of aeroelastic root locus plots as a function of either dynamic pressure or velocity, or velocity-damping-frequency (V-g-f) plots. These plots were generated rapidly and provided significant amount of insight regarding the aeroelastic mechanisms involved.

With the subsequent development of CFD methods, the analysis of complex, nonlinear flows, and their effect on the aeroelastic response, became a reality. While CFD tools are quite powerful and provide significant insight regarding flow physics, the significant increase in computational cost (time and CPU dollars) has had an effect on how aeroelastic analyses are performed. One side-effect of the increase in computational cost is the desire to keep the number of compute iterations, and the total number of solutions generated, at a minimum. Results are, therefore, computed for a small number of dynamic pressures (per Mach number) with only a few cycles computed per dynamic pressure. A second side-effect is that the resultant time histories of each generalized coordinate cannot be directly used to identify the governing aeroelastic mechanisms at work, as was the case for the classical linear methods. The recent development of reduced-order modeling (ROM) methods [2–4], provides a tool for the rapid generation of traditional aeroelastic tools such as the root-locus plots.

The origin of this method started with the author's PhD dissertation [5] and related publications [6,7]. An important conceptual development first presented in these references consists of the realization that unsteady aerodynamic impulse responses do, in fact, exist and can be computed using CFD methods. This concept is an important point that is claimed to be not realizable in some of the classic aeroelastic references. The reason for this discrepancy is actually quite simple as it relates to the difference between a continuous-time and a discrete-time impulse function.

For a continuous-time system, it is well known that the impulse input function is the Dirac delta function. This function serves the continuous domain well, in particular in the solution of ordinary and partial differential equations. However, its application to a discrete-time system such as a CFD-based solution, is not clear, thus the belief that an impulse input cannot be applied to a CFD code. Therefore, if an impulse input cannot be applied to a CFD code, then an unsteady aerodynamic response cannot be identified or realized.

An important contribution by the author [5] is the realization that in order to properly identify the unsteady aerodynamic impulse response using a CFD code, a discrete-time impulse input, also known as the unit sample input in discrete-time theory, is the proper function to use and not the Dirac delta function. The theory of Digital Signal Processing (DSP) demonstrates that a unit sample input is much simpler to apply and less complex to interpret than the Dirac delta function. These results proved the existence and realizability of a unit unsteady aerodynamic impulse (sample) response via a CFD code.

In the world of structural dynamics and modal identification, the concept of a structural dynamic impulse response is clear and well understood. As a result, various modal identification techniques consist of the identification of these responses and a subsequent realization of a system that captures the structural dynamic system of interest. Having familiarity with one of these methods by the name of Eigensystem Realization Algorithm (ERA) [8]/System Observer Controller Identification Toolbox (SOCIT) [9], the author applied the modal identification technique, previously limited to structural dynamic systems, to that of identifying an unsteady aerodynamic system via the identification of the unsteady aerodynamic impulse responses. Once the concept of a discrete-time unsteady aerodynamic impulse response was mathematically validated, the application of ERA/SOCIT became quite logical [10]. These results [10] represent the first time that the ERA/SOCIT algorithms were used for the identification of unsteady aerodynamic systems. It is valuable to point out that this method is now being applied at several organizations around the world [11–16]. In the area of fluid modal decompositions using, primarily, the Proper Orthogonal Decomposition (POD), the application of the ERA algorithm has become standard, with an initial appearence in the literature by Ma, Ahuja, and Rowley [17].

Following these fundamental advances, linearized, unsteady aerodynamic state-space models using the CFL3Dv6 [18] code were introduced [19]. The unsteady aerodynamic state-space models were coupled with a structural model within a MATLAB/SIMULINK™ environment for rapid calculation of aeroelastic responses, including the prediction of flutter. A comparison of the aeroelastic responses computed using the aeroelastic simulation ROM with the aeroelastic responses computed using the CFL3Dv6 code showed excellent correlation.

Initially, the excitation of one structural mode at a time was used to generate the unsteady aerodynamic state-space model [19]. However, the one-mode-at-a-time method becomes prohibitively expensive for more realistic cases where there exist a large number of modes. Methods based on the simultaneous application of structural modes as CFD input [20] have been proposed, greatly reducing the computational cost for a large number of structural modes. The method developed by Silva [2] enables the simultaneous excitation of the structural modes using orthogonal functions. These methods require only a single CFD solution and are, therefore, independent of the number of structural modes.

Static and matched-point aeroelastic solutions, using a ROM, have also been developed [2,21] and implemented in the FUN3D [22–25] CFD code. Methods for generating root locus plots of the ROM aeroelastic system have also been developed [3]. Applications of these methods include fixed-wing and launch vehicle configurations [4]. This paper will discuss the application of these methods to three configurations: a low-boom configuration, a full-span wind-tunnel model, and the AGARD 445.6 wing.

The AEROM software was granted a patent (November, 2011), Patent No. 8,060,350. The software has been distributed to the Air Force Research Laboratory, the Boeing Corporation, and the CFD Research Corporation.

2. Computational Methods

2.1. FUN3D Code

The FUN3D CFD code is the NASA-developed, RANS unstructured mesh solver used for this study. The code solves the Euler or Navier-Stokes equations with various turbulence models. Due to the differences between structural and CFD meshes, an interpolation between the two domains is required. Mode shape displacements are used to compute physical deformations that are then used to deform the mesh within FUN3D. Pressures are computed at each time step and then projected onto each mode shape to provide generalized aerodynamic forces (GAFs). These GAFs are then used by the linear state-space structural model (within FUN3D) to compute the next set of modal deformations for the next iteration.

2.2. System Identification Method

The development of algorithms such as the ERA [8] and the Observer Kalman Identification (OKID) [26] Algorithm have enabled the realization of discrete-time state-space models, used up to this point, primarily for structural dynamic modal identification. These algorithms use the Markov parameters (discrete-time impulse responses) of the systems of interest to perform the required state-space realization. The SOCIT contains these algorithms and others related to this methodology.

The PULSE algorithm is used to extract individual input/output impulse responses from simultaneous input/output responses. For a four-input/four-output system, for example, the PULSE algorithm is used to extract the individual sixteen (all combinations of four inputs and four outputs) impulse responses that define this input/output system. The individual sixteen impulse responses are then processed via the ERA in order to generate a four-input/four-output, discrete-time, state-space model. A brief summary of the basis of this algorithm follows.

A finite dimensional, discrete-time, linear, time-invariant dynamical system can be defined as

$$x(k+1) = Ax(k) + Bu(k) \tag{1}$$

$$y(k) = Cx(k) + Du(k) \tag{2}$$

where x is an n-dimensional state vector, u an m-dimensional control input, and y a p-dimensional output or measurement vector with k being the discrete time index. The transition matrix, A, characterizes the dynamics of the system. The goal of system realization is to generate constant matrices (A, B, C, D) such that the output responses of a given system due to a particular set of inputs is reproduced by the discrete-time state-space system described above.

The time-domain values of the discrete-time impulse responses of the system are also known as the Markov parameters and are defined as

$$Y(k) = CA^{k-1}B + D \tag{3}$$

with A an (n × n) matrix, B an (n × m) matrix, C a (p × n) matrix, and D an (p × m) matrix. The ERA algorithm generates the generalized Hankel matrix, consisting of the discrete-time impulse responses and then uses the singular value decomposition (SVD) to compute the (A, B, C, D) matrices. This is the method by which the ERA is applied to unsteady aerodynamic impulse responses to construct unsteady aerodynamic state-space models.

2.3. Simultaneous Excitation Input Functions

Clearly, the nonlinear unsteady aerodynamic responses of a flexible vehicle comprise a multi-input/multi-output (MIMO) system with respect to the modal inputs and generalized aerodynamic outputs. In the situation where the goal is the simultaneous excitation of such a MIMO system, system identification techniques [27–29] indicate that the input excitations must be properly defined in order to generate stable and accurate input/output models of the system. A critical requirement is that these input excitations be different, mathematically, from each other. If identical excitation inputs are applied simultaneously, the task of separating the various contributions of each input becomes nearly impossible. This effect makes it practically impossible for a system identification algorithm to extract the individual impulse responses for each input/output pair. It is essential that the individual impulse responses for each input/output pair be properly identified so that an accurate model can be generated.

These unsteady aerodynamic impulse responses are the time-domain generalized aerodynamic forces (GAFs), critical to understanding unsteady aerodynamic behavior. The Fourier-transformed version of these GAFs are the frequency-domain GAFs, that provide an important link to more traditional frequency-domain-based unsteady aerodynamic analyses.

The question is how different should these excitation inputs be from each other and how can we quantify a level of difference between them? Orthogonality (linear independence) is the most precise mathematical method for guaranteeing the difference between signals. Using orthogonal functions as the excitation inputs provides a mathematical guarantee of the desired difference between inputs.

In a previous paper [2], four families of functions were investigated to efficiently identify a CFD-based unsteady aerodynamic state-space model. For the present paper, the Walsh family of orthogonal functions [30] are used, shown in Figure 1 for four modes. These functions are orthogonal and therefore provide a benefit in the system identification process as discussed above. Also, this family of functions consists of a combination of step functions, which have been shown to be well-suited for the identification of CFD-based unsteady aerodynamic ROMs.

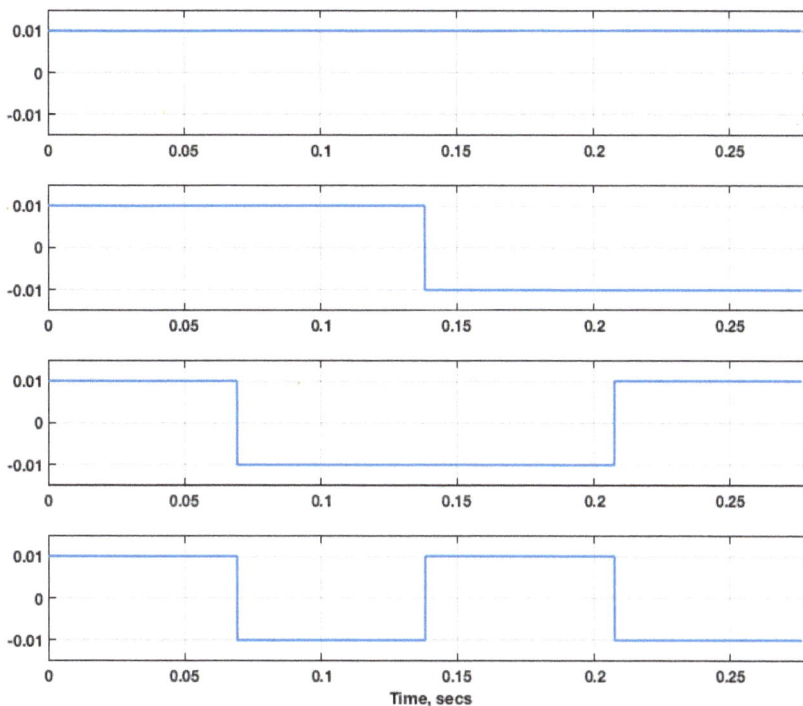

Figure 1. Walsh functions.

3. ROM Development Processes

The ROM development process consists of two parts: the creation of the unsteady aerodynamic ROM and the creation of the structural dynamic ROM. The combination of the unsteady aerodynamic ROM with the structural dynamic ROM yields what is referred to as the aeroelastic simulation ROM.

The original unsteady aerodynamic ROM development process consisted of the excitation of one structural mode at a time per CFD solution. That approach is not practical for realistic configurations with a large number of modes. As mentioned above, an improved method has been developed and is described below.

3.1. Improved ROM Development Process

The improved simultaneous modal excitation ROM development process:

1. Create as many orthogonal functions as the number of structural modes of interest;
2. Starting from the restart of a steady rigid CFD solution, execute a single CFD solution using the orthogonal excitation inputs simultaneously, resulting in GAF responses due to these inputs;
3. Identify the individual impulse responses from the responses computed in Step 2 using the PULSE algorithm;
4. Using the ERA, convert the impulse responses from Step 3 into an unsteady aerodynamic state-space model;
5. Using full-solution CFD results, compare with solutions generated using the model generated in Step 4;

Steps 1–4 of the improved process are presented in Figure 2.

Using generalized mass, modal frequencies, and modal dampings from the finite element model (FEM), a state-space model of the structure is generated, referred to as the structural dynamic ROM (Figure 3). The unsteady aerodynamic and structural dynamic ROMs are combined to form an aeroelastic simulation ROM (see Figure 4). Root locus plots are then extracted from the aeroelastic simulation ROM.

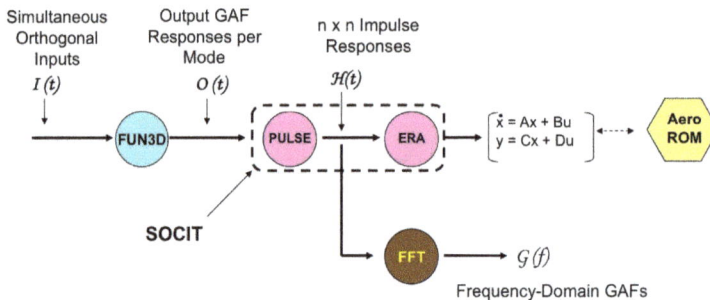

Figure 2. Improved process for generation of an unsteady aerodynamic reduced-order modeling (ROM) (Steps 1–4).

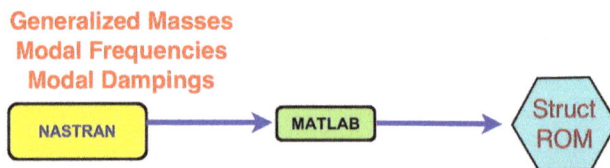

Figure 3. Process for generation of a structural dynamic state-space ROM.

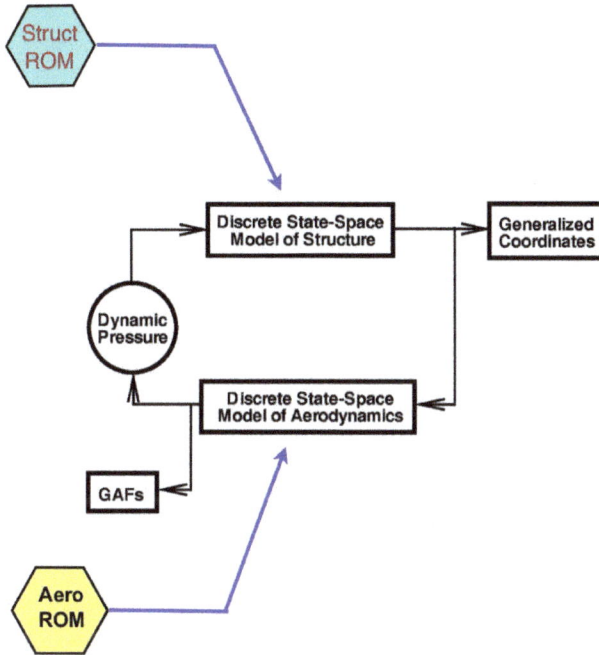

Figure 4. Process for generation of an aeroelastic simulation ROM consisting of an unsteady aerodynamic ROM and a structural dynamic state-space ROM.

For the original ROM process, the unsteady aerodynamic ROM was typically generated about a pre-selected static aeroelastic condition. Once the CFD-based converged static aeroelastic solution was obtained, the development of the unsteady aerodynamic ROM process was performed about that static aeroelastic condition. This approach would tend to limit the applicability of the unsteady aerodynamic ROM to the local neighborhood of that static aeroelastic condition.

In the early days, no method had been defined to enable the computation of a static aeroelastic solution using a ROM. These ROMs, based on a particular static aeroelastic condition were, therefore, limited to the prediction of dynamic responses about that condition. This included the methods by Raveh [31] and by Kim et al. [20]. The improved ROM method described above, however, consists of a ROM generated directly from a steady, rigid solution. Therefore, these improved ROMs can be used to predict static and dynamic aeroelastic solutions at any dynamic pressure [21]. All responses for the present results were computed from the restart of a steady, rigid FUN3D solution, bypassing the need (and the additional computational expense) of a static aeroelastic solution using FUN3D.

3.2. Error Minimization

Error minimization consists of error quantification and error reduction. Error quantification is defined as the difference (error) between the full FUN3D solution due to the orthogonal input functions used (Walsh) and the unsteady aerodynamic ROM solution due to the same orthogonal input functions. This was identified in Step 5 in the previous subsection and is shown schematically in Figure 5. The outputs shown are GAF responses per mode. Within the system identification algorithms, there are parameters that can then be used to reduce the error (error reduction). These parameters include number of states and the record length of the identified pulse responses, for example. The maximum error is the largest error encountered per mode. Using the maximum error as the figure of merit, the parameters are varied until an acceptable ROM has been obtained.

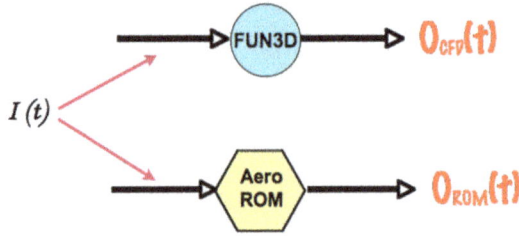

Figure 5. Error defined as difference between the FUN3D solution and the unsteady aerodynamic ROM solution due to input of orthogonal functions.

4. Sample Results

A brief summary of results for three configurations is presented in this section. These configurations are the Lockheed-Martin N+2 low-boom supersonic configuration, the Royal Institute of Technology (KTH) generic fighter wind-tunnel model, and the AGARD 445.6 wing.

4.1. Low-Boom N+2 Configuration

An artist's rendering of the Lockheed-Martin N+2 low-boom supersonic configuration is presented in Figure 6. This configuration has been used extensively as part of a NASA research effort to address the technologies required for a low-boom aircraft, including aeroelastic effects. Presented in Figure 7 is a comparison of the dynamic aeroelastic responses of the time histories of the first mode generalized displacements from a full FUN3D aeroelastic solution and the ROM aeroelastic solution at a Mach number of 1.7 and a dynamic pressure of 2.149 psi. Presented in Figure 8 is a comparison of the dynamic aeroelastic responses of the time histories of the second mode generalized displacements from a full FUN3D aeroelastic solution and the ROM aeroelastic solution at the same condition. As can be seen, the results indicate an excellent level of correlation between the full FUN3D solutions and the ROM solutions. Similar results are obtained for all the other modes, indicating good confidence in the ROM.

Figure 6. Artist's concept of the Lockheed-Martin N+2 configuration.

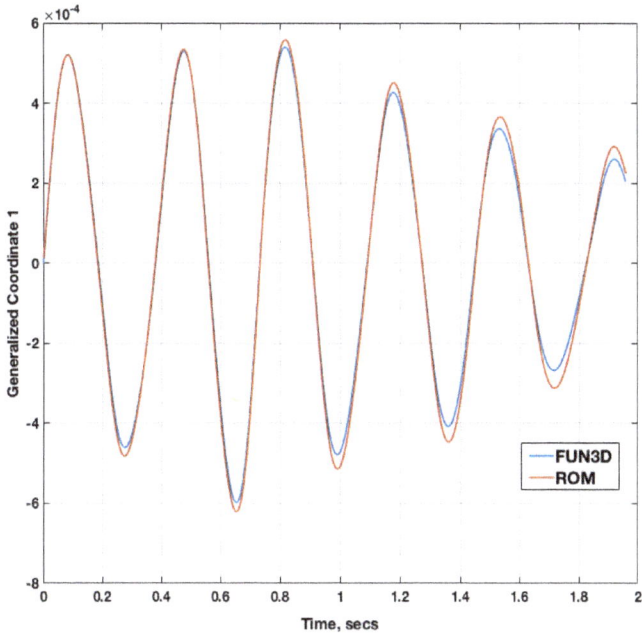

Figure 7. Comparison of full FUN3D aeroelastic response and ROM aeroelastic response for the first mode of the N+2 configuration at M = 1.7 and a dynamic pressure of 2.149 psi.

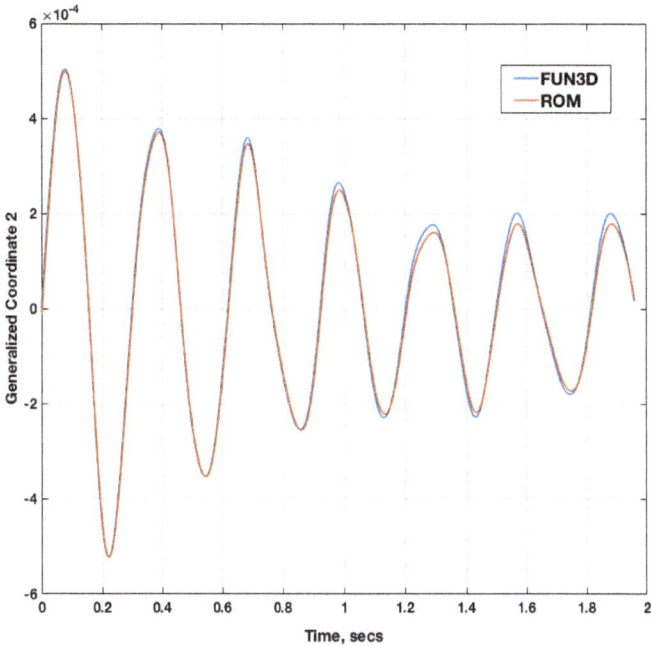

Figure 8. Comparison of full FUN3D aeroelastic response and ROM aeroelastic response for the second mode of the N+2 configuration at M = 1.7 and a dynamic pressure of 2.149 psi.

This ROM technology has the ability to rapidly generate an aeroelastic root locus plot that can reveal the aeroelastic mechanisms occurring at that flight condition. The aeroelastic root locus plot for the low-boom N+2 configuration at M = 1.7 is presented in Figure 9, revealing the aeroelastic mechanisms that affect this configuration. Each symbol represents the aeroelastic roots at a specific dynamic pressure, corresponding to a 2 psi increment in dynamic pressure.

In lieu of a ROM and its root locus solutions, multiple, expensive, and time consuming full FUN3D solutions would be required for each dynamic pressure of interest, with each solution requiring about two days. A full FUN3D analysis, at each dynamic pressure, requires two full FUN3D solutions: a static aeroelastic (~10 h) and a dynamic aeroelastic (~18 h) solution. Full FUN3D solutions for 20 dynamic pressures would require ~560 h of compute time.

The ROM solutions, on the other hand, consist of one full FUN3D solution that is used to generate the ROM at that Mach number. For this particular example, a full FUN3D solution of 2400 time steps ran for three hours. This ROM solution is then used to rapidly generate a ROM that can then be used to generate all the aeroelastic responses at all dynamic pressures and the corresponding root locus plots.

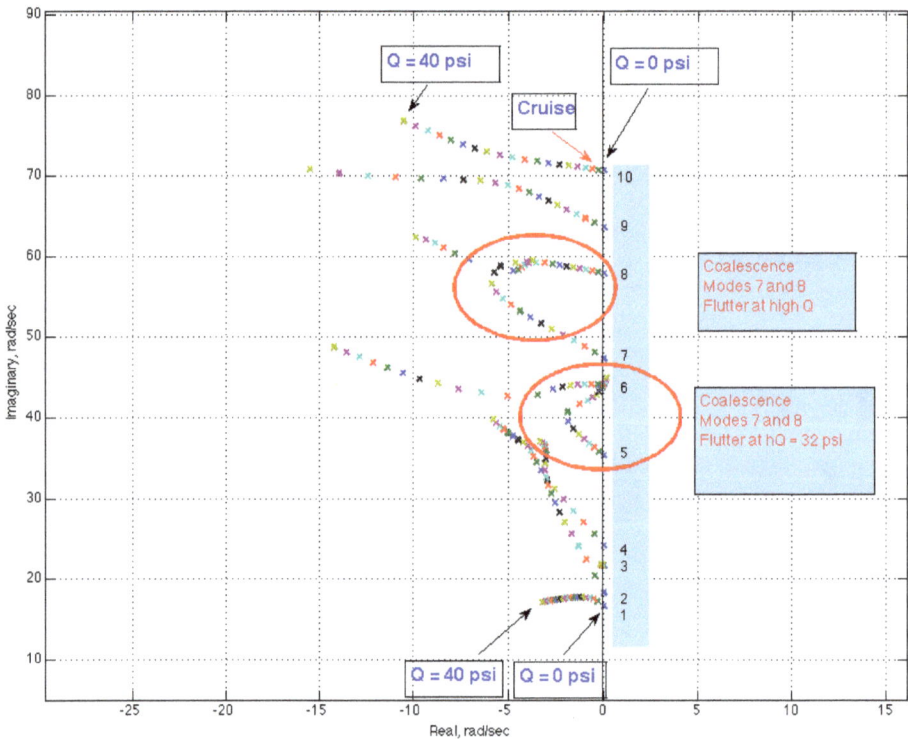

Figure 9. Aeroelastic root locus plot for the low-boom N+2 configuration at M = 1.7 with each colored marker indicating an increment of 2 psi in dynamic pressure for a given mode.

4.2. KTH Generic Fighter

Two wind-tunnel models of the Saab JAS 39 Gripen were designed, built, and tested in the NASA Transonic Dynamics Tunnel (TDT) for flutter clearance in 1985 and 1986. A stability model was designed to be stiff, while incorporating proper scaling of both the mass and geometry. The other model, the flutter model, was also designed for proper scaling of structural dynamics, and was used for flutter testing with various external stores attached.

A generic fighter flutter-model version of these earlier models was selected for the collaborative wind-tunnel testing campaign between the KTH and NASA. Shown in Figure 10 is the new model, with a similar outer mold line (OML) to the Gripen, but modified into a more generic fighter configuration. Details regarding the design, fabrication, and instrumentation of the wind-tunnel model can be found in the reference paper [32]. Figure 11 shows the wind-tunnel model installed in the TDT.

Figure 10. The generic fighter aeroelastic wind-tunnel model tested in summer of 2016.

Figure 11. The generic fighter aeroelastic wind-tunnel model installed in the Transonic Dynamics Tunnel (TDT).

Using the AEROM software, aeroelastic root locus plots were generated for the KTH wind-tunnel model in air test medium for a free-air case and a solution accounting for the effects of the TDT test section via CFD modeling [33,34], as can be seen in Figure 12. There were three configurations tested: wing with tip stores (configuration 1), wing with tip and under-wing stores (configuration 2), and wing with tip and under-wing stores with added masses at tip stores (configuration 3). The third configuration exhibited flutter while configurations 1 and 2 did not. Presented in Figure 13 is the aeroelastic ROM root locus plot for the free-air configuration at M = 0.90. For this case, the roots clearly indicate a flutter mechanism at about 8100 N/m^2 (or 169 psf) via a coalescence of modes 5 and 6. Using the ROM, any dynamic pressure can be quickly evaluated to determine the aeroelastic response, consistent with the root locus plots. At this dynamic pressure, the ROM-based flutter prediction is above the experimental flutter dynamic pressure at M = 0.9. Typically, a conservative flutter result occurs when the analysis predicts a flutter condition below the experimental flutter result. This result, therefore, implies a non-conservative result, indicative of potentially significantly non-linear phenomena. All results presented are for zero structural damping. Using the ROM, the effect of structural damping can be quickly evaluated as well but is not pursued in the present discussion.

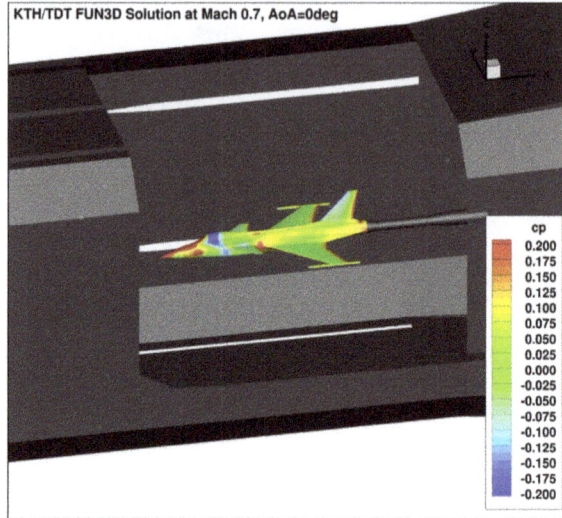

Figure 12. Pressure distributions at M = 0.7, AoA = 0 degrees on the KTH wind-tunnel model, as simulated inside the TDT using FUN3D code.

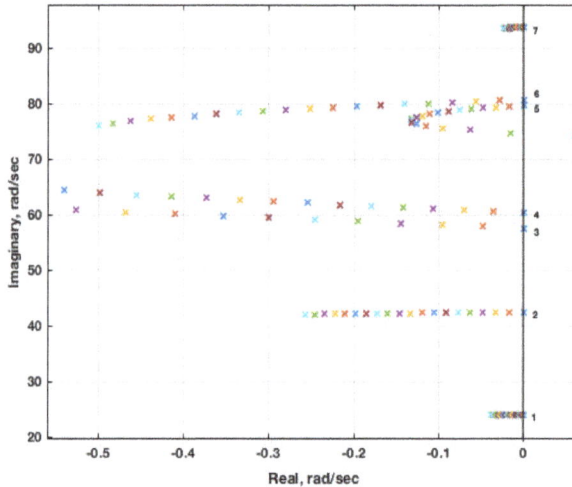

Figure 13. Root locus plot generated from ROM model indicating an aeroelastic instability at M = 0.90 in air test medium for the third configuration with each colored marker indicating an increment of 450 N/m² in dynamic pressure for a given mode.

Presented in Figures 14 and 15 are comparisons of the aeroelastic responses for modes 3, 4, 5, and 6 at M = 0.9 and Q = 7344 Pa for the FUN3D solution that includes the effect of the TDT and the ROM solution for the same configuration. As can be seen, the comparison is quite good with some variation in mode 6. Additional studies are currently underway to minimize these variations in order to reduce the error associated with the ROM.

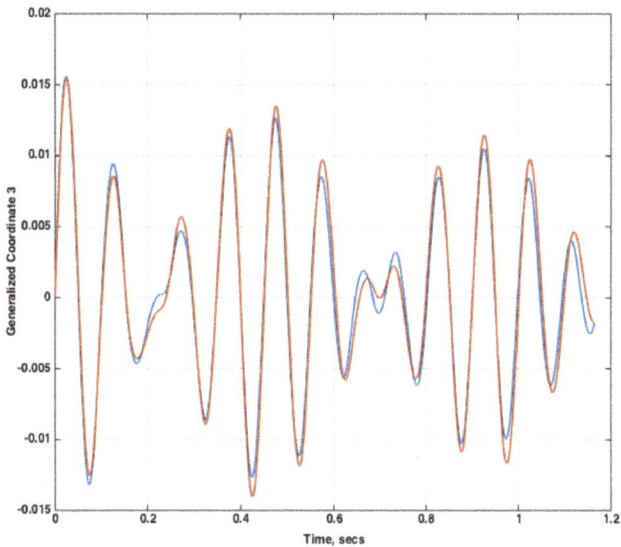

Figure 14. Aeroelastic response in mode 3 for the FUN3D (blue) and ROM (orange) solutions for the configuration including the Transonic Dynamics Tunnel (TDT).

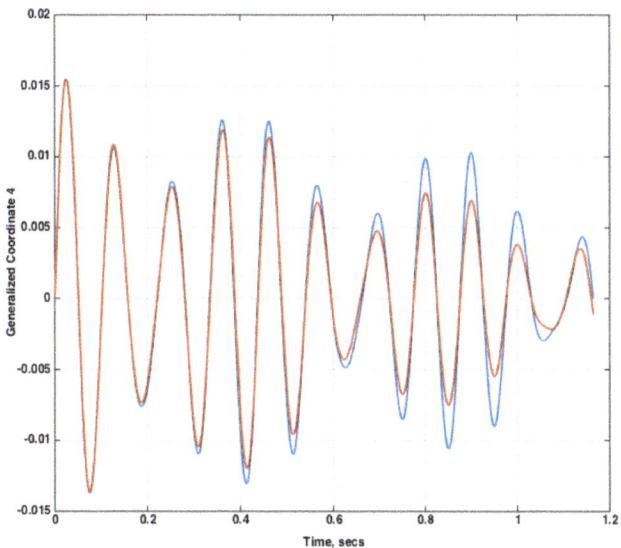

Figure 15. Aeroelastic response in mode 4 for the FUN3D (blue) and ROM (orange) solutions for the configuration including the TDT.

For the CFD model that included the TDT, the ROM solution required two days whereas the full solution (for only one dynamic pressure) required five days. The ROM solution could, of course, then be used to rapidly compute the aeroelastic response due to any dynamic pressure.

4.3. AGARD 445.6 Wing

Aeroelastic transients for the AGARD 445.6 aeroelastic wing [35] from the FUN3D full solution and from the aeroelastic ROM, for inviscid and viscous solutions, are presented in this section. The FUN3D full solution aeroelastic transients are presented for two Mach numbers (M = 0.96, M = 1.141) at various dynamic pressures. FUN3D full and ROM aeroelastic solutions are compared at specific dynamic pressures, including root locus plots.

4.3.1. Inviscid Results

Inviscid FUN3D full and ROM solutions are presented in this section. The aeroelastic root locus plot for M = 0.96 generated using the ROM method is presented in Figure 16. In these root locus plots, dynamic pressures vary from zero to 114 psf in twenty increments. These root locus plots can be generated for any number, and any increment, of dynamic pressures rapidly. A flutter mechanism dominated by the first mode with some coupling with the second mode is indicated at this Mach number, while the third and fourth modes are stable.

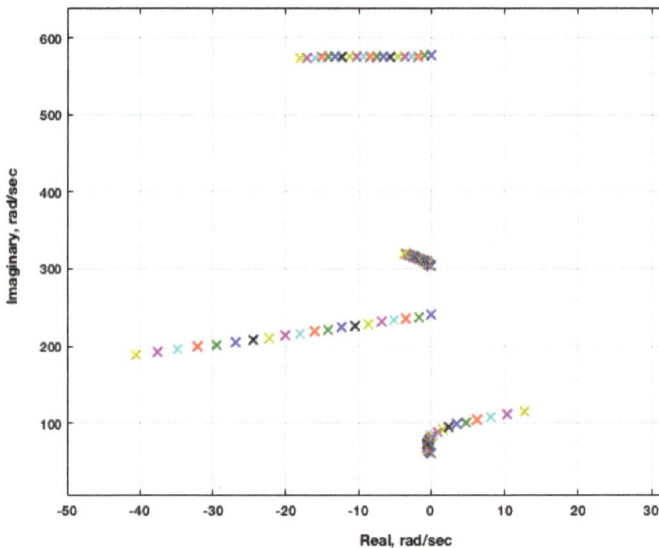

Figure 16. ROM aeroelastic root locus plot for M = 0.96, inviscid solution.

Figure 17 presents a close-up version of the root locus plot. The dynamic pressure for this root locus plot starts at 0 psf with an increment of 6 psf, resulting in a flutter dynamic pressure of approximately 30 psf, consistent with the full FUN3D solution flutter dynamic pressure [35]. The inviscid result does not compare well with the experimental flutter dynamic pressure at this Mach number. Inviscid solutions tend to have stronger shocks that are farther aft and, therefore, induce a stronger and earlier onset of flutter, so this discrepancy is not surprising. Including viscosity, the shock strength is reduced and the shock moves forward, yielding a higher flutter dynamic pressure.

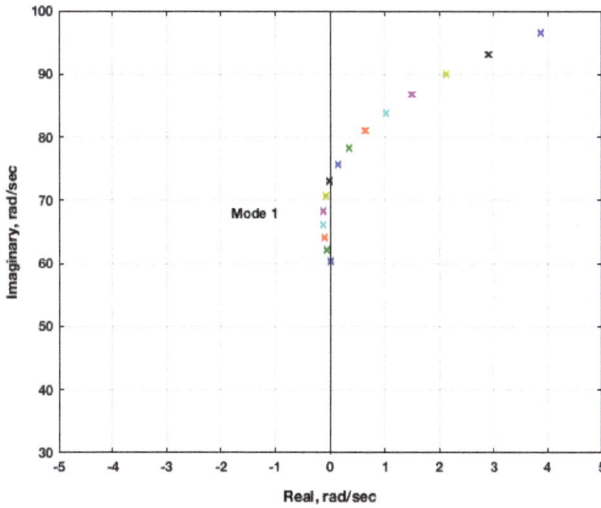

Figure 17. Detailed view of ROM aeroelastic root locus plot for M = 0.96, inviscid solution.

Figure 18 presents the ROM aeroelastic root locus plot for M = 1.141. There are two flutter mechanisms at this condition. The first flutter mechanism consists of a first-mode instability at a dynamic pressure of about 300 psf. The second flutter mechanism involves a third-mode instability that is always unstable. In order to validate the accuracy of this aeroelastic root locus plot, the generalized coordinates from a full FUN3D solution are analyzed.

A visual examination of the aeroelastic transients for the four modes at M = 1.141 and a dynamic pressure of 30 psf, presented in Figure 19, indicate that the first mode, with the largest amplitude, is clearly stable. The stability of the other three modes is harder to discern due to smaller and similar amplitudes. Figure 20 presents the generalized coordinate response of the third mode, clearly showing that this mode becomes unstable.

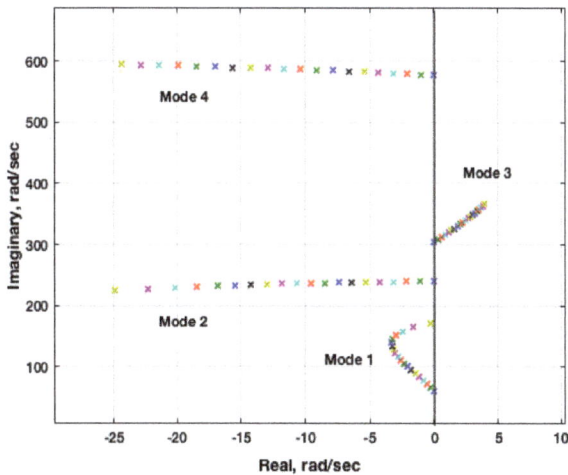

Figure 18. ROM aeroelastic root locus plot for M = 1.141, inviscid solution.

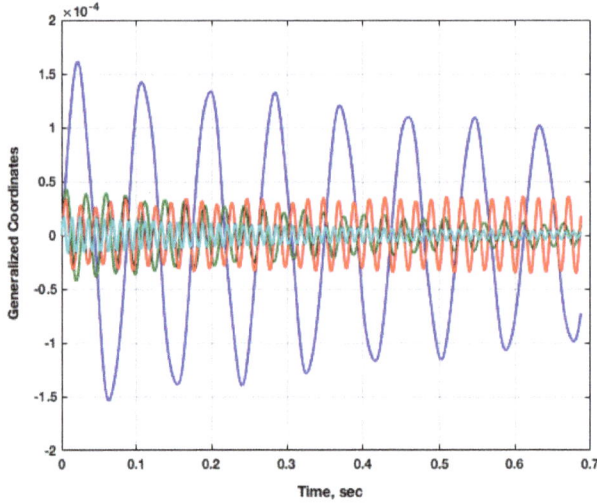

Figure 19. FUN3D full solution generalized coordinates at M = 1.141, Q = 30 psf, inviscid solution. Mode 1 = blue, Mode 2 = green, Mode 3 = red, Mode 4 = cyan.

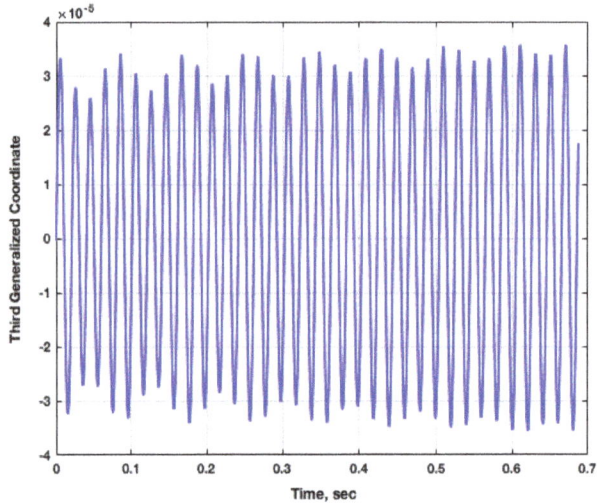

Figure 20. FUN3D full solution third generalized coordinate at M = 1.141, Q = 30 psf, inviscid solution.

Other publications on the flutter boundary of the AGARD 445.6 wing do not mention this third mode instability. It is not clear if this instability is present in all inviscid (Euler) solutions of the AGARD 445.6 wing. It is possible that the first mode instability dominated all inviscid analyses at supersonic conditions to date. If evaluation of stability was based on a visual examination of the generalized coordinates, it is understandable how the third mode instability might have been missed. Analyses performed in the early days of computational aeroelasticity would have consisted of fewer time steps (due to computational cost at the time), thereby making it difficult to visually notice the third mode instability presented in Figure 19. The authors have confirmed the existence of this third mode instability in previous solutions obtained using the NASA Langley CFL3D code as well.

4.3.2. Viscous Results

Viscous FUN3D full and ROM solutions are presented at M = 1.141 in this section. The results for FUN3D full and ROM viscous solutions at subsonic Mach numbers agree well with each other and with experiment and are not presented here.

The root locus plot generated using the FUN3D ROM viscous solution at M = 1.141, in dynamic pressure increments of 6 psf to 114 psf, is presented as Figure 21. It is clear that the inclusion of viscous effects has stabilized the third mode instability noticed in the inviscid solution.

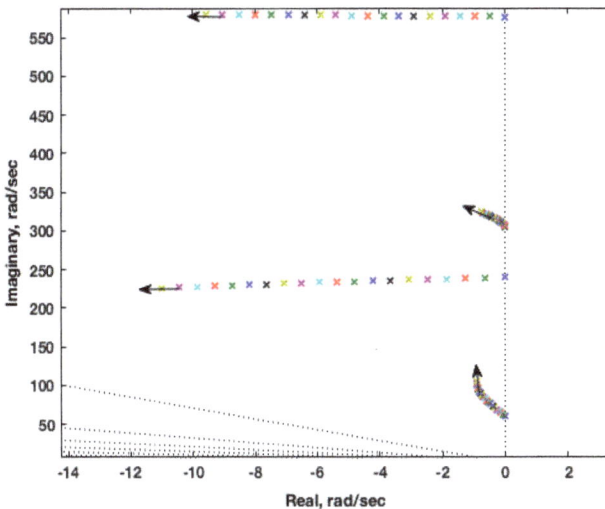

Figure 21. Viscous ROM root locus plot at M = 1.141.

A fundamental difference exists between a root locus plot and the post-processed analysis of generalized coordinates over a short period of time. By definition, a root locus plot reveals the roots of a system as time approaches infinity. The analysis of the initial transient response of a generalized coordinate over a short period of time, on the other hand, can be deceiving as the response can change if the response was viewed (or analyzed) over a longer period of time. This property of root locus plots is critical for the accurate evaluation of aeroelastic stability.

5. Conclusions

The origin, implementation, and applications of AEROM, the patented NASA reduced-order modeling software, have been presented. Recent applications of the software to analyze complex configurations, including computation of the aeroelastic responses of the Lockheed-Martin low-boom N+2 configuration, the KTH (Royal Institute of Technology, Stockholm, Sweden) generic fighter wind-tunnel model, and the AGARD 445.6 wing were presented. The results presented here demonstrate the computational efficiency and analytical capability of the AEROM software.

Conflicts of Interest: The author declares no conflict of interest.

References

1. Adams, W.M.; Hoadley, S.T. ISAC: A Tool for Aeroservoelastic Modeling and Analysis. In Proceedings of the 34th AIAA/ASME/ASCE/AHS/ASC Structures, Structural Dynamics, and Materials Conference, La Jolla, CA, USA, 19–22 April 1993; AIAA-1993-1421.
2. Silva, W.A. Simultaneous Excitation of Multiple-Input/Multiple-Output CFD-Based Unsteady Aerodynamic Systems. *J. Aircr.* **2008**, *45*, 1267–1274.
3. Silva, W.A.; Vatsa, V.N.; Biedron, R.T. Development of Unsteady Aerodynamic and Aeroelastic Reduced-Order Models Using the FUN3D Code. Presented at the International Forum on Aeroelasticity and Structural Dynamics, Seattle, WA, USA, 21–25 June 2009; IFASD Paper No. 2009-30.
4. Silva, W.A.; Vatsa, V.N.; Biedron, R.T. Reduced-Order Models for the Aeroelastic Analyses of the Ares Vehicles. Presented at the 28th AIAA Applied Aerodynamics Conference, Chicago, IL, USA, 28 June–1 July 2010; AIAA Paper No. 2010-4375.
5. Silva, W.A. Discrete-Time Linear and Nonlinear Aerodynamic Impulse Responses for Efficient CFD Analyses. Ph.D. Thesis, College of William & Mary, Williamsburg, VA, USA, December 1997.
6. Silva, W.A. Identification of Linear and Nonlinear Aerodynamic Impulse Responses Using Digital Filter Techniques. In Proceedings of the Atomospheric Flight Mechanics, New Orleans, LA, USA, 11–13 August 1997; AIAA Paper 97-3712.
7. Silva, W.A. Reduced-Order Models Based on Linear and Nonlinear Aerodynamic Impulse Responses. In Proceedings of the CEAS/AIAA/ICASE/NASA Langley International Forum on Aeroelasticity and Structural Dynamics, Williamsburg, VA, USA, 22–25 June 1999; pp. 369–379.
8. Juang, J.-N.; Pappa, R.S. An Eigensystem Realization Algorithm for Modal Parameter Identification and Model Reduction. *J. Guidance Control Dyn.* **1985**, *8*, 620–627.
9. Juang, J.-N. *Applied System Identification*; Prentice-Hall PTR: Upper Saddle River, NJ, USA, 1994.
10. Silva, W.A.; Raveh, D.E. Development of Unsteady Aerodynamic State-Space Models from CFD-Based Pulse Responses. Presented at the 42nd Structures, Structural Dynamics, and Materials Conference, Seattle, WA, USA, 16–19 April 2001; AIAA Paper No. 2001-1213.
11. Gaitonde, A.L.; Jones, D.P. Study of Linear Response Identification Techniques and Reduced-Order Model Generation for a 2D CFD Scheme. *J. Numer. Methods Fluids* **2006**, *52*, 1367–1402.
12. Gaitonde, A.L.; Jones, D.P. Calculations with ERA Based Reduced-Order Aerodynamic Models. In Proceedings of the 24th Applied Aerodynamics Conference, San Francisco, CA, USA, 5–8 June 2006.
13. Griffiths, L.; Jones, D.P.; Friswell, M.I. Model Updating of Dynamically Time Linear Reduced-Order Models. In Proceedings of the International Forum on Aeroelasticity and Structural Dynamics, Paris, France, 27–30 June 2011.
14. Fleischer, D.; Breitsamter, C. Efficient Computation of Unsteady Aerodynamic Loads Using Computational Fluid Dynamics Linearized Methods. *J. Aircr.* **2013**, *50*, 425–440.
15. Xiaoyan, L.; Zhigang, W.; Chao, Y. Aerodynamic Reduced-Order Models Based on Observer Techniques. In Proceedings of the 51st AIAA/ASME/ASCE/AHS/ASC Structures, Structural Dynamics, and Materials Conference, Orlando, FL, USA, 12–15 April 2010.
16. Song, H.; Qian, J.; Wang, Y.; Pant, K.; Chin, A.W.; Brenner, M.J. Development of Aeroelastic and Aeroservoelastic Reduced-Order Models for Active Structural Control. In Proceedings of the 56th AIAA/ASCE/AHS/ASC Structures, Structural Dynamics, and Materials Conference, Kissimmee, FL, USA, 5–9 January 2015.
17. Ma, Z.; Ahuja, S.; Rowley, C.W. Reduced-Order Models for Control of Fluid Using the Eigensystem Realization Algorithm. *Theor. Comput. Fluid Dyn.* **2011**, *25*, 233–247.
18. Krist, S.L.; Biedron, R.T.; Rumsey, C.L. *CFL3D User's Manual Version 5.0*; Technical Report; NASA Langley Research Center: Hampton, VA, USA, 1997.
19. Silva, W.A.; Bartels, R.E. Development of Reduced-Order Models for Aeroelastic Analysis and Flutter Prediction Using the CFL3Dv6.0 Code. *J. Fluids Struct.* **2004**, *19*, 729–745.
20. Kim, T.; Hong, M.; Bhatia, K.G.; SenGupta, G. Aeroelastic Model Reduction for Affordable Computational Fluid Dynamics-Based Flutter Analysis. *AIAA J.* **2005**, *43*, 2487–2495.

21. Silva, W.A. Recent Enhancements to the Development of CFD-Based Aeroelastic Reduced Order Models. In Proceedings of the 48th AIAA/ASME/ASCE/AHS/ASC Structures, Structural Dynamics, and Materials Conference, Honolulu, HI, USA, 23–26 April 2007; AIAA Paper No. 2007-2051.

22. Anderson, W.K. Bonhaus, D.L. An Implicit Upwind Algorithm for Computing Turbulent Flows on Unstructured Grids. *Comput. Fluids* **1994**, *23*, 1–21.

23. NASA LaRC. *FUN3D Manual, v12.9*; NASA LaRC: Hampton, VA, USA, 2015. Available online: http://fun3d.larc.nasa.gov (accessed on 21 September 2015).

24. Biedron, R.T.; Vatsa, V.N.; Atkins, H.L. Simulation of Unsteady Flows Using an Unstructured Navier-Stokes Solver on Moving and Stationary Grids. In Proceedings of the 23rd AIAA Applied Aerodynamics Conference, Toronto, ON, Canada, 6–9 June 2005; AIAA Paper 2005-5093.

25. Biedron, R.T.; Thomas, J.L. Recent Enhancements to the FUN3D Flow Solver for Moving-Mesh Applications. In Proceedings of the 47th AIAA Aerospace Sciences Meeting Including The New Horizons Forum and Aerospace Exposition, Orlando, FL, USA, 5–8 January 2009; AIAA Paper 2009-1360. Available online: https://arc.aiaa.org/doi/abs/10.2514/6.2009-1360 (accessed on 10 April 2009).

26. Juang, J.-N.; Phan, M.; Horta, L.G.; Longman, R.W. Identification of Observer/Kalman Filter Markov Parameters: Theory and Experiments. *J. Guid. Control Dyn.* **1993**, *16*, 320–329.

27. Eykhoff, P. *System Identification: Parameter and State Identification*; Wiley Publishers: Hoboken, NJ, USA, 1974.

28. Ljung, L. *System Identification: Theory for the User*; Prentice-Hall Publishers: Upper Saddle River, NJ, USA, 1999.

29. Zhu, Y. *Multivariable System Identification for Process Control*; Pergamon Publishers: Oxford, UK, 2001.

30. Pacheco, R.P.; Steffen, V., Jr. Using Orthogonal Functions for Identification and Sensitivity Analysis of Mechanical Systems. *J. Vib. Control* **2002**, *8*, 993–1021.

31. Raveh, D.E. Identification of Computational-Fluid-Dynamic Based Unsteady Aerodynamic Models for Aeroelastic Analysis. *J. Aircr.* **2004**, *41*, 620–632.

32. Silva, W.A.; Ringertz, U.; Stenfelt, G.; Eller, D.; Keller, D.F.; Chwalowski, P. Status of the KTH-NASA Wind-Tunnel Test for Acquisition of Transonic Nonlinear Aeroelastic Data. In Proceedings of the 15th Dynamics Specialists Conference, AIAA SciTech Forum, San Diego, CA, USA, 4–8 January 2016; No. AIAA 2016-2050.

33. Silva, W.A.; Chwalowski, P.; Wieseman, C.D.; Keller, D.F.; Eller, D.; Ringertz, U. Computational and Experimental Results for the KTH-NASA Wind-Tunnel Model Used for Acquistion of Transonic Nonlinear Aeroelastic Data. In Proceedings of the International Forum on Aeroelasticity and Structural Dynamics, Como, Italy, 26–28 June 2017.

34. Chwalowski, P.; Silva, W.A.; Wieseman, C.D.; Heeg, J. CFD Model of the Transonic Dynamics Tunnel with Applications. In Proceedings of the International Forum on Aeroelasticity and Structural Dynamics, Como, Italy, 26–28 June 2017.

35. Silva, W.A.; Chwalowski, P.; Perry, B. Evaluation of Linear, Inviscid, Viscous, and Reduced-Order Modeling Aeroelastic Solutions of the AGARD 445.6 Wing Using Root Locus Analysis. *Int. J. Comput. Fluid Dyn.* **2014**, *28*, 122–139.

MDPI

Article

A Hybrid Reduced-Order Model for the Aeroelastic Analysis of Flexible Subsonic Wings—A Parametric Assessment

Marco Berci [1] and Rauno Cavallaro [2,*]

[1] School of Mechanical Engineering, University of Leeds, Leeds LS2 9JT, UK; m.berci07@members.leeds.ac.uk
[2] Department of Bioengineering and Aerospace Engineering, Universidad Carlos III de Madrid, 28911 Leganés (Madrid), Spain
* Correspondence: rauno.cavallaro@uc3m.es

Received: 30 April 2018; Accepted: 9 July 2018; Published: 17 July 2018

Abstract: A hybrid reduced-order model for the aeroelastic analysis of flexible subsonic wings with arbitrary planform is presented within a generalised quasi-analytical formulation, where a slender beam is considered as the linear structural dynamics model. A modified strip theory is proposed for modelling the unsteady aerodynamics of the wing in incompressible flow, where thin aerofoil theory is corrected by a higher-fidelity model in order to account for three-dimensional effects on both distribution and deficiency of the sectional air load. Given a unit angle of attack, approximate expressions for the lift decay and build-up are then adopted within a linear framework, where the two effects are separately calculated and later combined. Finally, a modal approach is employed to write the generalised equations of motion in state-space form. Numerical results were obtained and critically discussed for the aeroelastic stability analysis of a uniform rectangular wing, with respect to the relevant aerodynamic and structural parameters. The proposed hybrid model provides sound theoretical insights and is well suited as an efficient parametric reduced-order aeroelastic tool for the preliminary multidisciplinary design and optimisation of flexible wings in the subsonic regime.

Keywords: hybrid reduced-order model; quasi-analytical; aeroelasticity; flexible wings; subsonic

1. Introduction

Efficient aeroelastic methods and tools [1] based on reduced complexity are increasingly sought for the preliminary multidisciplinary design and optimisation (MDO) [2,3] of flexible aircraft and unmanned air vehicles (UAV). Smart optimisation strategies and algorithms [4] still rely on effective and robust simulations, where the relevant aeroelastic issues and behaviours [5,6] are parametrically analysed in a large design variables space [7]. Fluid-structure interaction (FSI) [8] models coupling finite element methods (FEM) [9] and computational fluid dynamics (CFD) [10] have increasingly been proposed to enhance accuracy [11,12]. However, these high-fidelity tools are computationally expensive [13,14] and require special care [15,16] to ensure that the correct physics are reproduced in their coupling [17,18], especially at all boundaries and interfaces [19,20]. A large amount of time and efforts is then typically necessary for pre-processing the simulations and post-processing the results, which are key features for a reliable implementation of automated MDO routines [21].

A hybrid reduced-order model (ROM) [22–26] for the aeroelastic analysis of subsonic wings in unsteady incompressible flow is presented here within a generalised quasi-analytical formulation. A modified strip theory (MST) is adopted for the aerodynamic load [27,28], tuned (TST) and standard (SST) strip theories being readily resumed for comparison [29]. Thin aerofoil theory is employed for calculating the unsteady air load around each flexible wing section [30–35], where the lift deficiency function is corrected by a high-fidelity model in order to account for downwash effects [36]. First, the spanwise decay [37,38] and time-wise build-up [39–43] of the load are separately calculated for the rigid wing by means of a steady and unsteady simulation using the doublet lattice method (DLM) [44,45], as available in the commercial software Nastran [46]. They are then approximated via nonlinear curve-fitting [47] and re-combined a posteriori for use in the linear framework [48] of the proposed quasi-analytical model. A beam-like linear model is considered for the wing structural dynamics [49] and the principle of virtual work (PVW) [50] is used to derive the equilibrium equations. Ritz's method [51] is finally employed for solving the latter within a modal approach [52], where shape functions are assumed for the displacement [53]. The resulting hybrid ROM allows for arbitrary distributions of the wing properties, providing continuous deformations and loads [54]. Goland's wing is analysed first for the sake of a thorough validation [55]. The numerical results for both the divergence speeds and flutter frequency of a uniform rectangular flat wing are then shown and critically discussed with respect to the relevant aero-structural parameters, such as aspect and thickness ratios. Finally, Appendix A presents both steady [56–58] and unsteady [59,60] lifting line models which can serve as effective and inherently parametric semi-analytical tools [61].

2. Aeroelastic Problem Formulation

According to the closely-spaced rigid diaphragm assumption [62], a slender wing is considered flexible spanwise only and a beam-like model is then suitably employed [63]. Since the latter can be derived from a plate-like model [64], it may represent a physical ROM in itself where the chordwise dependency of the structural properties is dropped [65], and only the pitch and plunge rigid modes of the aerofoil section are retained to allow wing bending and torsion [66], respectively.

The wing has a chord $c(y)$, semi-span l and aspect ratio AR. The elastic axis (EA, where all loads are acting [67]) is modelled as a Rayleigh beam [68] and drawn by the locus of the shear centre of each chordwise section, with $x_{EA}(y) \equiv 0$ fixed for convenience [51,52], whereas the inertial axis $x_{CG}(y)$ is drawn by the locus of the sectional centre of gravity (CG, where the inertial load is applied [67]). Thus, there results a mass $m(y)$ as well as bending and torsion moments of inertia $\mu_\zeta(y)$ and $\mu_\vartheta(y)$ per unit length, an area moment of inertia $I(y)$ and a torsion factor $J(y)$, Young's and shear elastic modules $E(y)$ and $G(y)$ are distributed along the span $-l \leq y \leq +l$.

With $\zeta(y,t)$ and $\vartheta(y,t)$ being the vertical displacement and rotation of the EA, respectively, the wing deformation is given as $w = \zeta - x\vartheta$ directly. Neglecting gravity and concentrated loads, the PVW for the arbitrary virtual displacements $\delta\zeta(y,t)$ and $\delta\vartheta(y,t)$ then reads:

$$\int_0^l EI\zeta'' \delta\zeta'' dy + \int_0^l GJ\vartheta' \delta\vartheta' dy = \int_0^l \Delta L \delta\zeta dy + \int_0^l \Delta M \delta\vartheta dy$$
$$- \int_0^l m\ddot{w}_{CG} \delta w_{CG} dy - \int_0^l \mu_\zeta \ddot{w}'_{CG} \delta w'_{CG} dy - \int_0^l \mu_\vartheta \ddot{\vartheta} \delta\vartheta dy \tag{1}$$

where $w_{CG} = \zeta - x_{CG}\vartheta$ is the vertical displacement of the inertial axis, whereas $\Delta L(y,t)$ and $\Delta M(y,t)$ are the sectional unsteady aerodynamic force (positive upwards) and pitching moment (positive clockwise), respectively. The virtual displacement being arbitrary, the bending and torsion virtual work separate and are integrated by parts twice in order to give the linear system of coupled PDEs for the dynamic aeroelastic equilibrium of wing bending and torsion as:

$$\left(EI\zeta''\right)'' + m\left(\ddot{\zeta} - x_{CG}\ddot{\vartheta}\right) - \mu_\zeta\left(\ddot{\zeta} - x_{CG}\ddot{\vartheta}\right)' = \Delta L$$
$$\left(GJ\vartheta'\right)' - \mu_\vartheta\ddot{\vartheta} + x_{CG}\left[m\left(\ddot{\zeta} - x_{CG}\ddot{\vartheta}\right) - \mu_\zeta\left(\ddot{\zeta} - x_{CG}\ddot{\vartheta}\right)'\right] = -\Delta M \tag{2}$$

which are consistently completed by both geometrical and natural boundary conditions as:

$$\vartheta(0,t) = 0, \ \zeta(0,t) = 0, \ \zeta'(0,t) = 0$$

$$GJ\vartheta'\big|_l = 0, \ EI\zeta''\big|_l = 0, \ (EI\zeta'')'\big|_l - \mu_\zeta\ddot{\zeta}'\big|_l = 0 \tag{3}$$

Note that this standard problem formulation assumes an isotropic material and holds for swept wings too when a chordwise approach is employed [52,69]; if necessary, lumped masses, dampers or springs may easily be included using Dirac's delta function centred at their applicable location [27,63]. For a slender composite wing, the anisotropic material exhibits different mechanical characteristics in different directions and the resulting elastic coupling between bending and torsion may then be included using the applicable constitutive law, with a more complex calculation of the structural stiffness but no conceptual changes in the overall aeroelastic problem formulation [51].

Modal Solution Approach

Ritz's method [51] is employed and the beam displacement is then modally expressed as:

$$\zeta = \sum_{i=1}^{n_\zeta} \phi_i \varepsilon_i, \ \vartheta = \sum_{i=1}^{n_\vartheta} \varphi_i \eta_i, \ \delta\zeta = \sum_{i=1}^{n_\zeta} \phi_i \delta\varepsilon_i, \ \delta\vartheta = \sum_{i=1}^{n_\vartheta} \varphi_i \delta\eta_i \tag{4}$$

where the n_ζ functions, $\varepsilon_i(t)$ and n_ϑ functions $\eta_i(t)$ are the unknown generalised coordinates relative to the n_ζ mode shapes $\phi_i(y)$ for bending deformation and n_ϑ mode shapes $\varphi_i(y)$ for torsion deformation, respectively. All mode shapes shall satisfy the geometrical boundary condition for clamped-free beams [63] and may be either assumed [53] or obtained from FEM eigen-analysis or vibrations tests [70,71]. Unlike plate-like models (which require the clamping of the wing root along its entire chord [64]), beam-like models allow a linear torsion mode.

3. Generalised Aerodynamic Load

Within the modal formulation, the generalised unsteady sectional air load is implicitly given as:

$$F_i^\zeta = \int_0^l \Delta L \phi_i dy, \ F_i^\vartheta = \int_0^l \Delta M \varphi_i dy \tag{5}$$

whereas the unsteady lift, pitching moment and rolling moment of the wing are given by:

$$L = \int_0^l \Delta L dy, \ M_p = \int_0^l \Delta M dy, \ M_r = \int_0^l y \Delta L dy \tag{6}$$

Provided that the effect of the unsteady downwash is included in the indicial function for the wing load development [72], it can be reasonably assumed that steady effects on the spanwise lift distribution due to the wing-tip vortices and unsteady effects on the chordwise lift build-up due to the travelling wake may be considered separately and eventually assembled in a quasi-steady sense [28]; of course, the slower the wing motion (with respect to the aircraft speed) and the higher the aspect ratio, the more accurate the assumption [73]. MST is hence proposed as a physical ROM [27] for calculating the generalised unsteady aerodynamic load, where the lift distribution and evolution are independently combined a posteriori. Note that MTS holds also for compressible flows around arbitrary wings, as long as the coupling between both three-dimensional and compressible effects remains weak and the appropriate indicial functions are employed for the different types of wing motion [5,48,52,74].

Unsteady Modified Strip Theory

According to thin aerofoil theory for incompressible flow [75], the non-circulatory aerodynamic force and moment of each wing section act at its mid-chord (MC), whereas the circulatory ones act at both its aerodynamic centre (AC, where the pitching moment is independent of the angle of attack [76]) and its control point (CP, where the non-penetration boundary condition for the inviscid flow is imposed and the fluid-structure interaction hence enforced [33]). The AC and CP positions $x_{AC}(y)$ and $x_{CP}(y)$ falling at the first and last quarters of the chord [5], respectively, the sectional unsteady aerodynamic force and pitching moment due to the wing motion read as [30,54]:

$$\Delta L = \frac{1}{2}\rho c \left[\frac{\pi c}{2}\left(U\dot{\vartheta} - \ddot{w}_{MC} \right) + \kappa U C_{L/\alpha} \left(V_0 W + \int_0^t \frac{dV(\iota)}{d\iota} W(t-\iota)d\iota \right) \right]$$

$$\Delta M = -\frac{1}{2}\rho c \left[\frac{\pi c}{2}\left(\frac{c^2}{32}\ddot{\vartheta} + x_{CP}U\dot{\vartheta} - x_{MC}\ddot{w}_{MC} \right) + x_{AC}\kappa U C_{L/\alpha} \left(V_0 W + \int_0^t \frac{dV(\iota)}{d\iota} W(t-\iota)d\iota \right) \right]$$

(7)

where all terms involving the lift derivative $C_{L/\alpha}$ are of a circulatory nature, whereas all others are of a non-circulatory nature and include apparent inertia effects. Here, U and ρ are the speed and density of the reference airflow, $w_{AC} = \zeta - x_{AC}\vartheta$, then $w_{MC} = \zeta - x_{MC}\vartheta$ and $w_{CP} = \zeta - x_{CP}\vartheta$ are the instantaneous vertical displacements of AC, MC and CP, respectively, while $W(t)$ is the equivalent of Wagner's indicial-admittance function for the circulatory lift build-up due to a unit step in the angle of attack [30]; additional terms appear in the presence of sweep [69] or ailerons [33] (included by means of Heaviside's step function [52,77] centered at the applicable location). Due to its own motion, each wing section experiences an effective instantaneous angle of attack $\alpha_e(y,t)$ induced by the net vertical flow velocity $V(y,t)$, namely [30,54],

$$\alpha_e = \left(\frac{V_0}{U} \right) W + \int_0^t \frac{dV(\iota)}{Ud\iota} W(t-\iota)d\iota, \quad V = U\vartheta - \dot{w}_{CP}, \quad V_0 = V(y,0)$$

(8)

starting from an initial (rest) condition. Adopting the lift-curve slope $C_{L/\alpha} = 2\pi$ for a flat aerofoil [78], note that Wagner's and Theodorsen's formulations [30,33] represent a physical ROM in itself, where the load contribution from small chordwise deformations is neglected and only the pitch and plunge rigid modes are retained from Peters' general formulation for a morphing aerofoil [65,79].

Within MST [27], the scaling function $\kappa(y)$ is introduced in order to account for the span-wise influence of the wing-tip vortices on the sectional air load [37,38,80] and it is consistently derived from Kutta-Joukowsky's theorem [81,82] based on the steady lift distribution as:

$$\kappa_{MST} = \frac{2\Gamma}{UcC_L}, \quad \kappa_{TST} = \frac{\pi AR}{\pi AR + C_{L/\alpha}(1+)}, \quad \kappa_{SST} = 1$$

(9)

where $\Gamma(y)$ is the steady circulation distribution and is Oswald's efficiency factor [83], which embeds the downwash effect. In fact, MST considers the airflow around each wing section as quasi-independent, whereas TST treats it as fully independent and a global scaling factor is then applied to the wing lift; SST disregards all three-dimensional effects and is obtained for the limit of infinitely slender wings, with Wagner's function giving the lift-deficiency [30].

When the scaling function is based on lifting line theory (LLT, of which SST is the forcing term [27]; see Appendix A), MST may be regarded as a quasi-unsteady ROM for the unsteady LLT [84–86]; yet, the overall tuning concept is completely general [87,88] and such a function may then be derived based on the steady lift distribution obtained from any appropriate source [89–98], such as the vortex lattice method (VLM) [36], DLM [99], CFD [100] or experiments. In all cases, the proposed correction applies only to the circulatory load development of each wing section, whereas the non-circulatory load has an impulsive nature and remains uncorrected for three-dimensional effects [42,52]. From a theoretical point of view [80], the scaling function depends on both flow conditions and actual wing geometry, which in turn, depends on the total applied load and is not known a priori for flexible aircraft. In the presence of small deformations, the scaling function may be calculated for the undeformed wing only and then consistently used for the deformed wing [27]; furthermore, a unit angle of attack may be assumed without loss of generality when linear aerodynamic methods are used [5]. In the presence of large deformations [101] or in the case of non-planar wings [102], the scaling function would also include the geometrically nonlinear effects of the wing curvature on the local load; thus, the flying wing shape and actual angle of attack shall be considered, especially when nonlinear aerodynamic methods are used. However, note that a database of steady lift distributions may be available from previous aerodynamic design studies and also parametrically approximated in order to boost the overall design process efficiency via multi-fidelity surrogate models [103,104].

The calculated scaling function may then be approximated with Prandtl's expansion [56] as:

$$\kappa = \sum_{j=1}^{n_\kappa} \kappa_j \sin(j\psi) \tag{10}$$

where the n_κ coefficients can be obtained via either Fourier integrals or curve-fitting directly [105]; this scaling function would also apply to wind gust loading, regardless of the penetration effect [31]. Contrary to LLT, note that the proposed MST inherently prevents projection of all the structural modes on each of the assumed aerodynamic mode, thus reducing the problem size n_κ times [27].

4. Added Aerodynamic States

For two-dimensional unsteady incompressible potential flow, Wagner's lift-deficiency function accounts for the inflow generated by the travelling wake of a flat aerofoil [30,61]; it may be obtained from Theodorsen's function [33] and vice versa, due to reciprocal relations [106,107] and analytical continuation in the Laplace domain. For three-dimensional flow, the lift-deficiency function includes the unsteady downwash of the trailed wing-tip vortices [39] and is approximated for computational convenience with a series of exponential terms in the reduced time τ domain [108] (i.e., a series of rational terms in the reduced frequency k domain [109]), namely,

$$W = 1 - \sum_{i=1}^{n} A_i e^{-B_i \tau}, \quad \sum_{i=1}^{n} A_i = 1 - W_0, \quad \tau = 2\left(\frac{U}{c}\right)t \tag{11}$$

where all coefficients are obtained by best-fitting the reference curve for the specific wing shape with the exact constraint $W_0 = \lim_{\tau \to 0} W$ [105], whereas $\overline{c}$ is a reference chord (e.g., the wing root chord). In the case of SST, $A_1 = 0.165$ with $B_1 = 0.0455$ and $A_2 = 0.335$ with $B_2 = 0.3$ are commonly used [39]; two approximation terms are also typically sufficient for wings of industrial interest [110] and an added aerodynamic state $v(y,t)$ is hence introduced which evolves according to the linear ODE:

$$\ddot{v} + 2(B_1 + B_2)\left(\frac{U}{\overline{c}}\right)\dot{v} + 4B_1 B_2 \left(\frac{U}{\overline{c}}\right)^2 v = U\vartheta - \dot{\zeta} + x_{CP}\dot{\vartheta} \tag{12}$$

$$A_1 + A_2 = 1 - W_0$$

Note that the concept of indicial-admittance function is completely general [5,48,111] and the reference curve may be obtained from any appropriate analytical or numerical source, such as DLM [99], CFD [112,113] or experiments; still, when the lift-deficiency function is based on unsteady LLT (see Appendix A), MST may indeed be regarded as a quasi-unsteady ROM. From a theoretical point of view, the indicial aerodynamic function also depends on both flow conditions and actual wing geometry; therefore, the very same comments and assumptions as for the scaling function of the lift distribution apply. The lift-deficiency function may be calculated for the undeformed wing and then consistently be used for the deformed wing; a unit angle of attack step may be assumed without loss of generality whenever linear aerodynamic methods are used [5]. In fact, a database of lift-deficiency functions may be available from previous flight dynamics studies and parametrically approximated in order to significantly accelerate the overall design process [103,104]. Accounting for the penetration effect [31,35], the very same methodology and considerations would also apply to wind gust loading, due to linearity.

Unsteady Air Load

The generalised unsteady aerodynamic load per unit span now reads in the state space as:

$$\Delta L = \frac{\pi}{4}\rho c^2 \left(U\dot{\vartheta} - \ddot{\zeta} + x_{MC}\ddot{\vartheta}\right) + \frac{\kappa}{2}\rho U c C_{L/\alpha} W_0 \left(U\vartheta - \dot{\zeta} + x_{CP}\dot{\vartheta}\right)$$
$$+ \kappa\rho U c C_{L/\alpha} \left[(A_1 B_1 + A_2 B_2)\left(\frac{U}{c}\right)v + 2(A_1 + A_2)B_1 B_2 \left(\frac{U}{c}\right)^2 v\right] \tag{13}$$

$$\Delta M = -\frac{\pi}{4}\rho c^2 \left[\frac{c^2}{32}\ddot{\vartheta} + x_{CP}U\dot{\vartheta} - x_{MC}\left(\ddot{\zeta} - x_{MC}\ddot{\vartheta}\right)\right] - \frac{\kappa}{2}\rho U x_{AC} c C_{L/\alpha} W_0 \left(U\vartheta - \dot{\zeta} + x_{CP}\dot{\vartheta}\right)$$
$$- \kappa\rho U x_{AC} c C_{L/\alpha} \left[(A_1 B_1 + A_2 B_2)\left(\frac{U}{c}\right)v + 2(A_1 + A_2)B_1 B_2 \left(\frac{U}{c}\right)^2 v\right] \tag{14}$$

and can also be expanded in terms of the assumed modal base as $\Delta L = \sum\limits_{i=1}^{n_\zeta} \phi_i \Delta L_i^\varepsilon + \sum\limits_{i=1}^{n_\theta} \varphi_i \Delta L_i^\eta$ and $\Delta M = \sum\limits_{i=1}^{n_\zeta} \phi_i \Delta M_i^\varepsilon + \sum\limits_{i=1}^{n_\theta} \varphi_i \Delta M_i^\eta$, with $v = \sum\limits_{i=1}^{n_\zeta} \phi_i v_i^\varepsilon + \sum\limits_{i=1}^{n_\theta} \varphi_i v_i^\eta$; therefore, taking advantage of linear superposition, the modal unsteady air load $\Delta L_i^\varepsilon(t), \Delta M_i^\varepsilon(t), \Delta L_i^\eta(t), \Delta M_i^\eta(t)$ and the added states $v_i^\varepsilon(t)$, $v_i^\eta(t)$ are also eventually found in terms of the generalised coordinates, with each resulting term of strip theory projected onto the mode shapes ϕ_i and φ_i, respectively.

5. Analytical Aeroelastic Analysis

By substituting the modal expansions in the PVW, the aeroelastic equilibrium PDEs eventually become a linear system of ODEs for the generalised coordinates, namely [114]:

$$\mathbf{M}^s\ddot{\chi} + \mathbf{C}^s\dot{\chi} + \mathbf{K}^s\chi = \mathbf{F}^a, \quad \mathbf{F}^a = \mathbf{M}^a\ddot{\chi} + \mathbf{C}^a\dot{\chi} + \mathbf{K}^a\chi \tag{15}$$

with generalised structural mass $\mathbf{M}^s$, damping $\mathbf{C}^s$ and stiffness $\mathbf{K}^s$ matrices, aerodynamic load vector $\mathbf{F}^a(t)$, aerodynamic mass $\mathbf{M}^a$, damping $\mathbf{C}^a$ and stiffness $\mathbf{K}^a$ matrices, all depending on the wing shape and properties; $\chi(t)$ is the unknown vector of generalised coordinates (including added aerodynamic states), which drives the aeroelastic dynamic response. It is worth noting that a change of variables is always possible, as long as a rigorous transformation matrix can be defined and all aero-structural matrices are then consistently projected onto the new modal base [51,79].

The aeroelastic response and stability analysis of the subsonic wing are then governed by:

$$\mathbf{M}\ddot{\chi} + \mathbf{C}\dot{\chi} + \mathbf{K}\chi = 0, \quad \det\left(\mathbf{M}\lambda^2 + \mathbf{C}\lambda + \mathbf{K}\right) = 0 \tag{16}$$

respectively, or their equivalent first-order forms [114]:

$$\left\{\begin{array}{c}\ddot{\chi}\\\dot{\chi}\end{array}\right\} = \left[\begin{array}{cc}-\mathbf{M}^{-1}\mathbf{C} & -\mathbf{M}^{-1}\mathbf{K}\\\mathbf{I} & 0\end{array}\right]\left\{\begin{array}{c}\dot{\chi}\\\chi\end{array}\right\}, \quad \det\left[\begin{array}{cc}\mathbf{I}\lambda + \mathbf{M}^{-1}\mathbf{C} & \mathbf{M}^{-1}\mathbf{K}\\-\mathbf{I} & \mathbf{I}\lambda\end{array}\right] = 0 \tag{17}$$

where $\mathbf{M} = \mathbf{M}^s - \mathbf{M}^a$, $\mathbf{C} = \mathbf{C}^s - \mathbf{C}^a$ and $\mathbf{K} = \mathbf{K}^s - \mathbf{K}^a$ are the generalised aeroelastic mass, damping and stiffness matrices, which depend parametrically on the aircraft speed. In particular, flutter occurs at the lowest flow speed U_F, hence the real part of at least one of the complex eigenvalues λ_i becomes positive (i.e., the aeroelastic dynamic behaviour becomes unstable through a Hopf bifurcation [55,66], where a couple of conjugate eigenvalues cross the imaginary axis and leave the response undamped for unsteady flow), two or more generalised aeroelastic modes coupling at the flutter frequency f_F. Note that real and imaginary parts of the complex eigenvalue are related to the effective modal damping and vibration frequency of the wing [115], respectively, with its natural vibration modes being correctly recovered in the absence of air [51,68]. Finally, the static divergence speed is the lowest flow speed U_D making at least one of the eigenvalues cross the imaginary axis on the real axis, and then, the aeroelastic stiffness matrix and static response become singular [55,66] (i.e., structural and aerodynamic forces do not find a stable equilibrium for steady flow).

6. Numerical Aeroelastic Analysis

The commercial aeroelastic solver Nastran [46] was here used as the full-order model (FOM), following the good practice shown in the chapter, "Dynamic Aeroelastic Response Analysis" of the Aeroelastic Analysis User's Guide [116]. A beam-like FEM and a lifting-surface DLM were built and coupled via splines interface [117] for transferring both loads and displacements, resulting in the numerical aeroelastic model. Note that the latter needed to be re-generated for each and every case to be parametrically investigated, including pre-processing and post-processing.

The beam was modelled using CBEAM elements and RBE2 rigid elements were added to each finite element node in order to support splining; the node lying at the wing root was then clamped. The aerodynamic lifting surface was aligned with the freestream and modelled as a flat plate divided into an appropriate number of CAERO panels placed along the wing span and chord. The RB2 elements were designed to match the leading and trailing edge of the aerodynamic surface and provide a natural support for splining; in particular, surface splines SPLINE1 were used and the infinite plate spline (IPS) [118] was selected among the available options.

For the natural vibration analysis, shear deformation was neglected and Rayleigh beam theory [68] was used, with PBEAM defining the properties (i.e., inertia and stiffness) of the beam element and SPC1 defining the single-point constraint for the clamped root. The vibration analysis was then performed using Lanczos' method [119], which was among those available in EIGRL, normalising the modes to unit values of the generalized mass.

For the aerodynamic analysis, the matrix of complex aerodynamic influence coefficients (AIC) [44] was generated (in the physical space) at several reduced frequencies specified in MKAERO1; which includes coupling terms between unsteady wake inflow, vortices downwash, fluid compressibility and apparent inertia, as the DLM is formulated for subsonic potential flow [114]. Load symmetry with respect to the vertical plane was always imposed at the wing root. Well-established guidelines on results accuracy and robustness relate the highest reduced frequency to the number of panels placed along the chord [120]; no additional correction [121] was implemented.

For the steady aerodynamic load analysis and scaling function derivation, a unit angle of attack was specified for the clamped rigid wing in TRIM. Taking advantage of SPLINE2, the strip-wise normal force and pitching moment coefficients (at the strip leading edge) distribution was finally obtained along the wing span using the straightforward MONCNCM.

For the unsteady aerodynamic load analysis and lift-deficiency function derivation, the plunge motion was released to impose a unit step change in the angle of attack. The latter was prescribed as a time-dependent dynamic excitation in TLOAD1, in terms of an enforced velocity motion in SPCD, within the framework of a transient response analysis; the full-time history of the dynamic excitation (i.e., a positive square wave followed by an equal and opposite one) was specified in TABLED1. A very stiff scalar spring element was then defined at the EA root in CELAS2, in order to avoid any significant vertical displacement of the rigid wing resulting from the imposed dynamic excitation; a rigid-body spline SPLINRB was used for interpolating the spring motion and forces. The dynamic excitation is automatically Fourier-transformed and the transient response problem is solved in the frequency domain, based on the frequency range and resolution specified in FREQ1. Fourier inverse transform is then used to obtain the solution back in the time domain, using the time-step interval specified in TSTEP. The wing lift was monitored at the EA root, where the stiff spring was located.

For the static aeroelastic divergence analysis, two different yet still equivalent approaches were adopted: a complex eigenvalue analysis using Lanczos' block method [119] as defined in EIGC with Mach numbers specified in DIVERG, a dynamic aeroelastic divergence analysis at zero frequency.

For the dynamic aeroelastic divergence analysis, the *p-k* method [122] as available in FLUTTER was used with a non-iterative frequency-sweeping technique, where the unsteady AIC matrix is first projected onto the modal space and mode-tracking is then performed via the eigenvectors correlation matrix. Note that the *p-k* method uses only real matrix terms for computing the flutter solution [114], meaning that any imaginary terms in any of the matrices are ignored and the imaginary part of the AIC matrix is added as a real matrix to the viscous damping matrix. Finally, flutter analyses were also performed with two-dimensional aerodynamics (PAERO4 and CAERO4) for direct validation.

7. Results and Discussion

Goland's wing [123,124] was considered first, since it is widely used as a fundamental reference for validation as well as an ideal prototype for investigating new methods and concepts. It is a flat thin uniform rectangular cantilevered wing with $c = 1.829$ m and $l = 6.096$ m; the wing root is clamped at its elastic axis with stiffness $EI = 9{,}772{,}200$ Pa·m^4 and $GJ = 987{,}600$ Pa·m^4 at 33% of the wing chord, while the inertial axis with mass $m = 35.72$ kg/m and $\mu_\theta = 7.452$ kg·m lays at 43% of the latter. The wing is aligned with the horizontal reference airflow, which is assumed to be incompressible [123]; then the aerodynamic center and control point are consistently at 25% and 75% of the wing chord, respectively. Goland's wing exhibits the prototypical flutter mechanism coupling its fundamental bending and torsion modes, the uncoupled natural vibration frequencies of which were found at $f_\zeta = 7.9$ Hz and $f_\theta = 13.9$ Hz, respectively, as expected [125]. Figure 1 shows the evolution with the airspeed of the real and imaginary parts of the relative complex eigenvalues calculated by the hybrid ROM, using unsteady SST with Wagner's function approximation [39] for the aeroelastic stability analysis in a standard atmosphere at sea level ($\rho = 1.225$ kg/m^3 [126]). Due to the torsion mode becoming unstable and extracting energy from the coupled bending mode, flutter is found at $U_F = 137.4$ m/s and $f_F = 11.1$ Hz, which is in excellent agreement with previous results [123–128]; then, the flutter Mach number $M_F = 0.40$ confirms that the incompressible flow was correctly assumed [129]. When using unsteady MST for the aeroelastic stability analysis at about 6100 ft altitude ($\rho = 1.02$ kg/m^3 [126]), flutter is found at $U_F = 163.7$ m/s and $f_F = 11.3$ Hz, which is still in remarkable agreement with the existing results [130]. In this case, five sinusoidal and two exponential terms were respectively employed to approximate the lit distribution and build-up as obtained by Nastran's DLM [45] with a grid of 48 spanwise and 12 chordwise panels and 14 reduced frequencies in the range $0 < k < 0.8$, which largely covers the reduced flutter frequency. In all cases, employing the first two bending and torsion modes (correctly found at $f_1 = 7.7$ Hz, $f_2 = 15.2$ Hz, $f_3 = 38.8$ Hz, $f_4 = 55.3$ Hz, also when using Nastran's FEM [46]) granted convergence of the aeroelastic stability analysis [130].

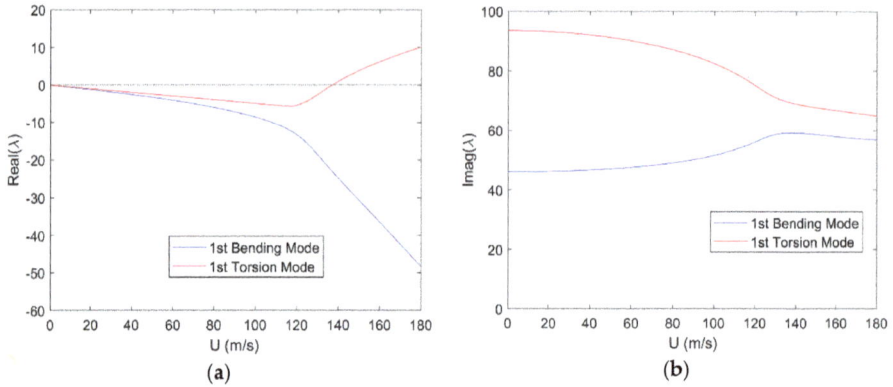

Figure 1. Real (**a**) and imaginary (**b**) parts of the first two eigenvalues for Goland's wing flutter.

Following Goland's concept, a flat rectangular cantilevered wing of uniform material and chord $c = 1$ m is now considered; the wing root is still clamped at the elastic axis. Flexural and torsional stiffness of the wing are respectively given by [67]:

$$EI = \frac{Ech^3}{12(1-v^2)}, \quad GJ = \frac{Ech^3}{6(1+v)}\left(1 - \frac{3h}{5c}\right) \tag{18}$$

for the rectangular cross-section of the beam-like model, where $h(y)$ is the section thickness, while the mass and second moments of inertia per unit area read [67]:

$$m = \rho_s hc, \quad \mu_\zeta = \frac{mh^2}{12}, \quad \mu_\theta = \frac{m}{12}\left(h^2 + c^2\right) \tag{19}$$

where $\rho_s(y)$ is the material density. Both elastic and inertial axes coincide with the symmetry $\hat{y}$ axis, namely $x_{CG} = x_{MC} = 0$ m; $x_{AC} = -0.25$ m and $x_{CP} = 0.25$ m for incompressible flow [33].

Neglecting structural damping without loss of generality, divergence speed, flutter speed and flutter frequency are parametrically investigated with respect to both aspect and thickness ratios for a uniform wing, considering $\rho_s = 2700$ kg/m^3, $E = 70^9$ Pa and $v = 0.35$ as in previous studies [131]. Nine aeroelastic configurations resulted from combining three aspect and thickness ratios, namely:

$$AR \equiv \frac{2l}{c} = \begin{bmatrix} 4 & 6 & 8 \end{bmatrix}, \quad TR \equiv \frac{h}{c} = \begin{bmatrix} 0.006 & 0.008 & 0.010 \end{bmatrix} \tag{20}$$

ranging from relatively low to relatively high values in order to investigate their physical role and influence on structural dynamic, aerodynamic and aeroelastic behaviours of the flat wing. This is particularly useful in multidisciplinary optimisation studies for preliminary design [7].

7.1. Structural FEM and Aerodynamic DLM for Numerical Simulations

A discrete aeroelastic model was built and used in Nastran for every configuration, based on rigorous convergence studies (not shown) on the aeroelastic results. Table 1 presents the selected number of nodes and elements of the structural FEM along with the number of spanwise and chordwise panels of the aerodynamic DLM. Figure 2 shows structural FEM and aerodynamic DLM arrangements for all wings with $AR = 6$, as an example: the wing EA is clamped at the root mid chord, while the RBE2 elements realise the wing torsion and transfer the chordwise pressure loads.

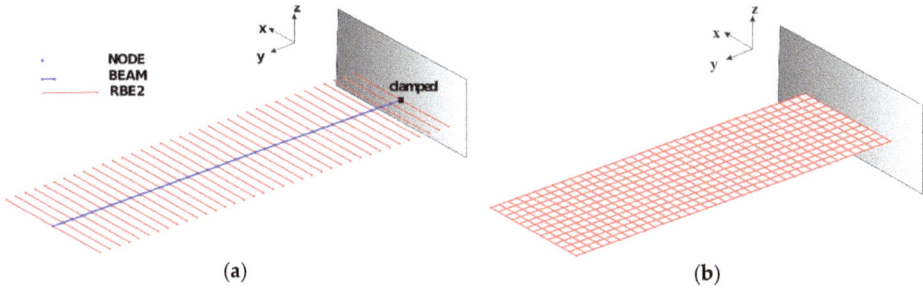

Figure 2. FEM (**a**) and DLM (**b**) numerical arrangement, for the wings with $AR = 6$.

Table 1. FEM and DLM numerical discretisation, for all considered wings.

AR	FEM: Nodes, Elements	DLM: Spanwise, Chordwise
4	75, 24	24, 12
6	108, 36	36, 12
8	144, 48	48, 12

Following the generation of the panels, the unsteady AIC matrix was calculated at 14 reduced frequencies [0.001, 0.005, 0.01, 0.02, 0.03, 0.04, 0.05, 0.06, 0.08, 0.1, 0.2, 0.4, 0.6, 0.8] and automatically interpolated elsewhere using cubic splines [46,116].

7.2. Natural Vibration Modes

In the case of uniform beams, the exact natural vibration mode shapes and frequencies are well known for both bending and torsion as [51,52]:

$$\phi_i = \cosh\left(\gamma_i \frac{y}{l}\right) - \cos\left(\gamma_i \frac{y}{l}\right) - \left(\frac{\cosh\gamma_i + \cos\gamma_i}{\sinh\gamma_i + \sin\gamma_i}\right)\left[\sinh\left(\gamma_i \frac{y}{l}\right) - \sin\left(\gamma_i \frac{y}{l}\right)\right]$$

$$f_{\phi_i} = \frac{\gamma_i^2}{2\pi l^2}\sqrt{\frac{EI}{m}}, \ \gamma_1 = 0.597\pi, \ \gamma_i \approx \left(i - \frac{1}{2}\right)\pi \tag{21}$$

$$\varphi_i = \sin\left(\theta_i \frac{y}{l}\right), \ f_{\varphi_i} = \frac{\theta_i}{2\pi l}\sqrt{\frac{GJ}{\mu_\theta}}, \ \theta_i = \left(i - \frac{1}{2}\right)\pi \tag{22}$$

and form the modal base for the analytical model and generalised solutions. Figure 3 shows the first four FEM modes for the wings with $AR = 6$, whereas Figure 4 shows the first two bending and torsion exact modes (which hold for all aero-structural configurations). Table 2 shows the typology of both FEM and the exact natural vibration modes, independent of the thickness ratio. The relative natural frequencies do depend on the latter and are shown in Figure 5, where exact agreement is found between numerical and analytical results: this cross-validates both and demonstrates that FOM and ROM are equivalent as far as structural dynamics are concerned.

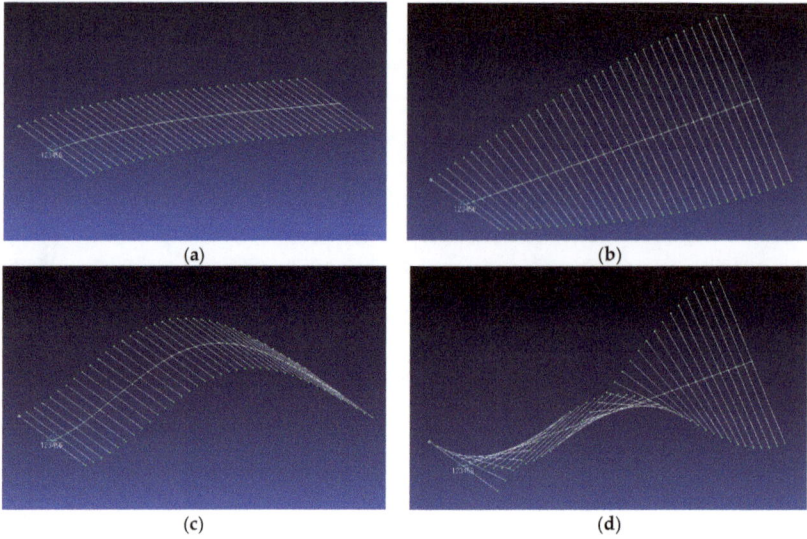

Figure 3. First (**a**), second (**b**), third (**c**) and fourth (**d**) FEM vibration modes, for the wings with *AR* = 6.

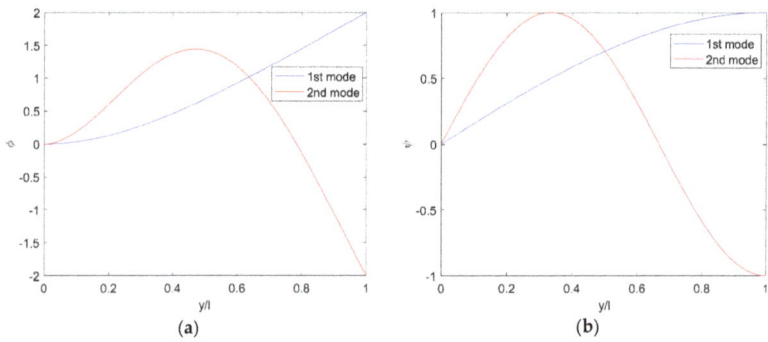

Figure 4. First two bending (**a**) and torsion (**b**) exact natural vibration modes.

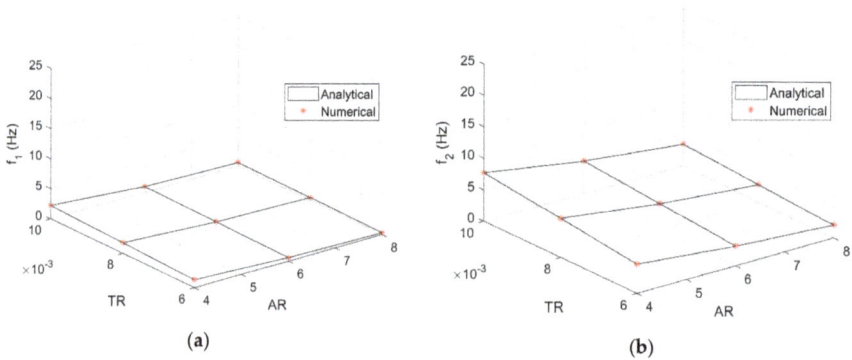

Figure 5. *Cont.*

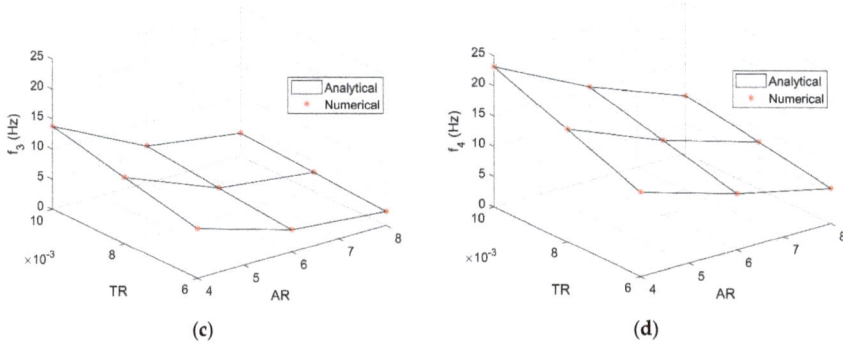

Figure 5. First (**a**), second (**b**), third (**c**) and fourth (**d**) vibration frequencies for all considered wings.

Table 2. FEM and analytical natural vibration modes, for all considered wings.

AR	1st Mode	2nd Mode	3rd Mode	4th Mode
4	1st Bending	1st Torsion	2nd Bending	2nd Torsion
6	1st Bending	1st Torsion	2nd Bending	2nd Torsion
8	1st Bending	2nd Bending	1st Torsion	3rd Bending

Note the switch between bending and torsion in the second, third and fourth modes for wings with $AR = 8$, since the natural frequency of the bending modes decreases more rapidly than that of the torsion modes with an increase in the wing span.

7.3. Steady and Unsteady Air Load

The air load steady distribution and unsteady evolution were calculated by the DLM with a unit angle of attack for the undeformed wings and then approximated to derive the analytical ROM. As the subsonic DLM formulation inherently tends to infinity at the start of the indicial response for the case of incompressible flow [5] (where the exact solution exhibits a Dirac delta), the initial value was then set as the theoretical one for the circulatory contribution (see Appendix A). As shown in Figure 6, the first 5 (odd) sinusoidal terms granted excellent approximation of the (symmetric) normalised lift distribution, whereas 2 exponential terms gave an excellent approximation of the lift-deficiency function due to a unit step in angle of attack; still, the three-dimensional coupling between unsteady wing-tip downwash and wing-wake inflow is only enforced a posteriori in the MST-based ROM.

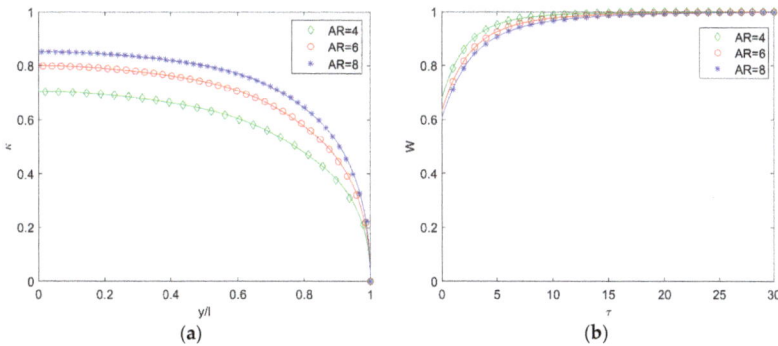

Figure 6. Lift decay (**a**) and build-up (**b**) for flat rectangular wings with different aspect ratio: symbols = numerical solution, lines = analytical approximation.

Note that a parametric database of steady lift distributions and unsteady lift evolutions results from the DLM simulations, which may then analytically be approximated and form an aerodynamic ROM where all curve-fitting coefficients are generally expressed as function of the wing AR [132].

7.4. Divergence and Flutter Analysis

The dependence of the divergence speed, the flutter speed and frequency on aspect and thickness ratios were then investigated. Figure 7 shows both divergence and flutter speeds as calculated by the aeroelastic models when SST is employed for the unsteady aerodynamics and exact agreement is found between the numerical and analytical results. The same is true for the related flutter frequency and reduced frequency shown in Figure 8, which cross-validate both numerical and analytical results. As further proof of the rigorous validation, note that the results for the divergence speed exactly reproduce the theoretical solutions derived in previous studies [27].

Figure 9 shows both divergence and flutter speeds as calculated by the aeroelastic FOM and ROM when MST and DLM are employed for the unsteady aerodynamics, respectively; excellent agreement is found between numerical and analytical results, with no appreciable difference in the divergence speed. In spite of the DLM compressible formulation, very good agreement is also found for the related flutter frequency and reduced frequency shown in Figure 10, where the discrepancy decreases with increasing the wing aspect ratio as the flow becomes progressively two-dimensional and quasi-steady; thus, coupling between apparent fluid inertia, tip-vortices downwash and wake inflow becomes gradually weaker and differences between FOM and ROM reduce accordingly.

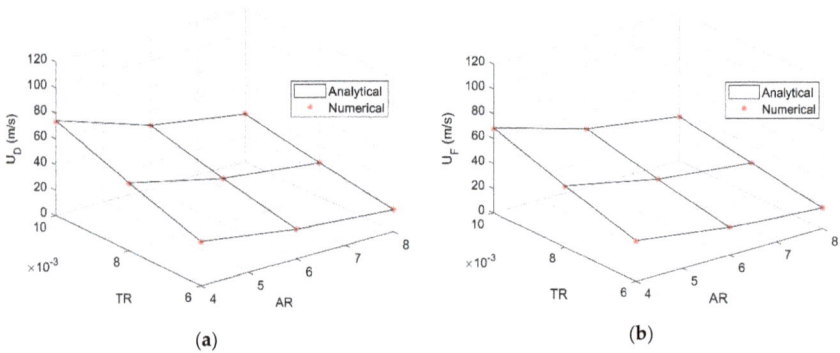

Figure 7. Divergence (**a**) and flutter (**b**) speeds according to SST.

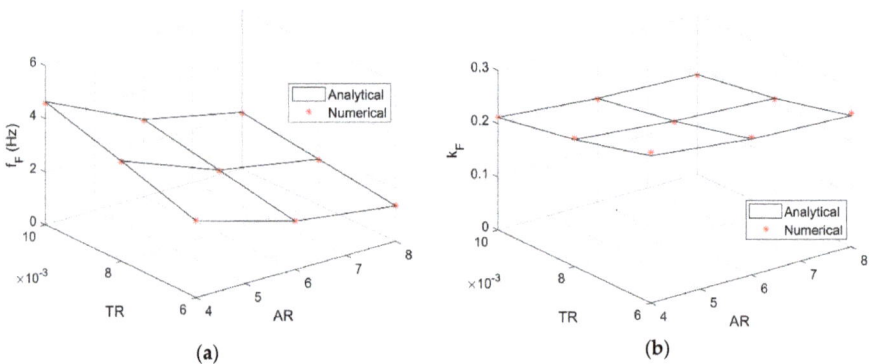

Figure 8. Flutter frequency (**a**) and reduced frequency (**b**) speeds according to SST.

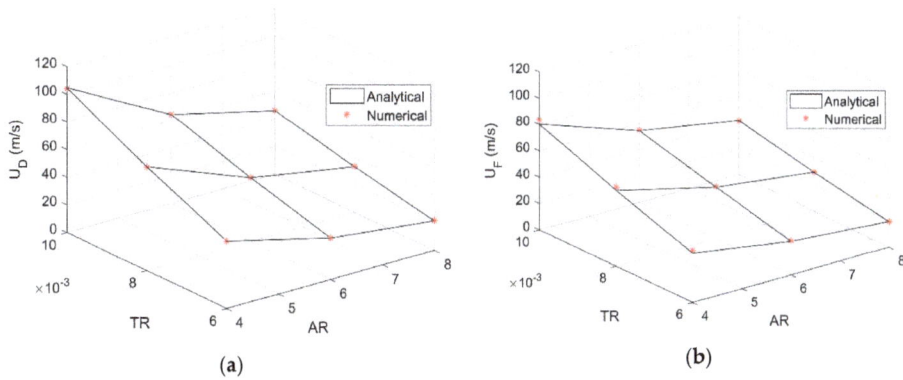

Figure 9. Divergence (a) and flutter (b) speeds according to MST and DLM.

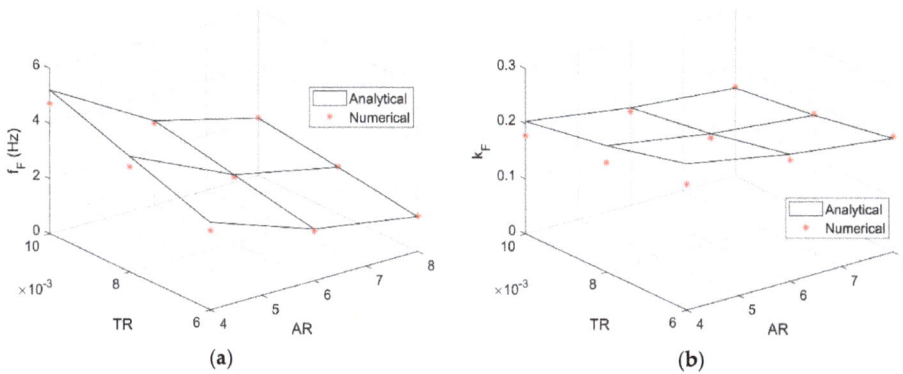

Figure 10. Flutter frequency (a) and reduced frequency (b) speeds according to MST and DLM.

As expected, the SST-based results are more conservative than the MST-based ones, as in the latter case the aerodynamic load builds up but also decays more rapidly towards the wing tip, which is the softer area. Indeed, the resulting bending moment at the wing root is lower in the MST-based cases. Note that both divergence speeds and flutter frequency consistently decrease with decreasing the thickness ratio and increasing the wing aspect ratio; yet, the reduced flutter frequency increases with a decrease in the thickness ratio, as the decrement in the flutter speed is higher than that in the flutter frequency along this dimension of the design variables space. It is also notable that all reduced flutter frequencies fall in the frequency range chosen for the numerical simulations.

The first 5 bending and torsion natural vibration modes granted ROM convergence in all cases. As the flutter phenomenon couples first bending and torsion, the switch between bending and torsion in the third and fourth modes for $AR = 8$ did not create discontinuity in the results; however, additional simulations showed that hump modes start to develop at higher aspect ratios and lower thickness ratios (i.e., when decreasing the stiffness), as also found in previous works [131].

8. Conclusions

A computationally efficient hybrid ROM for the aeroelastic stability analysis of flexible wings in subsonic flow has been presented. A new modified strip theory was formulated where the unsteady aerodynamic load provided by thin aerofoil theory is corrected by a higher-fidelity model to account for three-dimensional downwash effects on both distribution and build-up of the sectional pressure forces.

A slender beam model being coupled for the structural dynamics, the generalised aeroelastic equations were derived by means of the principle of virtual works and then solved using a modal approach that takes full advantage of the implicit projection concept embedded in the aerodynamic scaling function. The proposed FSI ROM allows an arbitrary distribution of the wing's physical properties and calculates a continuous solution for displacements and loads, which is ideal for parametric optimisation studies over a large design space within aircraft preliminary MDO. Numerical results were obtained using MSC NASTRAN and then compared for both divergence speeds and flutter frequency of a flat rectangular homogeneous wing, given different aspect and thickness ratios. The presented results offer sound insight into the aeroelastic stability of flexible subsonic wings and thus, may be used to assess high-fidelity FOMs. The proposed hybrid modified strip theory demonstrated excellent accuracy at low computational costs with respect to the classic DLM and it is therefore suggested as a general and efficient aerodynamic ROM for the MDO of flexible wings in subsonic flow, especially at the preliminary stage where fast and robust semi-analytical aero-structural tools are highly sought for best computing performance.

Author Contributions: M.B. derived the analytical model and results, whereas R.C. performed the numerical simulations; the authors then wrote the respective parts of the manuscript.

Funding: This research received no external funding.

Conflicts of Interest: The authors declare no conflict of interest.

Nomenclature

A	aerodynamic gain coefficient
AR	wing aspect ratio
B	aerodynamic pole coefficient
c	section chord
C_L	section lift
$C_{L/\alpha}$	section lift derivative
$C_{L/\alpha}^{3D}$	wing lift derivative
$\mathbf{C}$	generalised damping matrix
e	elliptic integral of the second kind
E	section Young's elastic modulus
f	angular frequency
$\mathbf{F}$	generalised aerodynamic load vector
G	section shear elastic modulus
h	section thickness
I	section flexural area moments of inertia
J	section torsional mass moments of inertia
k	reduced frequency
$\mathbf{K}$	generalised stiffness matrix
l	wing semi-span
ΔL	section aerodynamic force
m	section mass
ΔM	section aerodynamic moment
$\mathbf{M}$	generalised mass matrix
n	number of expansion terms
t	time
U	horizontal air speed
V	vertical air speed
w	section vertical displacement
W	aerodynamic indicial-admittance function
x	chordwise coordinate
y	spanwise coordinate
α	angle of attack
Γ	section circulation
Γ	section circulation

ε	flexural generalised coordinate
ζ	section flexural displacement
η	torsional generalised coordinate
ϑ	section torsional displacement
κ	aerodynamic load-scaling function
λ	eigenvalue
μ_ζ	section flexural mass moments of inertia
μ_ϑ	section torsional mass moments of inertia
ν	Poisson ratio
o	Oswald's efficiency factor
ρ	reference air density
τ	reduced time
υ	added aerodynamic state
ϕ	flexural assumed mode shape
φ	torsional assumed mode shapes
χ	generalised coordinates vector
ψ	spanwise Glauert angle

Appendix A. Lifting Line Models for Rectangular Straight Wings

Lifting line theory [56] accounts for the downwash angle induced by the tip vortex and is very powerful for slender straight wings. However, it is generally conservative as the distance between aerodynamic centre lines (where the bound circulation lays) and control points line (where the non-penetration boundary condition for the potential flow is enforced) is neglected when applying Helmholtz's theorem and Biot-Savart law [36]. Therefore, a correction [57] should be considered in the presence of a small aspect ratio.

The first three (odd) sinusoidal terms granted convergence of the (symmetric) normalised lift distribution shown in Figure A1, where the lift-deficiency function due to a unit step in angle of attack is also shown. These analytical results show good agreement with the numerical ones and provide sound comparisons.

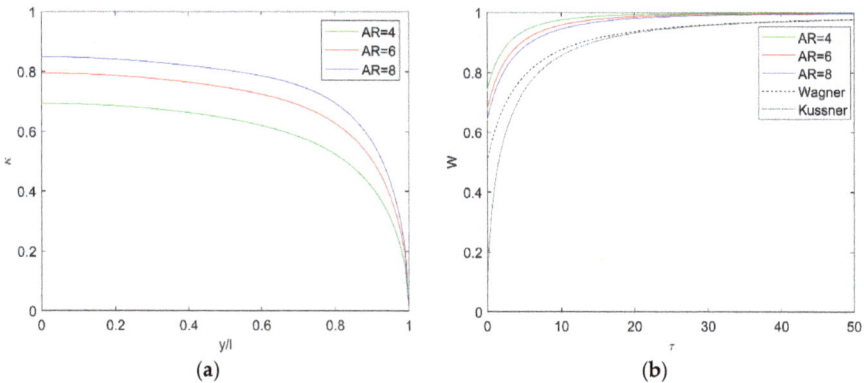

Figure A1. Lift-scaling (**a**) and lift-deficiency (**b**) functions for flat rectangular wings of different aspect ratios.

Appendix A.1. Steady Lift Distribution

Prandtl's equations for the sectional flow circulation $\Gamma(y)$ and downwash angle $\alpha_i(y)$ are generalised as [57,58]:

$$\Gamma\sqrt{1+\left(\frac{2}{AR}\right)^2}+\frac{c}{4}\int_{-l}^{l}\left(\frac{d\Gamma}{d\zeta}\right)\frac{d\zeta}{y-\zeta}=\frac{c}{2}UC_L, \alpha_i=\frac{1}{4\pi U}\int_{-l}^{l}\left(\frac{d\Gamma}{d\zeta}\right)\frac{d\zeta}{y-\zeta} \tag{A1}$$

respectively, where ζ is a dummy integration variable running along the wing span, whereas AR represents the ratio between the latter and mean chord for trapezoidal and elliptical planforms. This refined lifting-line model then gives a correct estimate of the downwash towards the wing root but underestimates the lift decay towards the wing tips, where VLM and DLM prescribe a stronger vortex effect. Inthis respect, note that tapering the wing increases the aspect ratio while decreasing the wing downwash and hence reduces the challenges to the MST [27].

Due to Glauert's integral (in principal value) [76] and being $\Gamma(\pm l)=0$, adopting Prandtl's expansion for the circulation gives [56]:

$$\Gamma=lU\sum_{j=1}^{n_\Gamma}\Gamma_j\sin(j\psi), \alpha_i=\sum_{j=1}^{n_\Gamma}j\Gamma_j\frac{\sin(j\psi)}{4\sin\psi} \tag{A2}$$

and results in a modified system of linear algebraic equations for the lifting-line model [57,58]:

$$\sum_{j=1}^{n_\Gamma}\left(\sin(j\psi)\sqrt{1+\left(\frac{2}{AR}\right)^2}+j\frac{cC_{L/\alpha}}{8l}\frac{\sin(j\psi)}{\sin\psi}\right)\Gamma_j=\frac{cC_L}{2l} \tag{A3}$$

where Prandtl's original equations [56] are asymptotically resumed for slender wings. Odd and even Fourier terms still give symmetric and antisymmetric circulation distributions, respectively; the n_Γ coefficients Γ_j are then found by solving such linear system in a least-squares sense [133], on N wing sections at various spanwise stations $y=l\cos(\psi)$, with $0\le\psi\le\pi$ running from tip to tip along the span. Note that $\Gamma=0$ for $c=0$, while the singularities at $\psi=j\pi$ can be lifted by multiplying both sides of the equation by $\sin\psi$. Thus, the scaling function coefficients are derived from:

$$\sum_{j=1}^{n_\kappa}\left(\sin(j\psi)\sqrt{1+\left(\frac{2}{AR}\right)^2}+j\frac{cC_{L/\alpha}}{8l}\frac{\sin(j\psi)}{\sin\psi}\right)\kappa_j=1 \tag{A4}$$

Oswald's efficiency factor [83] is finally used to write the lift coefficient derivative for the entire wing; the unified LLT [86] shall be employed for slender wings with significant sweep angle. Within a proper orthogonal decomposition (POD) perspective [26], scaling function and aerofoil differential pressure coefficient distribution may be interpreted as the dominant (normalised) modes of the load in the spanwise and chordwise directions, respectively, with the angle of attack driving their amplitude.

Appendix A.2. Unsteady Lift Development

For finite wings, the influence of the tip vortices on the unsteady lift distribution along the wing span may be calculated based on lifting line model as a function of the wing aspect ratio [29] (which introduces the dependency of the air load on the finite spanwise dimension) and the aerodynamic derivatives of the wing may be obtained from the ones of its airfoil section [39].

As an effective, simplified approach, a single vortex-ring is considered for modeling the total (lumped) wing circulation [59]. The bound vortex is placed at the AC line, as per thin airfoil theory, while the wing-tip vortices are trailed parallel to the free-stream; a single CP for the total downwash is then consistently placed at the wing's root, where the flow's non-penetration boundary condition is satisfied. All vortex lines have the same (lumped) intensity and the shed vorticity travels towards infinity with half the reference speed from half the wing's root chord behind the control point, hence stretching the vortex-ring and increasing the wake length; when the wake eventually approaches infinity, its influence fades away and the steady condition is asymptotically obtained. The influence of both tip vortices and unsteady wake on the wing lift is therefore calculated using the simplest implementation of unsteady lifting line theory [84–86] and the load build-up is obtained as a function of the aspect ratio.

Considering all contributions due to bound, trailed and shed vortices of the vortex-ring, the wing lift-deficiency coefficient $\hat{C}_{L/\alpha}^{3D}(\tau)$ from a unit step in the angle of attack is calculated based on Kutta-Joukowsky theorem and Biot-Savart law as [59]:

$$\hat{C}_{L/\alpha}^{3D} = \frac{ARC_{L/\alpha}}{\sqrt{1+AR^2} + \frac{2}{2+\tau}\sqrt{\left(1+\frac{\tau}{2}\right)^2 + AR^2}} \tag{A5}$$

with the initial (step-like) and asymptotic (steady) behaviours respectively given by:

$$\lim_{\tau \to 0} \hat{C}_{L/\alpha}^{3D} = \frac{ARC_{L/\alpha}}{2\sqrt{1+AR^2}}, \quad \lim_{\tau \to \infty} \hat{C}_{L/\alpha}^{3D} = \frac{ARC_{L/\alpha}}{1+\sqrt{1+AR^2}} \tag{A6}$$

Garrick's approximation [106] of Wagner's function for thin aerofoil is correctly resumed in the limit of infinitely slender wing [30]. Nevertheless, due to the inherent limitations of the vortex-system employed, the initial and final values of the lift coefficient are not very accurate and shall rather be provided by other higher-fidelity sources, with a suitable general expression being [39,83]:

$$C_{L0}^{3D} = \frac{\pi}{e}, \quad C_{L\infty}^{3D} = \frac{ARC_{L/\alpha}}{AR + 2(1+o)} \tag{A7}$$

where e is the elliptic integral giving the ratio of the semi-perimeter to the span for an elliptical planform with the same aspect ratio, whereas $o = 0$ only when Prandtl's original equations [56] are considered. Using linear mapping [77], the lift-deficiency coefficient $C_{L/\alpha}^{3D}(\tau)$ from a unit step in the angle of attack may finally be approximated as:

$$C_{L/\alpha}^{3D} = C_{L0}^{3D} + \left(\frac{C_{L\infty}^{3D} - C_{L0}^{3D}}{\hat{C}_{L\infty}^{3D} - \hat{C}_{L0}^{3D}}\right)\left(\hat{C}_{L/\alpha}^{3D} - \hat{C}_{L0}^{3D}\right), \quad W \equiv \frac{C_{L/\alpha}^{3D}}{C_{L\infty}^{3D}} \tag{A8}$$

where both asymptotic and initial conditions are automatically satisfied. It is worth stressing that this unsteady aerodynamic model was originally derived for slender wings with significant sweep angle and taper ratio [59].

Finally, in order to estimate the lift-deficiency function from a unit sharp-edge gust within the standard "frozen" approach [134], that from a unit step in the angle of attack shall be multiplied by the ratio between Kussner's [31] and Wagner's functions (introducing the two-dimensional effect of the gust penetration); all wing sections encountering the gust at the same time, Kussner's function for thin airfoils is then automatically resumed in the limit of infinite wings. Note that this is roughly equivalent to convolving the lift-deficiency coefficient from a unit step in the angle of attack with a fictitious angle of attack derived from the Laplace transform of the ratio between Sears' [34,35] and Theodorsen's functions (representing a delay function for the two-dimensional flow). Of course, the wind gust penetration delays the circulation growth and hence reaches the asymptotic (steady) lift. In the general case of swept wings [59], the gust-entry delay relative to each section is geometrically known and shall also be considered in order to obtain the lift build-up due to a unit a sharp-edge gust normal to the reference airflow, which is purely circulatory and acts at the AC.

Aerospace **2018**, *5*, 76

References

1. Cavagna, L.; Ricci, S.; Travaglini, L. NeoCASS: An Integrated Tool for Structural Sizing, Aeroelastic Analysis and MDO at Conceptual Design Level. *Prog. Aerosp. Sci.* **2011**, *47*, 621–635. [CrossRef]
2. Alexandrov, N.M.; Hussaini, M.Y. *Multidisciplinary Design Optimization: State of the Art*; Proceedings in Applied Mathematics Series, 80; SIAM: Philadelphia, PA, USA, 1997.
3. Martins, J.R.R.A.; Lambe, A.B. Multidisciplinary Design Optimization: A Survey of Architectures. *AIAA J.* **2013**, *51*, 2049–2075. [CrossRef]
4. Vanderplaats, G.N. *Numerical Optimization Techniques for Engineering Design: With Applications*; Series in Mechanical Engineering; McGraw Hill: New York, NY, USA, 1984.
5. Pike, E.C. *Manual on Aeroelasticity*; AGARD-R-578-71; AGARD: Neuilly sur Seine, France, 1971.
6. Livne, E. The Future of Aircraft Aeroelasticity. *J. Aircr.* **2003**, *40*, 1066–1092. [CrossRef]
7. Kesseler, E.; Guenov, M. *Advances in Collaborative Civil Aeronautical Multidisciplinary Design Optimization*; Progress in Astronautics and Aeronautics Series, 233; AIAA: Reston, VA, USA, 2010.
8. Bungartz, H.J.; Schafer, M. *Fluid-Structure Interaction: Modelling, Simulation, Optimization*; Lecture Notes in Computational Science and Engineering, 53; Springer: Berlin, Germany, 2006.
9. Dhatt, G.; Lefrancois, E.; Touzot, G. *Finite Element Method*; Numerical Methods Series; Wiley: Hoboken, NJ, USA, 2013.
10. Chung, T.J. *Computational Fluid Dynamics*; Cambridge Press: Cambridge, UK, 2002.
11. Cavagna, L.; Quaranta, G.; Ghiringhelli, G.L.; Mantegazza, P. Efficient Application of CFD Aeroelastic Methods Using Commercial Software. In Proceedings of the 11th IFASD, Munich, Germany, 28 June–1 July 2005.
12. Romanelli, G.; Serioli, E.; Mantegazza, P. *A "Free" Approach to Computational Aeroelasticity*; AIAA-2010-176; AIAA: Reston, VA, USA, 2010.
13. Sucipto, T.; Berci, M.; Krier, J. Gust Response of a Flexible Typical Section via High- and (Tuned) Low-Fidelity Simulations. *Comput. Struct.* **2013**, *122*, 202–216. [CrossRef]
14. Berci, M.; Mascetti, S.; Incognito, A.; Gaskell, P.H.; Toropov, V.V. Dynamic Response of Typical Section Using Variable-Fidelity Fluid Dynamics and Gust-Modeling Approaches—With Correction Methods. *J. Aerosp. Eng.* **2014**, *27*, 04014026. [CrossRef]
15. Farhat, C.; Lesoinne, M.; Le Tallec, P. Load and Motion Transfer Algorithms for Fluid/Structure Interaction Problems with Non-Matching Discrete Interfaces: Momentum and Energy Conservation, Optimal Discretization and Application to Aeroelasticity. *Comput. Methods Appl. Mech. Eng.* **1998**, *157*, 95–114. [CrossRef]
16. Cizmas, P.G.A.; Gargoloff, J.I. Mesh Generation and Deformation Algorithm for Aeroelasticity Simulations. *J. Aircr.* **2008**, *45*, 1062–1066. [CrossRef]
17. Heil, M.; Hazel, A.L.; Boyle, J. Solvers for Large-Displacement Fluid–Structure Interaction Problems: Segregated Versus Monolithic Approaches. *Comput. Mech.* **2008**, *43*, 91–101. [CrossRef]
18. Sheldon, J.P.; Miller, S.T.; Pitt, J.S. Methodology for Comparing Coupling Algorithms for Fluid-Structure Interaction Problems. *World J. Mech.* **2014**, *4*, 54. [CrossRef]
19. Kloppel, T.; Popp, A.; Wall, W.A. Interface Treatment in Computational Fluid-Structure Interaction. In Proceedings of the 3rd ECCOMAS COMPDYN, Island of Corfu, Greece, 26–28 May 2011.
20. Farhat, C.; Lakshminarayan, V. An ALE Formulation of Embedded Boundary Methods for Tracking Boundary Layers in Turbulent Fluid-Structure Interaction Problems. *J. Comput. Phys.* **2014**, *263*, 53–70. [CrossRef]
21. Berci, M.; Toropov, V.V.; Hewson, R.W.; Gaskell, P.H. Multidisciplinary Multifidelity Optimisation of a Flexible Wing Aerofoil with Reference to a Small UAV. *Struct. Multidiscip. Optim.* **2014**, *50*, 683–699. [CrossRef]
22. Quarteroni, A.; Rozza, G. *Reduced Order Methods for Modeling and Computational Reduction*; MS&A, 9; Springer International Publishing: Cham, Switzerland, 2014.
23. Qu, Z.Q. *Model Order Reduction Techniques with Applications in Finite Element Analysis*; Springer: London, UK, 2004.
24. Ghoreyshi, M.; Jirasek, A.; Cummings, R.M. Reduced Order Unsteady Aerodynamic Modeling for Stability and Control Analysis Using Computational Fluid Dynamics. *Prog. Aerosp. Sci.* **2014**, *71*, 167–217. [CrossRef]
25. Gennaretti, M.; Mastroddi, F. Study of Reduced-Order Models for Gust-Response Analysis of Flexible Fixed Wings. *J. Aircr.* **2004**, *41*, 304–313. [CrossRef]
26. Ripepi, M.; Verveld, M.J.; Karcher, N.W.; Franz, T.; Abu-Zurayk, M.; Görtz, S.; Kier, T.M. *Reduced Order Models for Aerodynamic Applications, Loads and MDO*; DLRK-2016-420057; DLR: Braunschweig, Germany, 2016.
27. Berci, M. Semi-Analytical Static Aeroelastic Analysis and Response of Flexible Subsonic Wings. *Appl. Math. Comput.* **2015**, *267*, 148–169. [CrossRef]

28. Sitaraman, J.; Baeder, J.D. Computational-Fluid-Dynamics-Based Enhanced Indicial Aerodynamic Models. *J. Aircr.* **2004**, *41*, 798–810. [CrossRef]

29. Anderson, J.D. *Fundamentals of Aerodynamics*; Series in Aeronautical and Aerospace Engineering; McGraw-Hill: New York, NY, USA, 2007.

30. Wagner, H. Uber die Entstenhung des Dynamischen Auftriebes von Tragflugeln. *Z. Angew. Math. Mech.* **1925**, *5*, 17–35. [CrossRef]

31. Kussner, H.G. Zusammenfassender Beritch uber den Instationaren Auftrieb von Flugeln. *Luftfahrtforsch* **1936**, *13*, 410–424.

32. Kayran, A. Küssner's Function in the Sharp Edged Gust Problem—A Correction. *J. Aircr.* **2006**, *43*, 1596–1599. [CrossRef]

33. Theodorsen, T. *General Theory of Aerodynamic Instability and the Mechanism of Flutter*; NACA 496; NACA: Washington, DC, USA, 1935.

34. Von Karman, T.; Sears, W.R. Airfoil Theory for Non-Uniform Motion. *J. Aeronaut. Sci.* **1938**, *5*, 379–390. [CrossRef]

35. Sears, W.R. Operational Methods in the Theory of Airfoils in Non-Uniform Motion. *J. Frankl. Inst.* **1940**, *230*, 95–111. [CrossRef]

36. Katz, J.; Plotkin, A. *Low Speed Aerodynamics*; Cambridge Aerospace Series; Cambridge University Press: Cambridge, UK, 2001.

37. Diederich, F.W. *Approximate Aerodynamic Influence Coefficients for Wings of Arbitrary Plan Form in Subsonic Flow*; NACA TN 2092; NACA: Washington, DC, USA, 1950.

38. Diederich, F.W.; Zlotnick, M. *Calculated Spanwise Lift Distributions, Influence Functions and Influence Coefficients for Unswept Wings in Subsonic Flow*; NACA 1228; NACA: Washington, DC, USA, 1955.

39. Jones, R.T. *The Unsteady Lift of a Wing of Finite Aspect Ratio*; NACA 681; NACA: Washington, DC, USA, 1940.

40. Jones, W.P. *Aerodynamic Forces on Wings in Simple Harmonic Motion*; ARC-RM-2026; HM Stationery Office: London, UK, 1945.

41. Jones, W.P. *Aerodynamic Forces on Wings in Non-Uniform Motion*; ARC-RM-2117; HM Stationery Office: London, UK, 1945.

42. Reissner, E. *Effect of Finite Span on the Airload Distributions for Oscillating Wings—Part I: Aerodynamic Theory of Oscillating Wings of Finite Span*; NACA TN-1194; Massachusetts Institute of Technology: Cambridge, MA, USA, 1947.

43. Reissner, E.; Stevens, J.E. *Effect of Finite Span on the Airload Distributions for Oscillating Wings—Part II: Methods of Calculation and Examples of Application*; NACA TN-1195; NACA: Washington, DC, USA, 1947.

44. Albano, E.; Rodden, W.P. A Doublet-Lattice Method for Calculating the Lift Distribution of Oscillating Surfaces in Subsonic Flows. *AIAA J.* **1969**, *7*, 279–285. [CrossRef]

45. Rodden, W.P.; Harder, R.L.; Bellinger, E.D. *Aeroelastic Addition to NASTRAN*; NASA CR-3094; NASA: Washington, DC, USA, 1979.

46. Quick Reference Guide. In *MSC Nastran*; MSC Software Corporation: Newport Beach, CA, USA, 2018.

47. Holmes, R.B. *A Course on Optimization and Best Approximation*; Lecture Notes in Mathematics, 257; Springer: Berlin, Germany, 1972.

48. Leishman, J.G. *Principles of Helicopter Aerodynamics*; Cambridge Aerospace Series; Cambridge University Press: Cambridge, UK, 2006.

49. Bisplinghoff, R.L.; Ashley, H. *Principles of Aeroelasticity*; Dover: Mineola, NY, USA, 2013.

50. Reddy, J.N. *Energy Principles and Variational Methods in Applied Mechanics*; Wiley: Hoboken, NJ, USA, 2002.

51. Hodges, D.H.; Pierce, G.A. *Introduction to Structural Dynamics and Aeroelasticity*; Cambridge Aerospace Series; Cambridge University Press: Cambridge, UK, 2002.

52. Bisplinghoff, R.L.; Ashley, H.; Halfman, R.L. *Aeroelasticity*; Dover: Mineola, NY, USA, 1996.

53. Demasi, L.; Livne, E. *Structural Ritz-Based Simple-Polynomial Nonlinear Equivalent Approach—An Assessment*; AIAA-2005-2093; AIAA: Reston, VA, USA, 2005.

54. Fung, Y.C. *An Introduction to the Theory of Aeroelasticity*; Dover: Mineola, NY, USA, 1993.

55. Dowell, E.H. *A Modern Course in Aeroelasticity*; Solid Mechanics and Its Applications, 217; Springer: Berlin, Germany, 2015.

56. Prandtl, L. *Applications of Modern Hydrodynamics to Aeronautics*; NACA TR-116; NACA: Washington, DC, USA, 1921.

57. Helmbold, H.B. Der Unverwundene Ellipsenflügel als Tragende Fläche. In *Jahrbuch der Deutschen Luftfahrtforschung*; Deutsche Akademie der Luftfahrtforschung: Munich, Germany, 1942; Volume I, pp. 111–113.

58. Diederich, F.W. *A Plan-Form Parameter for Correlating Certain Aerodynamic Characteristics of Swept Wings*; NACA TN 2335; NACA: Washington, DC, USA, 1951.

59. Queijo, M.J.; Wells, W.R.; Keskar, D.A. *Approximate Indicial Lift Function for Tapered, Swept Wings in Incompressible Flow*; NASA TP-1241; NASA: Washington, DC, USA, 1978.

60. Wells, W.R. *An Approximate Analysis of Wing Unsteady Aerodynamics*; AFFDL-TR-79-3046; University of Dayton Research Institute: Dayton, OH, USA, 1979.

61. Jones, R.T. *Classical Aerodynamic Theory*; NASA RP 1050; NASA: Washington, DC, USA, 1979.

62. Megson, T.H.G. *Aircraft Structures for Engineering Students*; Elsevier Aerospace Engineering Series; Elsevier: Oxford, UK, 2007.

63. Yang, B. *Strain, Stress and Structural Dynamics*; Elsevier: London, UK, 2005.

64. Amabili, M. *Nonlinear Vibrations and Stability of Shells and Plates*; Cambridge University Press: Cambridge, UK, 2008.

65. Berci, M.; Gaskell, P.H.; Hewson, R.W.; Toropov, V.V. A Semi-Analytical Model for the Combined Aeroelastic Behaviour and Gust Response of a Flexible Aerofoil. *J. Fluids Struct.* **2013**, *37*, 3–21. [CrossRef]

66. Wright, J.R.; Cooper, J.E. *Introduction to Aircraft Aeroelasticity and Loads*; Aerospace Series; Wiley: Chichester, UK, 2015.

67. Young, W.C.; Budynas, R.G. *Roark's Formulas for Stress and Strain*; McGraw-Hill: New York, NY, USA, 2011.

68. Han, S.M.; Benaroya, H.; Wei, T. Dynamics of Transversally Vibrating Beams Using Four Engineering Theories. *J. Sound Vib.* **1999**, *225*, 935–988. [CrossRef]

69. Marzocca, P.; Librescu, L.; Silva, W.A. Aeroelastic Response and Flutter of Swept Aircraft Wings. *AIAA J.* **2002**, *40*, 801–812. [CrossRef]

70. Craig, R.R.; Bampton, M.C.C. Coupling of Substructures for Dynamic Analyses. *AIAA J.* **1968**, *6*, 1313–1319.

71. Allemang, R.J. The modal assurance criterion—Twenty years of use and abuse. *Sound Vib.* **2003**, *37*, 14–23.

72. Drischler, J.A. *Approximate Indicial Lift Function for Several Wings of Finite Span in Incompressible Flow as Obtained from Oscillatory Lift Coefficients*; NACA TN-3639; NACA: Washington, DC, USA, 1956.

73. Mateescu, D.; Seytre, J.F.; Berhe, A.M. Theoretical Solutions for Finite-Span Wings of Arbitrary Shapes Using Velocity Singularities. *J. Aircr.* **2003**, *40*, 450–460. [CrossRef]

74. Lomax, H.; Heaslet, M.A.; Fuller, F.B.; Sluder, L. *Two- and Three-Dimensional Unsteady Lift Problems in High-Speed Flight*; NACA 1077; NACA: Washington, DC, USA, 1950.

75. Gulcat, U. *Fundamentals of Modern Unsteady Aerodynamics*; Springer: Berlin, Germany, 2011.

76. Glauert, H. *The Elements of Aerofoil and Airscrew Theory*; Cambridge Science Classics; Cambridge University Press: Cambridge, UK, 1983.

77. Quarteroni, A.; Sacco, R.; Saleri, F. *Numerical Mathematics*; Texts in Applied Mathematics, 37; Springer: Berlin, Germany, 2007.

78. Abbott, I.H.; von Doenhoff, A.E. *Theory of Wing Sections: Including a Summary of Aerofoil Data*; Dover: New York, NY, USA, 1959.

79. Peters, D.A.; Hsieh, M.C.A.; Torrero, A. A State-Space Airloads Theory for Flexible Airfoils. *J. Am. Helicopter Soc.* **2007**, *52*, 329–342. [CrossRef]

80. Pettit, G.W. Model to Evaluate the Aerodynamic Energy Requirements of Active Materials in Morphing Wings. Master's Thesis, Virginia Polytechnic Institute and State University, Blacksburg, VA, USA, 2001.

81. Kutta, M.W. Auftriebskräfte in Strömenden Flüssigkeiten. *Illustrierte Aeronautische Mitteilungen* **1902**, *6*, 133–135.

82. Joukowski, N.E. Sur les Tourbillons Adjionts. *Traraux de la Section Physique de la Societé Imperiale des Amis des Sciences Naturales* **1906**, *13*, 2.

83. Nita, M.F. Contributions to Aircraft Preliminary Design and Optimization. Ph.D. Thesis, Politehnica University of Bucharest, Bucharest, Romania, 2012.

84. Van Holten, T. Some Notes on Unsteady Lifting-Line Theory. *J. Fluid Mech.* **1976**, *77*, 561–579. [CrossRef]

85. Sclavounos, P.D. An Unsteady Lifting-Line Theory. *J. Eng. Math.* **1987**, *21*, 201–226. [CrossRef]

86. Guermond, J.L.; Sellier, A. A Unified Unsteady Lifting-Line Theory. *J. Fluid Mech.* **1991**, *229*, 427–451. [CrossRef]

87. Wieseman, C.D. *Methodology for Matching Experimental and Computational Aerodynamic Data*; NASA TM-100592; NASA: Washington, DC, USA, 1988.

88. Palacios, R.; Climent, H.; Karlsson, A.; Winzell, B. Assessment of Strategies for Correcting Linear Unsteady Aerodynamics Using CFD or Test Results. In Proceedings of the 9th IFASD, Madrid, Spain, 5–7 June 2001.

89. Munk, M.M. *Elements of the Wing Section Theory and of the Wing Theory*; NACA 191; NACA: Washington, DC, USA, 1924.

90. Lippisch, A. *Method for the Determination of the Spanwise Lift Distribution*; NACA TM 778; NACA: Washington, DC, USA, 1935.

91. Pearson, H.A. *Span Load Distribution for Tapered Wings with Partial-Span Flaps*; NACA 585; NACA: Washington, DC, USA, 1937.

92. DeYoung, J. *Theoretical Additional Span Loading Characteristics of Wings with Arbitrary Sweep, Aspect Ratio, and Taper Ratio*; NACA TN 1491; NACA: Washington, DC, USA, 1947.

93. DeYoung, J. *Theoretical Antisymmetric Span Loading for Wings of Arbitrary Plan Form at Subsonic Speeds*; NACA TN 2140; NACA: Washington, DC, USA, 1950.

94. Multhopp, H. *Methods for Calculating the Lift Distribution of Wings (Subsonic Lifting-Surface Theory)*; ARC R&M 2884; Aeronautical Research Council: London, UK, 1955.

95. Simpson, R.W. *An Extension of Multhopp's Lifting Surface Theory*; Cranfield CoA Report 132; Cranfield University: Bedford, UK, 1960.

96. Brebner, G.G.; Lemaire, D.A. *The Calculation of the Spanwise Loading Sweptback Wings with Flaps or All-Moving Tips at Subsonic Speeds*; ARC R&M 3487; Aeronautical Research Council: London, UK, 1967.

97. Woodward, F.A. *An Improved Method for the Aerodynamic Analysis of Wing-Body-Tail Configurations in Subsonic and Supersonic Flow*; NASA CR-2228; NASA: Washington, DC, USA, 1973.

98. Morino, L. *A General Theory of Unsteady Compressible Potential Aerodynamics*; NASA CR-2464; NASA: Washington, DC, USA, 1974.

99. Blair, M. *A Compilation for the Mathematics Leading to the Doublet Lattice Method*; WL-TR-92-3028; USAF: Washington, DC, USA, 1992.

100. Fossati, M. Evaluation of Aerodynamic Loads via Reduced-Order Methodology. *AIAA J.* **2015**, *53*, 2389–2405. [CrossRef]

101. Thwapiah, G.; Campanile, L.F. Nonlinear Aeroelastic Behavior of Compliant Aerofoils. *Smart Mater. Struct.* **2010**, *19*, 253–262. [CrossRef]

102. Cone, C.D. *The Theory of Induced Lift and Minimum Induced Drag of Non-Planar Lifting Systems*; NASA TR R-139; NASA: Washington, DC, USA, 1962.

103. Forrester, A.J.; Keane, A.J. *Engineering Design via Surrogate Modelling: A Practical Guide*; Wiley: Chichester, UK, 2008.

104. Berci, M.; Gaskell, P.H.; Hewson, R.W.; Toropov, V.V. Multifidelity Metamodel Building as a Route to Aeroelastic Optimization of Flexible Wings. *J. Mech. Eng. Sci.* **2011**, *225*, 2115–2137. [CrossRef]

105. Tiffany, S.H.; Adams, W.M. *Nonlinear Programming Extensions to Rational Function Approximation Methods for Unsteady Aerodynamic Forces*; NASA-TP-2776; NASA: Washington, DC, USA, 1988.

106. Garrick, L.E. *On Some Reciprocal Relations in the Theory of Nonstationary Flows*; NACA-629; NACA: Washington, DC, USA, 1938.

107. Heaslet, M.A.; Spreiter, J.R. *Reciprocity Relations in Aerodynamics*; NACA-1119; NACA: Washington, DC, USA, 1953.

108. Beddoes, T.S. Practical Computation of Unsteady Lift. *Vertica* **1984**, *8*, 55–71.

109. Dowell, E.H. *A Simple Method for Converting Frequency Domain Aerodynamics to the Time Domain*; NASA-TM-81844; NASA: Washington, DC, USA, 1980.

110. Leishman, J.G.; Nguyen, K.Q. State-Space Representation of Unsteady Airfoil Behavior. *AIAA J.* **1990**, *28*, 836–844. [CrossRef]

111. Tobak, M. *On the Use of the Indicial-Function Concept in the Analysis of Unsteady Motions of Wings and Wing-Tail Combinations*; NACA 1188; NACA: Washington, DC, USA, 1954.

112. Silva, W. Discrete-Time Linear and Nonlinear Aerodynamic Impulse Responses for Efficient Use of CFD Analyses. Ph.D. Thesis, College of William & Mary, Williamsburg, VA, USA, 1997.

113. Ghoreyshi, M.; Jirásek, A.; Cummings, R.M. Computational Investigation into the Use of Response Functions for Aerodynamic-Load Modeling. *AIAA J.* **2012**, *50*, 1314–1327. [CrossRef]

114. Rodden, W.P. *Theoretical and Computational Aeroelasticity*; Crest Pub.: Columbus, OH, USA, 2011.

115. Van Zyl, L.H. Aeroelastic Divergence and Aerodynamic Lag Roots. *J. Aircr.* **2001**, *38*, 586–588. [CrossRef]

116. Aeroelastic Analysis User's Guide. In *MSC Nastran*; MSC Software Corporation: Newport Beach, CA, USA, 2018.

117. Appa, K. Finite-Surface Spline. *J. Aircr.* **1989**, *26*, 495–496. [CrossRef]

118. Harder, R.L.; Desmarais, R.N. Interpolation Using Surface Splines. *J. Aircr.* **1972**, *9*, 189–191. [CrossRef]

119. Lanczos, C. *An Iteration Method for the Solution of the Eigenvalue Problem of Linear Differential and Integral Operators*; United States Governm. Press Office: Los Angeles, CA, USA, 1950; Volume 45.

120. Bellinger, D.; Pototzky, T. A Study of Aerodynamic Matrix Numerical Condition. In Proceedings of the 3rd MSC Worldwide Aerospace Conference and Technology Showcase, Toulouse, France, 24–26 September 2001.

121. Giesing, J.P.; Kalman, T.P.; Rodden, W.P. *Correction Factor Techniques for Improving Aerodynamic Prediction Methods*; NASA CR-144967; NASA: Washington, DC, USA, 1976.

122. Lawrence, A.J.; Jackson, P. *Comparison of Different Methods of Assessing the Free Oscillatory Characteristics of Aeroelastic Systems*; ARC CP 1084; HM Stationery Office: London, UK, 1970.

123. Goland, M. The Flutter of a uniform Cantilever Wing. *J. Appl. Mech.* **1945**, *12*, A197–A208.

124. Goland, M.; Luke, Y.L. The Flutter of a Uniform Wing with Tip Weights. *J. Appl. Mech.* **1948**, *15*, 13–20.

125. Wang, I. Component Modal Analysis of a Folding Wing. Ph.D. Thesis, Duke University, Durham, NC, USA, 2011.

126. NASA. *U.S. Standard Atmosphere*; NASA TM-X-74335; NASA: Washington, DC, USA, 1976.

127. Banerjee, J.R. Flutter sensitivity studies of high aspect ratio aircraft wings. *WIT Trans. Built Environ.* **1993**, *2*, 374–387.

128. Sotoudeh, Z.; Hodges, D.H.; Chang, C.S. Validation Studies for Aeroelastic Trim and Stability Analysis of Highly Flexible Aircraft. *J. Aircr.* **2010**, *47*, 1240–1247. [CrossRef]

129. Qin, Z. Vibration and Aeroelasticity of Advanced Aircraft Wings Modeled as Thin-Walled Beams—Dynamics, Stability and Control. Ph.D. Thesis, Virginia Polytechnic Institute and State University, Blacksburg, VA, USA, 2001.

130. Palacios, R.; Epureanu, B. *An Intrinsic Description of the Nonlinear Aeroelasticity of Very Flexible Wings*; AIAA-2011-1917; AIAA: Reston, VA, USA, 2011.

131. Dunning, P.D.; Stanford, B.K.; Kim, H.A.; Jutte, C.V. Aeroelastic Tailoring of a Plate Wing with Functionally Graded Materials. *J. Fluids Struct.* **2014**, *51*, 292–312. [CrossRef]

132. Berci, M. Semi-Analytical Reduced-Order Models for the Unsteady Aerodynamic Loads of Subsonic Wings. In Proceedings of the 5th ECCOMAS COMPDYN, Crete, Greece, 25–27 May 2015.

133. Hildebrand, F.B. *A Least-Squares Procedure for the Solution of the Lifting-Line Integral Equation*; NACA TN 925; NACA: Washington, DC, USA, 1944.

134. Hoblit, F.M. *Gust Loads on Aircraft: Concepts and Applications*; AIAA Education Series; AIAA: Reston, VA, USA, 1988.

aerospace

MDPI

Article

A Multi-Fidelity Approach for Aerodynamic Performance Computations of Formation Flight

Diwakar Singh, Antonios F. Antoniadis *, Panagiotis Tsoutsanis, Hyo-Sang Shin, Antonios Tsourdos, Samuel Mathekga and Karl W. Jenkins

School of Aerospace, Transport and Manufacturing, Cranfield University, Cranfield, Bedfordshire MK43 0AL, UK; d.singhaliasangel@cranfield.ac.uk (D.S.); panagiotis.tsoutsanis@cranfield.ac.uk (P.T.); h.shin@cranfield.ac.uk (H.-S.S.); a.tsourdos@cranfield.ac.uk (A.T.); mathekga.ms@gmail.com (S.M.); k.w.jenkins@cranfield.ac.uk (K.W.J.)
* Correspondence: a.f.antoniadis@cranfield.ac.uk; Tel.: +44-123-475-4691

Received: 9 May 2018; Accepted: 7 June 2018; Published: 15 June 2018

Abstract: This paper introduces a multi-fidelity computational framework for the analysis of aerodynamic performance of flight formation. The Vortex Lattice and Reynolds Averaged Navier–Stokes methods form the basis of the framework, as low- and high-fidelity, respectively. Initially, the computational framework is validated for an isolated wing, and then two rectangular NACA23012 wings are considered for assessing the aerodynamic performance of this formation; the optimal relative position is through the multi-fidelity framework based on the total drag reduction. The performance estimates are in good agreement with experimental measurements of the same configuration. Total aerodynamic performance of formation flight is also assessed with respect to attitude variations of the lifting bodies involved. The framework is also employed to determine the optimal position of blended-wing-body unmanned aerial vehicles in tandem formation flight.

Keywords: multi-fidelity; aerodynamic performance; formation; VLM; RANS

1. Introduction

It is widely known that aeroplanes flying in formation yield fuel savings. Fuel savings are achieved due to the reduced induced drag on the following aeroplane, when it is positioned within the upwash created by the leader's wingtip vortices. It has been established experimentally and computationally that birds flying in formation results in energy savings [1,2]. Lissaman and Shollenberg [3] showed that at the optimal distance between the leader and the followers in a flock of twenty-five birds could increase the flight range by 70% compared to their isolated flights. Hainsberg [4] used fifty-five geese against the model of Lissaman and Schollenberger which reflected the power savings of 36%. Hummel [5] devised a theory to predict the optimal wing tip spacing ($WTS_{OPT} = 0.5b(1 - 0.89)$) for the maximum drag reduction and concluded that it is a negative value which is only possible in the V-shaped formation configurations. Use of lifting line theory to study formation and control has also been of keen interest for researchers [6].

Unmanned Aerial Vehicles (UAVs) have been incorporated in civilian and military fields for surveillance, reconnaissance, search and rescue missions and UAV swarms have been proved to be more performance efficient in formations [7,8].

Many in-flight tests, wind tunnel experiments and computational analyses were performed in recent decades to study formation flight. Hummel and Beukenberg [5] conducted an experiment using two Dornier Do-28 aircraft and established that the maximum obtainable power savings are of about 15%. Pahle and Berger [9] tested several flights for large transport class vehicle C-17 aircraft with maximum fuel savings of 7–8%. Inasawa et al. [10] used two rectangular wings of aspect ratio (AR) 5 in a wind tunnel while Blake and Gingras [11] tested two delta wings and compared against the analysis obtained from Vortex Lattice Method (VLM). Recently, Slotnick et al. [12] performed aerodynamic analysis using a multi-fidelity approach. The analysis performed used low-fidelity VLM along with high-fidelity hybrid RANS solver and established the total drag benefits of around 25%, which was validated against flight test data with good agreement.

Formation flights involving more than just one follower have also been the prime focus of research to quantify individual savings and the resulting overall drag reduction of the aerodynamic system [13]. Blake and Multhopp [14] showed that 10% wing span overlap between the leader and the follower could result in 60% increase in the flight range for a cluster of five aircraft in V formation while Maskew et al. [15] found the flight range augmentation of 46–67% for three to five aircraft. It was further noted that power savings were more in the case of second follower than the first. Wagner et al. [16] confirmed increased savings of 17.5% for the second wingman compared to first which yielded 15% savings [17]. Recently, Ivan and Roberts [18] conducted wing tunnel tests using multiple wings configurations with each wing having low aspect ratio of two; a total power savings of 14% and 24% in two and three wings configurations was obtained. In correspondence to finding an optimum spacing between the various units in formation, numerous studies also shed light on effect of leader's shape and size on the follower [19,20].

The present study involves the development of flexible and robust multi-fidelity numerical framework for studying formation flight. The numerical framework accuracy is assessed with grid/panel sensitivity analysis and comparison with experimental data. A computational analysis of two identical NACA 23012 rectangular wings in formation using the multi-fidelity approach of RANS and VLM methodologies is compared against Inasawa et al. [10] experimental data. Furthermore, to demonstrate the flexibility of the numerical framework, two blended-wing aircraft in formation flight are modelled. Section 2 introduces the governing equations for VLM and RANS along with mesh generation strategy and solver attributes. Section 3 expands on the aerodynamic performance analysis of an isolated NACA23012 wing and also in two wings of the same configuration in tandem formation, the local angle of incident and the relative position is in order to maximise performance gains. The computational analysis is further augmented by altering the angle of attack of the lead wing and its effect on the trail wing is assessed. Furthermore, two blended-wing-body aircraft in formation are introduced and the aerodynamic performance is assessed within the multi-fidelity approach. Finally, the conclusions are presented in Section 4 comprising of the inferences obtained for both computational techniques and their applicability to analysis.

2. Computational Methodology

Multi-fidelity strategies incorporating different levels of uncertainty for aerodynamic performance estimations are becoming a standard tool for design, parameterisation and optimisation tasks for aircraft systems. Inexpensive methods based on panels, lifting line theory and vortex lattice methods can generate large data sets for design of experiment phases and multi-objective optimisation analysis. The limitations of these methods with respect to the inherited physical assumptions must be considered, as their applicability range has to be tailored to produce realistic aerodynamic performance indicators particularly in terms of drag estimation. Higher fidelity approaches such as Navier–Stokes solvers can be used to complement the predictions of the lower fidelity approaches.

The computational approach adopted for this study is based on the Vortex Lattice Method solving the Laplace's equation, where each discrete element corresponds to the vortex line solution of the incompressible potential flow equation. The open-source VLM code Tornado [21] is employed as the foundation of the low-fidelity framework. Additional modules are developed enabling individual trim conditions and Cartesian coordinates to be set for multiple surfaces representing additional lift-generating bodies. High-fidelity aerodynamic estimations are obtained with a second-order RANS solver that accounts for three-dimensionality of the flow, turbulence, viscous drag and boundary layer effects. The developed low-fidelity framework enables the construction of extensive simulation data sets for large flight formation configurations to run in a fraction of the computational cost compared with the RANS approach.

2.1. Low-Fidelity: Vortex Lattice Method

Vortex Lattice Method assumes that the flow is incompressible, inviscid and irrotational. The resulting velocity field is assumed to be a conservative vector field which essentially means that a scalar velocity potential ϕ whose gradient will produce a velocity vector field combined with the free stream velocity as given in Equation (1).

$$V = V_\infty + \nabla\phi \qquad (1)$$

The velocity field has to satisfy the incompressible continuity equation written as

$$\nabla \cdot V = 0 \qquad (2)$$

the assumed conservative vector field is the velocity field expressed in terms of scalar potential ϕ that results in it being satisfied, given by

$$\nabla^2\phi = 0 \qquad (3)$$

The scalar potential ϕ satisfies the Laplace's equation which further implies that if ϕ_1 and ϕ_2 are two potential solutions then their linear combination such as $c_1\phi_1 + c_2\phi_2$ is also a solution for any values of c_1 and c_2. This forms the potential flow theory in which such solutions are combined together to represent a lifting surface. VLM uses one such elementary potential solution to form a vortex sheet and represent a lifting surface.

The thickness of the lifting surface is ignored and the planform of a cambered surface is divided into a number of quadrilateral panels as shown in Figure 1. The number of panels to be incorporated for a particular analysis is subjected to the required accuracy. The lifting surface is replaced with a vortex sling on each panel such that the bound vortex part of the sling is placed at $1/4$ chord line. Solution of the velocity field at each panel is computed, based on the attributed vortex sheet of the unknown strength Γ_j on the corresponding panel. A normal vector n at every panel is calculated respective collocation points accounting for camber of the lifting surface. For a problem consisting of N number of panels, the perturbation velocity at ith collocation point is given in Equation (4).

$$\nabla_{\phi_i} = +\sum_{j=i}^{N} w_{ij}\Gamma_j \qquad (4)$$

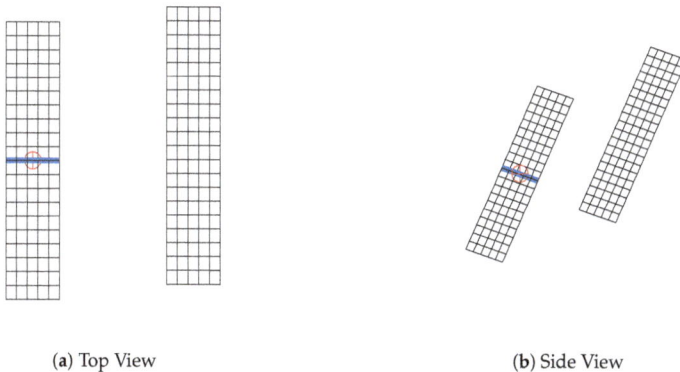

(a) Top View   (b) Side View

Figure 1. Panel distribution on rectangular wings.

The free-stream velocity components are given in terms of angle of attack (α) and angle of side slip (β), as given in Equation (5), and are depicted in Figure 2a,b.

$$V_\infty = V_\infty \begin{bmatrix} cos(\alpha)cos(\beta) \\ -sin(\beta) \\ sin(\alpha)cos(\beta) \end{bmatrix} \tag{5}$$

Note that for the present study the angle of side slip (β) is zero. For a problem with N panels, the perturbation velocity at collocation point i is given by summing the contributions of all the horseshoe vortices in terms of an Aerodynamic Influence Coefficient (AIC) matrix w_{ij}. A physical boundary condition of no normal flow component to each panel of the lifting surface is applied and is expressed by Equation (6) at the collocation points.

$$V_i \cdot n_i = \left(V_\infty + \sum_{j=i}^{N} w_{ij}\Gamma_j \right) \tag{6}$$

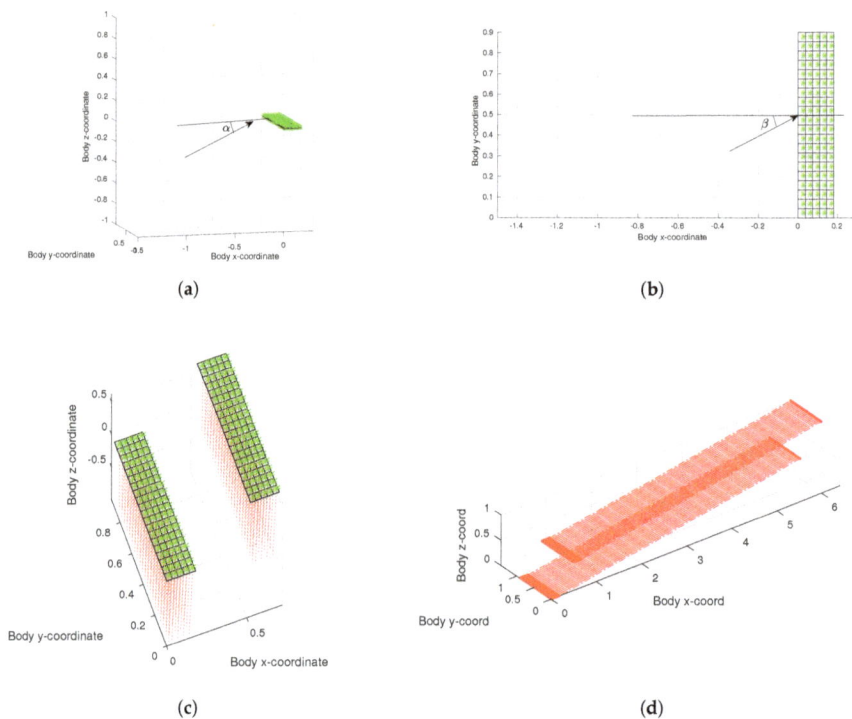

(a)

(b)

(c)

(d)

Figure 2. (**a**) Normal vectors (in red) and collocation points on wing surface (green); (**b**) free-stream wake of the lifting surface; (**c**) schematic of angle of attack (α) with arrow showing the direction of free-stream velocity (isometric view); and (**d**) schematic of angle of side-slip (β) with arrow showing the direction of free-stream velocity (plan-view).

This is demonstrated in Figure 2c,d. This is also known as flow tangency condition. After evaluating the dot products, a new normal-wash aerodynamic influence coefficient (AIC) matrix is formed given by $a_{ij} = w_{ij} \cdot n_i$, as given in Equation (7)

$$
\begin{bmatrix}
a_{11} & a_{12} & . & . & . & a_{1n} \\
a_{21} & . & . & . & . & . \\
. & . & . & . & . \\
. & . & . & . & . \\
a_{n1} & a_{n2} & . & . & . & a_{nn}
\end{bmatrix}
\begin{bmatrix}
\Gamma_1 \\
\Gamma_2 \\
. \\
. \\
\Gamma_n
\end{bmatrix}
=
\begin{bmatrix}
b_1 \\
b_2 \\
. \\
. \\
b_n
\end{bmatrix}, \tag{7}
$$

the right hand side of the equation is formed by a free-stream velocity and the two aerodynamic angles given below (8).

$$
b_i = V_\infty \left[cos(\alpha)cos(\beta), -sin(\beta), sin(\alpha)cos(\beta) \right] \cdot n_i \tag{8}
$$

A horseshoe vortex is imparted to the panel which starts from several lengths downstream and moves forward to the panel and crosses it at the quarter chord line and then again runs to the far down stream. All the vortices create downwash on each panel on the lifting surface. V_{ind} is the induced velocity at the centre of the panel which is calculated once the vortex strength is evaluated using Biot–Savarts Law given by Equation (9).

$$
dV_{ind} = \frac{\Gamma}{4\pi} \frac{dl \times r}{|r|^3} \tag{9}
$$

The induced flow is used to get the force acting on the panel by using Kutta–Jukovski theorem given by Equation (10)

$$
F_i = \rho \Gamma_i (V_\infty + V_{ind}) \times l_i \tag{10}
$$

where F_i is the force contribution from the ith panel, ρ is the air density, l_i is the vortex transverse segment vector (bound vortex) of that panel and r_i is the position vector of the segment's centre. The wake is assumed to be flat and in the position in the free-stream direction, as shown in Figure 2b.

The VLM code is based on the Tornado solver. Several modules are created that provide the control of the non-dimensionalised lateral (l/s), vertical (h/c) and longitudinal (x/c) coordinates for n number of wings, as shown in Figure 3 for a rectangular wing of the NACA23012 configuration. The modules include the development of the connectivity of multiple wings based upon the number of panels associated with the respective wing. The right hand side term of the boundary condition given in Equation (7) is split. The value of α in the term $-V_\infty sin(\alpha)$ is the angle of attack. Splitting the right hand side of the boundary condition is achieved by assigning a different value of α for the lead and trail wing depending upon the number of panels associated with that wing. The resulting vector can then be obtained using the same procedure discussed within this section; the method first solves for the aerodynamic forces on the nth wing in locally altered coordinates. Their effect on the other members of the formation is analysed by comparing the aerodynamic coefficients attributed for each wing. Simulations are run to find an optimized solution by constructing a matrix for longitudinal and lateral overlapping positions of the wings. The resulting matrix of the aerodynamic coefficients is plotted for all the lateral and longitudinal positions of the trail wing at a particular stream-wise distance and is discussed in detail in Section 3.3. The developed VLM framework also enables flight dynamics of attitude parameters including pitch, yaw and roll angles to be set for each individual aircraft or wing.

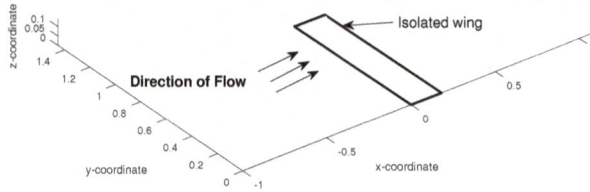

Figure 3. Schematic of isolated wing NACA23012 and spatial domain within the VLM solver.

2.2. High-Fidelity: Reynolds Averaged Navier-Stokes

The high-fidelity approach considered is the RANS method in three spatial dimension. The RANS equations describe time-averaged motion of fluid flow through a control volume. The instantaneous physical quantities are decomposed into their time-averaged and fluctuating parts. The flow considered in the present computations involves the NACA23012 configuration at a Reynolds number of $Re = 0.24 \times 10^6$ based on the wing's chord length, corresponding to a fully turbulent regime. As the flow test cases considered for this work are below the compressibility limit, steady state solutions are obtained by solving the incompressible RANS equations; the system of equations expressing mass and momentum conservation within a finite controlled volume are written as in Equation (11).

$$\frac{\partial u_i}{\partial x_i} = 0$$
$$\frac{\partial(\rho u_i u_j)}{\partial x_j} = -\frac{\partial p}{\partial x_i} + \frac{\partial \tau_{ij}}{\partial x_j} \tag{11}$$

where u_i, τ_{ij} and p are the velocity vector, the shear stress tensor and the pressure, respectively. The SST $\kappa - \omega$ turbulence model is considered for modelling the viscous turbulent stresses. The model is based on the shear–stress–transport (SST) formulation which blends the $\kappa - \omega$ [22] and $\kappa - \epsilon$ [23] turbulence models. The $\kappa - \omega$ model is well suited to model the flows inside the viscous sub-layer and and $\kappa - \epsilon$ model is understood to be good at predicting flows in the regions away from the wall [24].

The CFD model is completed with the generation of three-dimensional grids with frontal radius and downstream length of 200 and 500 chord lengths, respectively, as shown in Figure 4. Hybrid meshes are generated comprising of a quadrilateral dominant surface grid for both wings. A hexahedral inflation layer is wrapped around each wing to capture the boundary layer up to the sub-viscous region maintaining a $y^+ \approx 1$. Figure 5 demonstrates the different focus angles of the cut-section volume grid close to the wing as well as the refinement region in the wake. The far-field fluid domain is filled with unstructured tetrahedral cells with varying spatial resolution to capture the flow exhibiting strong gradients near and around the surfaces of wings.

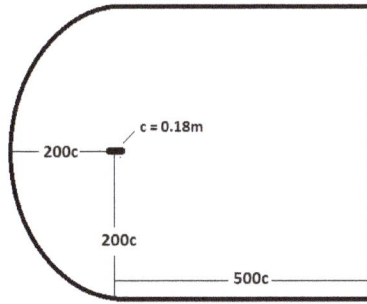

Figure 4. Topology and dimensions of spatial domain for RANS computations of NACA23012 case.

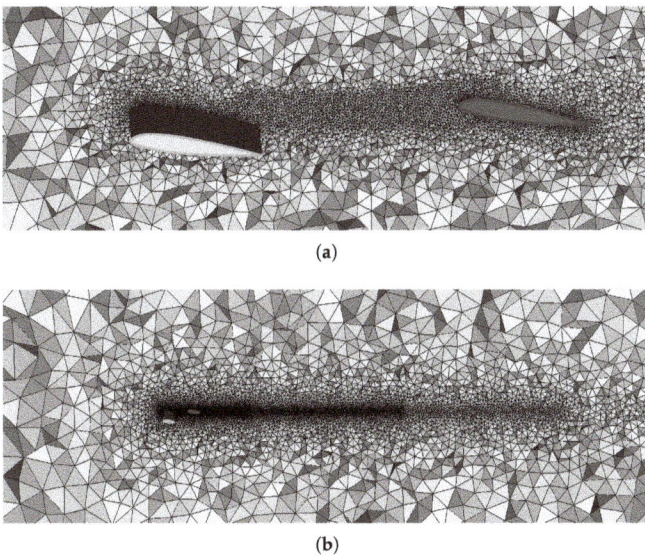

(a)

(b)

Figure 5. Cut-section of the 3D grids: (**a**) close-up of both wings; and (**b**) overview of both wings and grid refinement in the wake region.

3. Application of the Multi-Fidelity Framework

This section is divided in four subsections. Sections 3.1 and 3.2 contain the verification procedure of the multi-fidelity framework, where RANS and VLM calculations are carried out for an isolated NACA23012 wing and for the same wing in double tandem configuration at 8° angle of attack. The computed solutions are compared against wind-tunnel data. An analysis is carry-out where we investigate the effect of the leading wing's angle of attack on the overall formation flight aerodynamic performance within Section 3.3. Finally, we demonstrate the flexibility aspects of the developed framework for blended wing formation configurations in Section 3.4.

3.1. Isolated Wing

Initially, an isolated wing configuration is considered which consists of a single NACA23012 wing at angle of attack of $AOA = 8°$, which is shown in Figure 3 at corresponding Reynolds number of $Re = 0.24 \times 10^6$ and a free-stream velocity of 20 m/s. The main objective of this analysis is to assess the accuracy and uncertainty levels of each of the computational fidelity approaches, i.e., VLM and RANS, and compare with experiment.

A numerical simulation is considered to be consistent when the numerical solution approaches the exact solution as the grid spacing tends to zero [25]. To study the effect of grid refinement for the RANS simulation on the solution accuracy, four meshes with increasing element count are generated with 3, 5, 8 and 10 million cells, for the isolated wing case. Similarly, the spatial resolution is also assessed for the VLM method, by computing the solution from 20 to 100 panels; for 100 panels, 5 in chordwise and 20 in spanwise direction are set. Lift (C_{LO}) and drag (C_{DO}) coefficients are plotted against the number of cells and number of panels for RANS and VLM in Figure 6. In the case of RANS, the change in C_{LO} and C_{DO} obtained from the 8 and 10 millions element meshes is less than 0.1%.

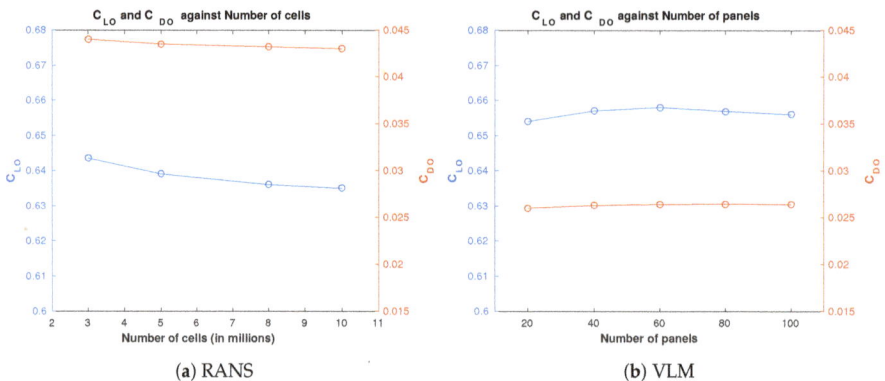

Figure 6. Spatial resolution study for the isolated NACA23012 wing at 8° angle of attack, in terms of lift (C_{LO}) and drag (C_{DO}) coefficients for RANS and VLM methods.

The lift coefficient (C_{LO}) and drag coefficient (C_{DO}) for an isolated NACA23012 wing are plotted against angle of attacks AOA ranging from −4° to 14° and compared against the experimental data from Inasawa et al. [10] in Figure 7. The C_{LO} values obtained from VLM maintain a linear trend throughout the polar and superimpose the experimental data up to around 8° incident. This demonstrates that the VLM method performs adequately well for small angles. Slight over-predictions become apparent with respect to the lift for angles greater than 8° by the VLM estimations. With respect to the drag, VLM solutions are under-predicted throughout the polar; this is expected to some extent as the method is based on the inviscid, irrotational potential flow theory and accounts just for the lift-induced drag. RANS solutions are clearly more accurate and with smaller deviations from the experiment in terms of both drag and lift. The experimental lift to drag ratio (L/D) at 5° angle of attack is 16.5, whereas, for the RANS, the L/D value turns out to be 14.93 with an error of 9.5%. When compared against the VLM, RANS is conclusively more accurate at predicting the drag coefficient.

Both approaches demonstrated to be effective and reliable techniques for performing aerodynamic analysis with an error of under 5% for C_{LO} and of 11% in the case of C_{DO} up to 11° angle incident. The computational effort required to run one VLM simulation with 200 panels is 30 s, as compared to 2.5 h for the RANS computation using open-foam, on a Intel Core i5-7600K processor, 4 CPUs, 12 GB of RAM, with the 8 million grid.

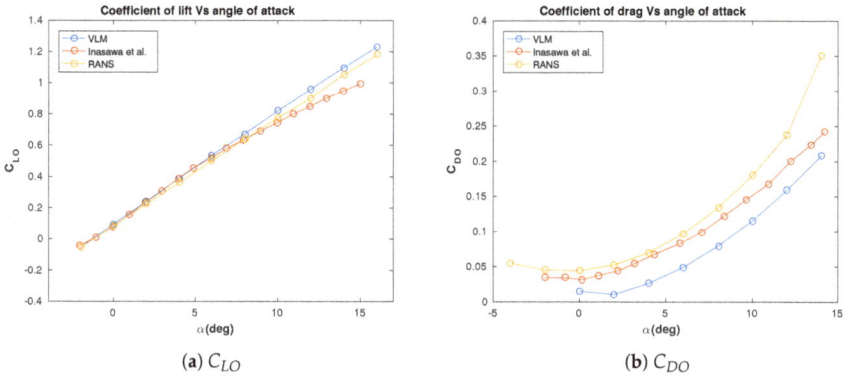

Figure 7. Polar analysis of isolated NACA23012 wing plotting for coefficient of lift (C_{LO}) and drag (C_{DO}); results include VLM, RANS and experiment from Inasawa et al. [10].

3.2. Two Wings in Tandem Configuration

An additional rectangular NACA23012 wing is introduced to the isolated wing configuration discussed in the previous section. For this section, the isolated wing is now referred to as trail wing in the two wing configuration, as depicted in Figure 8. This is done to directly compare the aerodynamic performance of the isolated wing to the same isolated wing in the wake region of the lead wing. By incorporating a second wing, the relative positioning and flight dynamics parameters (yaw, pitch and roll) are varied of each body to obtain the best aerodynamic performance in terms of total quantities, e.g., total drag reduction and lift over drag ratio. The position of the lead wing is held constant while the trail wing is allowed to move in the y-z plane at different stream-wise positions. Both the lead and the trail wings are set at incident angle of $AOA = 8°$; the freestream velocity is set to 20 m/s and a Reynolds number of $Re = 0.24 \times 10^6$.

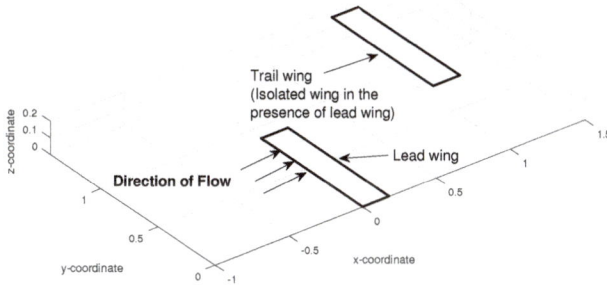

Figure 8. Schematic of two NACA23012 wings in tandem configuration.

The displacement of the trail wing behind the lead wing is monitored by the given transformed coordinates as depicted in Figure 8 where the lateral displacement corresponds to (l/s), the vertical to (h/c) and the stream-wise direction at (x/c). Note that the l/s is positive when the lead and the trail wing are in overlapping position. The aerodynamic performance gains are assessed by calculating the percentage change in the lift and drag coefficient of the trail wing (C_L and C_D) in the presence of the lead wing compared against the lift and drag coefficient of an isolated configuration (C_{LO} and

C_{DO}). The aerodynamic performance is evaluated at the trail wing by VLM and RANS, and compared with experiment, it is appropriate to evaluate the percentage's variation than the absolute values; thus the percentage change in lift and drag coefficients are calculated as $\%\Delta C_L = (C_L - C_{LO})/C_{LO}$ and $\%\Delta C_D = (C_{DO} - C_D)/C_{DO}$. Note that the change between the isolated and trail wings is monitored in percentages. This implies that at the optimum position, the trail wing is expected to have maximum increase in lift and maximum decrease in drag. Moreover, it is further to be noted that, in the expressions for $\%\Delta C_L$ and $\%\Delta C_D$, the terms representing the respective quantities for the isolated and the trail wings are switched. This is to keep consistency with the expressions detailed in the experimental analysis.

Figure 9 demonstrates the performance of the percentage variations of lift and drag, $\%\Delta C_L$ and $\%\Delta C_D$, at five lateral overlapping positions $l/s = -0.05$, 0, 0.05, 0.01 and 0.15, solutions from VLM and RANS are compared with experiment. RANS and VLM over-predict the $\%\Delta C_L$ values for $l/s = 0.05$ with an error of 13% and 18%, respectively, compared with reference data.

The RANS computations suggest a negative overlapping position ($l/s = -0.05$) where the ΔC_D attains the minimum value of 5% as shown in Figure 9b. As the wing overlap becomes positive ($l/s = 0$ to 0.05), the ΔC_D value increases from 9% to maximum value of 11% with an error of 17%. By plugging in a span value of 0.9 units in the formula given by Hummel [5] discussed in Section 1, the optimum wing tip overlap value (l/s) turns out to be 0.0495 units which translates to around $l/s = 0.05$. This overlapping position in fact marks the position for maximum decrease in ΔC_D values for both the experiment (8%) and the RANS (9.3%) with an error of 16.25%. Same degree of error is observed in deeper overlapping positions i.e., at $l/s = 0.1$ and 0.15 where both the ΔC_L and ΔC_D values start to decrease as compared to the optimum lateral overlapping position of $l/s = 0.05$. This is attributed to the fact that, along with the upwash, the downwash from the lead wing vortex also acts on the trail wing thereby offsetting the upwash, resulting in decreased aerodynamic performance.

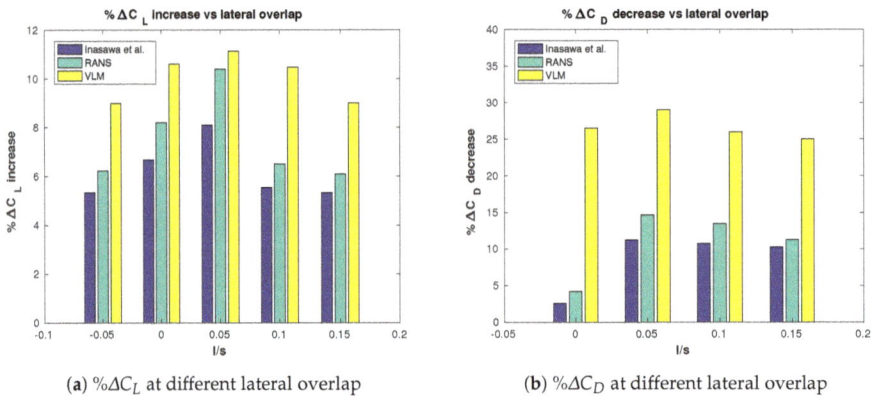

(a) $\%\Delta C_L$ at different lateral overlap

(b) $\%\Delta C_D$ at different lateral overlap

Figure 9. Percentage variations of drag ($\%\Delta C_D$) and lift ($\%\Delta C_L$) at five different displacements; solutions are shown for RANS and VLM and compared with the experimental data [10].

For the VLM solutions, drag over-predictions are observed at all the discrete locations by almost 20–25% when compared against experimental values. Maximum value of ΔC_L is also observed at $l/s = 0.05$ position, as shown in Figure 10a,b, where the absolute values of lift and drag of trailwing are plotted as a response surface of the h/c and l/s displacement variables. It is also noted that the maximum drag reduction predicted by the VLM occurs at the same location and is as large as 27% in magnitude as compared to 10% in the case of experiment. Such high values of drag reductions are often cited by the various researchers undertaking computational analysis based on potential flow theory. Iglesis and Mason [26] used the discrete vortex method and concluded that the drag reductions of 30% for formation of three aircraft is achievable with central aircraft as the leader.

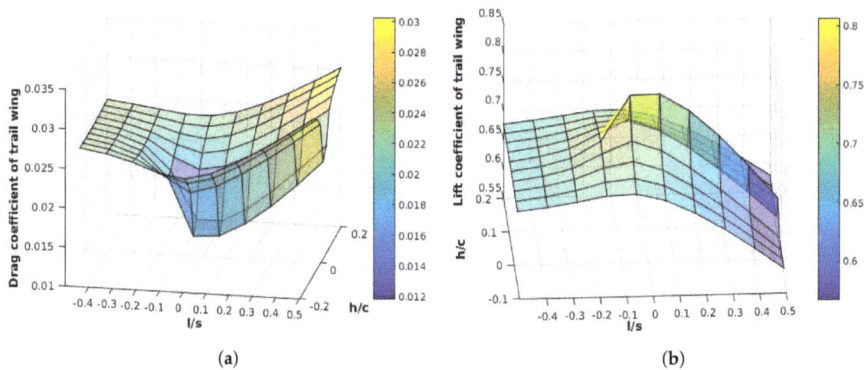

Figure 10. Three-dimensional response surfaces of absolute quantities against relative positions: (**a**) drag coefficient; and (**b**) lift coefficient, for VLM solutions.

3.3. Effect of the Lead Wing's AOA

Increasing the angle of attack can be favourable as the induced lift is increased, stronger pressure gradient would be present in the vicinity of wingtip vortices. This is attributed to the higher pressure difference between the upper and lower surfaces of the wing, which provides a more energetic upwash for the trailing wing. Therefore, this section is devoted to the assessment of the effect of the leading wing's angle of incident to the overall aerodynamic performance of the system. The angle of the lead wing is varied from 6° to 10° keeping the trail wing angle constant at 8°. When both wings are set at 8° angle, the configuration is referred to as baseline configuration (BSL). The setup for these computations is identical as described in the previous Section 3.2 with the exception that lead wing's AOA is varied. Results obtained from RANS and VLM are compared against the baseline configuration prediction of the previous section. The analysis is restricted within the linear range as higher angles might lead to larger discrepancies particularly for the low-fidelity solver as boundary layer separation might be present [27].

It is observed in Figure 11 that for both techniques at the trail wing's ΔC_L values share a linear relationship with the lead wing's angle of attack when compared against the results obtained from the baseline configuration. In the case of the RANS predictions, there is an increase in ΔC_L value of 10% for the trail wing compared to 6% in the case of the experiment. The ΔC_L of the trail wing drops to 8% when lead wing's angle is at 6°, whereas it increases to 12% at 10°, as shown in Figure 11a. Figure 11c shows similar trend in the case of the VLM but with slight increase in the magnitude of ΔC_L values for all the incidence angles. The ΔC_D values are over predicted by almost twice, as observed in the case of RANS. It can been noted that the low-fidelity prediction for all angles, predict the expected increase in lift and also the increase in absolute values of ΔC_D are in comparable range as seen in Figure 11d.

The aerodynamic performance predictions obtained from the VLM solver are plotted in terms of ΔC_L and ΔC_D with colour contours at different lead wing's angles at 7, 8 and 9 in Figure 12. It can be noted that, for all three lead wing's angles, there is no substantial benefit from the upwash created by lead wing's vortices beyond 13% lateral overlap marked by contours in the blue colour region. As the lateral overlap (l/s) is further increased, ΔC_L values become negative indicating that an isolated wing has greater lift coefficient than the trail wing in the presence of the lead wing, as the trail wing is in the pronounced downwash region created by the lead wing. It is also noted that the maximum increase in ΔC_L and maximum decrease in ΔC_D values are observed between the overlapping positions $l/s = 0$ to $l/s = 0.05$. Furthermore, as shown in Figure 12a,c,e, the regions pronouncing the best aerodynamic performance "sweet spot" of the trail wing shift upwards as the angle is increased to 9° and drops down when is reduced to 7° when compared against the baseline configuration. This is mainly because the wake in the VLM is parallel to the free-stream direction, whereas, in the real flows, the self induced velocities cause the longitudinal position of the vortex to drop in longitudinal direction.

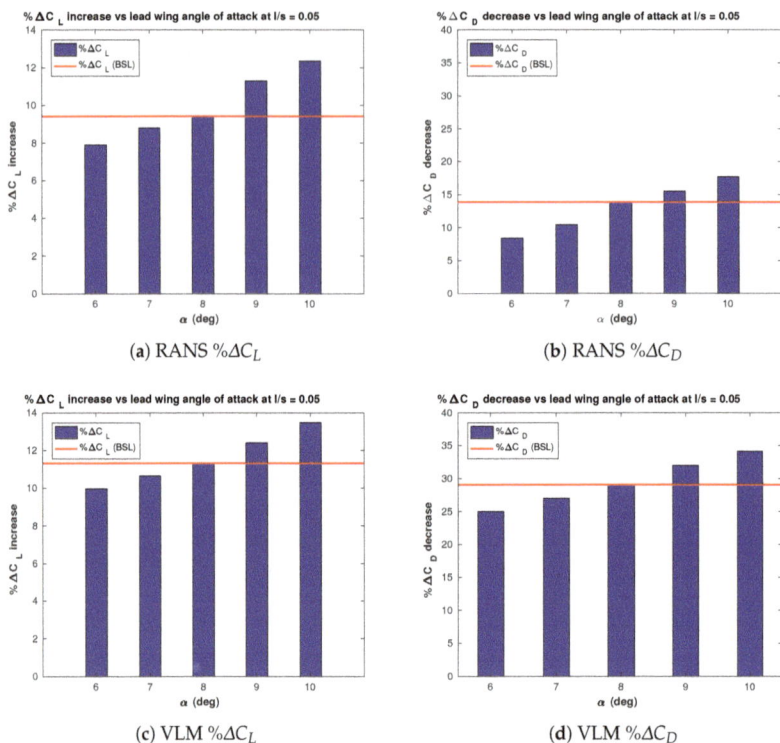

(a) RANS %ΔC_L

(b) RANS %ΔC_D

(c) VLM %ΔC_L

(d) VLM %ΔC_D

Figure 11. Effect of lead wing's AOA on aerodynamic coefficients of trail wing with red line corresponding to Baseline Configuration (Both wings $AOA = 8$): (**a**) percentage lift coefficient (%ΔC_L) for RANS; (**b**) percentage drag coefficient (%ΔC_D) for RANS; (**c**) percentage lift coefficient (%ΔC_L) for VLM; and (**d**) percentage drag coefficient (%ΔC_D) for VLM.

The qualitative analysis of ΔC_D contours shown in Figure 12b,d,f reflect somewhat the same position for the maximum drag reduction. As can be seen in Figure 11, VLM predicts as much as 30% increase in induced drag component at higher AOA. Even though the quantification of drag reduction does not yield reliable replication of the experimental values yet, it proves to be of immense importance when used in conjunction with the ΔC_L contours to study various variables in formation flights for relatively lower computational expense method. The multi-fidelity approach i.e employing VLM approach at early stages of analysis clearly proves prudent in deciphering the main flow features and general quantification of the aerodynamic coefficients at early stages of the analysis provide good insight of the design changes required for a particular requirement. This inexpensive approach helps in downsizing the simulation space for high-fidelity computation in the final stages of a computational analysis.

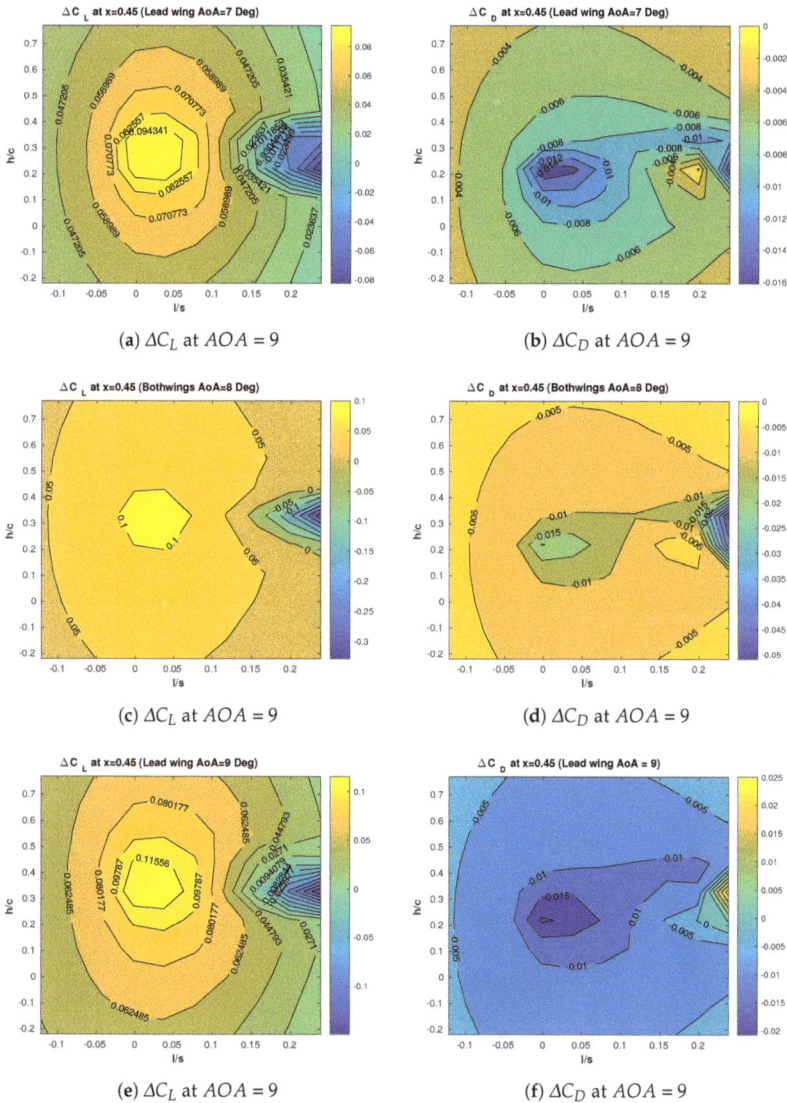

(**a**) ΔC_L at $AOA = 9$

(**b**) ΔC_D at $AOA = 9$

(**c**) ΔC_L at $AOA = 9$

(**d**) ΔC_D at $AOA = 9$

(**e**) ΔC_L at $AOA = 9$

(**f**) ΔC_D at $AOA = 9$

Figure 12. Contours of lift (ΔC_L) and drag coefficient (ΔC_D) percentage change for three lead wing incidence angles at streamwise distance of $x = 0.45$.

3.4. Blended-Wing-Body Aircraft in Formation

We extend the application of multi-fidelity approach to a realistic aircraft formation of a blended-wing-body aircraft configuration in formation. The main objective is to assess the flexibility of the developed framework. The flight conditions are set as steady level flight at a free-stream Mach number of 0.3, at an angle of attack of 3° with a corresponding Reynolds number of $(Re) = 40 \times 10^6$ based on the mean aerodynamic chord of $mac = 6.6$ m.

Initially, an isolated blended-wing-body configuration is modelled to determine the panel and grid sensitivity on the solutions. Figure 13 shows the lift and drag coefficient estimates for panel and

RANS approaches; it can be noted that monotone behaviours are observed and an obvious trend for grid/panel independence for both methods, at approximately 800 panels and 6 million elements for VLM and RANS, respectively. The setup of the RANS computations is similar to in the previous section including the meshing strategy, numerical scheme and turbulence model, whereas the incompressible steady state RANS equations are solved and discretised by a second-order upwind method and turbulence effects are accounted with the $\kappa - \omega \; SST$ model.

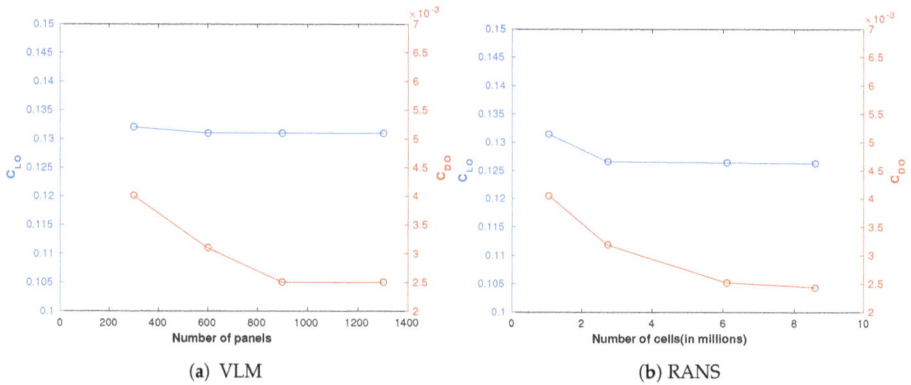

(a) VLM   (b) RANS

Figure 13. Grid/Panel dependence analysis for the isolated blended-wing-body aircraft configuration.

A two aircraft configuration is considered. The distribution of the quadrilateral panels and triangular surface elements for the configuration in depicted in Figure 14. Note, for the low-fidelity VLM approach, thin surfaces are considered.

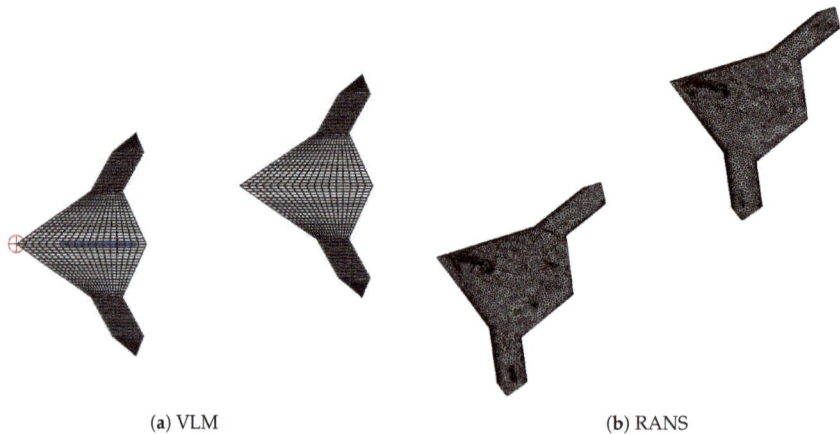

(a) VLM   (b) RANS

Figure 14. Panel distribution and surface gird distribution on the blended-wing-body two-aircraft configuration.

The procedure for determining the optimum position of the second aircraft to maximise the overall aerodynamic performance is set by fixing the lead aircraft and displacing the trailing aircraft in the Cartesian reference frame. With respect to y-z plane, a increment of $\delta y/c = 0.15$ and $\delta z/c = 0.075$ and three free-stream locations at $x/c = 1, 3$ and 4.5 are set, as shown in Figure 15.

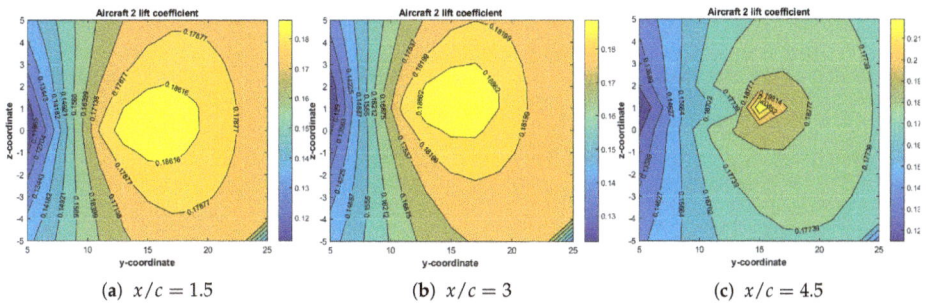

(a) $x/c = 1.5$ (b) $x/c = 3$ (c) $x/c = 4.5$

Figure 15. Lift coefficient contours of blended-wing-body configuration in double tandem formation flight of the trailing aircraft on y-z plane at three different stream-wise positions.

Figure 15 shows the lift of the second aircraft obtained by the VLM solver, with the second aircraft at three stream-wise (x/c) locations. The plots suggest that the highest lift is achieved at 11% wing-tip overlap of the aircraft and at lower x/c distance. Figure 16 shows the drag coefficient, in which it can be seen that the point of the highest drag coefficient drop, optimal position for aerodynamic improvement, is at the coordinates $x/c = 1.5$, $y/c = 2.5$ and $z/c = 0.15$. The drag coefficient is 0.0016 which suggests 30% reduction in drag as compared to an isolated aircraft. The overlap distance between the two wing-tips is about 10.2% of the aircraft span. When the overlapping positioning of the two aircraft increases, the drag also increases and the lift starts to drop. This is mainly because, in this region, the trail aircraft is under a direct influence of just the downwash created by the lead wing and hence no aerodynamic performance improvements are encountered.

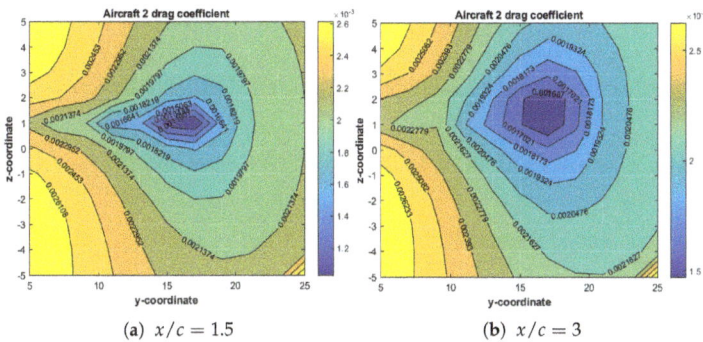

(a) $x/c = 1.5$ (b) $x/c = 3$

Figure 16. Drag coefficient contours of blended-wing-body configuration in double tandem formation flight of the trailing aircraft on y-z plane at two different stream-wise positions.

High-fidelity computations are performed at the optimum solution suggested by the VLM with the second aircraft positioned at $x/c = 1.5$, $y/c = 2.5$ and $z/c = 0.15$. The high-fidelity solutions predict a drag coefficient of 10% higher than the VLM estimation, as shown in Table 1. Note also that a 5% reduction in drag is obtained compared to the isolated aircraft at the same coordinate position. The "sweet spot", which marks the location for the maximum drag reduction for the trail wing behind the lead wing, is found to be around 10% wing-tip overlap for both VLM and the RANS.

Aerospace **2018**, *5*, 66

Table 1. Drag and lift coefficients for isolated and trail aircraft in formation for the blended-wing-configuration at optimal position.

Configuration	Aerodynamic Coefficients	VLM	RANS
Isolated aircraft	C_D	0.0022	0.0028
	C_L	0.1758	0.1290
Trail aircraft in formation	C_D	0.0016	0.0019
	C_L	0.1935	0.1450

4. Conclusions

A novel computational framework is proposed to determine flight formation and analyse the aerodynamic performance benefits using a multi-fidelity approach. The multi-fidelity approach for this study is based on a low-fidelity, Vortex Lattice Method and a high-fidelity, Reynolds Averaged Navier–Stokes solver. The developed modules for the low-fidelity solver are validated against the high-fidelity and experimental measurements on two distinct flow cases.

Initially, a rectangular wing NACA23012 is considered to validate the numerical framework; the main aim is to determine and to quantify the accuracy of each method. Two NACA23012 are considered in tandem configuration; the relative position of the second wing is altered at various longitudinal, horizontal and vertical positions. Computations are performed with both fidelity approaches and compared with wind-tunnel measurements [10]. The low-fidelity solver over-predicts lift and drag value compared with the available experiment, however the experiment and both fidelity solvers predict the same location of minimum drag and maximum lift. Moreover, the effect of lead's wing angle of attack on the two wing aerodynamic performance is studied by considering three angles within the linear range. It was found that no considerable variation of the lateral and vertical position occurs for maximum aerodynamic performance for these three angles. The last test case considered is the flow over a blended-wing-aircraft configuration; the multi-fidelity framework is employed to determine the best aerodynamic position of the trail aircraft, and a drag reduction of 30% is achieved, which agrees with similar studies. The developed multi-fidelity framework forms the basis for aerodynamic performance estimation of larger formation flights (<3 bodies) and the robust data generation tool for future multi-objective optimisation analyses.

Author Contributions: D.S., A.F.A., P.T., H.-S.S., A.T., S.M. and K.W.J. contributed equally to this manuscript.

Acknowledgments: The present research work was financially supported by the Centre for Computational Engineering Sciences at Cranfield University under Project Code EEB6001R. The authors would like to acknowledge the IT support and the use of the High Performance Computing (HPC) facilities at Cranfield University, U.K.

Conflicts of Interest: The authors declare no conflict of interest.

Abbreviations

The following abbreviations are used in this manuscript:

RANS	Reynolds Averaged Navier–Stokes
VLM	Vortex Lattice Method
AOA	Angle of Attack
AR	Aspect Ratio
Re	Reynolds Number
SIMPLE	Semi Implicit Pressure Linked Equations

References

1. Weimerskirch, H.; Martin, J.; Clerquin, Y.; Alexandre, P.; Jiraskova, S. Energy saving in flight formation. *Nature* **2001**, *413*, 697–698. [CrossRef] [PubMed]
2. Bajec, I.; Heppner, F. Organized flight in birds. *Anim. Behav.* **2009**, *78*, 777–789. [CrossRef]
3. Lissaman, P.; Shollenberger, C.A. Formation flight of birds. *Science* **1970**, *168*, 1003–1005. [CrossRef] [PubMed]
4. Hainsworth, F.R. Precision and dynamics of positioning by Canada geese flying in formation. *J. Exp. Biol.* **1987**, *128*, 445–462.
5. Beukenberg, M.; Hummel, D. Aerodynamics, Performance and Control of airplanes in formation flight. In Proceedings of the 17th Congress of the International Council of the Aeronautical Sciences, Stockholm, Sweden, 9–14 September 1990; Volume 2, pp. 1777–1794.
6. DeVries, L.D.; Paley, D.A. Wake estimation and optimal control for autonomous aircraft in formation flight. In Proceedings of the Conference on AIAA Guidance, Navigation, and Control (GNC), Boston, MA, USA, 19–22 August 2013; p. 4705.
7. Shin, H.; Antoniadis, A.; Tsourdos, A. Parametric Study on Formation Flying Effectiveness for a Blended-Wing UAV. *J. Intell. Robot. Syst.* **2018**, *10*.
8. Qiu, H.; Duan, H. Multiple UAV distributed close formation control based on in-flight leadership hierarchies of pigeon flocks. *Aerosp. Sci. Technol.* **2017**, *70*, 471–486. [CrossRef]
9. Pahle, J.; Berger, D.; Venti, M.; Duggan, C.; Faber, J.; Cardinal, K. An initial flight investigation of formation flight for drag reduction on the C-17 aircraft. In Proceedings of the AIAA Atmospheric Flight Mechanics Conference, Minneapolis, MN, USA, 13–16 August 2012; pp. 13–16.
10. Ayumu Inasawa, F.M.; Asai, M. Detailed Observation of Interaction of Wingtip Vortices in Close-formation Flight. *J. Aircr.* **2012**, *49*, 206–213. [CrossRef]
11. Blake, W.; Gingras, D.R. Comparison of predicted and measured formation flight interference effects. *J. Aircr.* **2004**, *41*, 201–207. [CrossRef]
12. Slotnick, J.P.; Clark, R.W.; Friedman, D.M.; Yadlin, Y.; Yeh, D.T.; Carr, J.E.; Czech, M.J.; Bieniawski, S.W. Computational aerodynamic analysis for the formation flight for aerodynamic benefit program. In Proceedings of the 52nd Aerospace Sciences Meeting, National Harbor, MD, USA, 13–17 January 2014; AIAA Paper; p. 1458.
13. Ning, A.; Flanzer, T.C.; Kroo, I.M. Aerodynamic performance of extended formation flight. *J. Aircr.* **2011**, *48*, 855–865. [CrossRef]
14. Blake, W.; Multhopp, D. Design, Performance and Modeling Considerations for Close Formation Fligh. In Proceedings of the 23rd Atmospheric Flight Mechanics Conference, Boston, MA, USA, 10–12 August 1998; AIAA Paper 98-4343.
15. Maskew, B. *Formation Flying Benefits Based on Vortex Lattice Calculations*; NASA CR-151974; NASA: Washington, DC, USA, 1977.
16. Wagner, E.; Jacques, D.; Blake, W.; Pachter, M. Flight test results of close formation flight for fuel saving. In Proceedings of the AIAA Atmospheric Flight Mechanics Conference and Exhibit, Monterey, CA, USA, 5–8 August 2002.
17. Wagner, H.E. An Analytical Study of T-38 Drag Reduction in Tight Flight Formation. Air Force Inst of Tech Wright-patterson Afb oh School of Engineering and Management. In Proceedings of the Atmospheric Flight Mechanics Conference, Denver, CO, USA, 14–17 August 2000.
18. Korkischko, I.; Konrath, R. Formation Flight of Low-Aspect-Ratio Wings at Low Reynolds Number. *J. Aircr.* **2016**, *54*, 1025–1034. [CrossRef]
19. Thien, H.; Moelyadi, M.; Muhammad, H. Effects of leaders position and shape on aerodynamic performances of V flight formation. *arXiv* **2008**, arXiv:0804.3879.
20. Gunasekaran, M.; Mukherjee, R. Behaviour of trailing wing(s) in echelon formation due to wing twist and aspect ratio. *Aerosp. Sci. Technol.* **2017**, *63*, 294–303.
21. Melin, T. A Vortex Lattice MATLAB Implementation for Linear Aerodynamic Wing Applications. Master's Thesis, Department of Aeronautics, Royal Institute of Technology (KTH), Stockholm, Sweden, 2000.
22. Wilcox, D. Formulation of the kappa-omega turbulence model revisited. *AIAA J.* **2008**, *46*, 2823–2838. [CrossRef]
23. Chien, K. Predictions of channel and boundary-layer flows with a low-reynolds-number turbulence model. *AIAA J.* **1982**, *20*, 33–38. [CrossRef]

24. Menter, F.R. Two-Equation Eddy-Viscosity Turbulence Models for Engineering Applications. *AIAA J.* **1994**, *32*, 1598–1605. [CrossRef]

25. Roache, P.J.; Ghia, K.N.; White, F.M. Editorial policy statement on the control of numerical accuracy. *J. Fluids Eng.* **1986**, *108*. [CrossRef]

26. Iglesias, S.; Mason, W. Optimum spanloads in formation flight. In Proceedings of the 40th AIAA Aerospace Sciences Meeting & Exhibit, Reno, NV, USA, 14–17 January 2002; AIAA Paper; Volume 258.

27. Ekaterinaris, J.; Platzer, M. Computational prediction of airfoil dynamic stall. *Prog. Aerosp. Sci.* **1998**, *33*, 759–846. [CrossRef]

aerospace

MDPI

Article

Simulation and Modeling of Rigid Aircraft Aerodynamic Responses to Arbitrary Gust Distributions

Mehdi Ghoreyshi [1,*], Ivan Greisz [2], Adam Jirasek [1] and Matthew Satchell [1]

[1]   High Performance Computing Research Center, U.S. Air Force Academy, Air Force Academy,
     CO 80840, USA; Adam.Jirasek@usafa.edu (A.J.); Matthew.Satchell@usafa.edu (M.S.)
[2]   Department of Mechanical and Aerospace Engineering, University of Colorado at Colorado Springs,
     Colorado Springs, CO 80918, USA; igreisz@uccs.edu
*   Correspondence: Mehdi.Ghoreyshi@usafa.edu

Received: 19 March 2018; Accepted: 14 April 2018; Published: 18 April 2018

Abstract: The stresses resulting from wind gusts can exceed the limit value and may cause large-scale structural deformation or even failure. All certified airplanes should therefore withstand the increased loads from gusts of considerable intensity. A large factor of safety will make the structure heavy and less economical. Thus, the need for accurate prediction of aerodynamic gust responses is motivated by both safety and economic concerns. This article presents the efforts to simulate and model air vehicle aerodynamic responses to various gust profiles. The computational methods developed and the research outcome will play an important role in the airplane's structural design and certification. COBALT is used as the flow solver to simulate aerodynamic responses to wind gusts. The code has a user-defined boundary condition capability that was tested for the first time in the present study to model any gust profile (intensity, direction, and duration) on any arbitrary configuration. Gust profiles considered include sharp edge, one minus cosine, a ramp, and a 1-cosine using tabulated data consisting of gust intensity values at discrete time instants. Test cases considered are a flat plate, a two-dimensional NACA0012 airfoil, and the high Reynolds number aero-structural dynamics (HIRENASD) configuration, which resembles a typical large passenger transport aircraft. Test cases are assumed to be rigid, and only longitudinal gust profiles are considered, though the developed codes can model any gust angle. Time-accurate simulation results show the aerodynamic responses to different gust profiles including transient solutions. Simulation results show that sharp edge responses of the flat plate agree well with the Küssner approximate function, but trends of other test cases do not match because of the thin airfoil assumptions made to derive the analytical function. Reduced order aerodynamic models are then created from the convolution integral of gust amplitude and the time-accurate responses to sharp-edge gusts. Convolution models are next used to predict aerodynamic responses to arbitrary gust profiles without the need of running time-accurate simulations for every gust shape. The results show very good agreement between developed models and simulation data.

Keywords: wind gust responses; computational fluid dynamics; convolution integral; sharp-edge gust; reduced order aerodynamic model

---

1. Introduction

Currently, the use of computational fluid dynamic (CFD) solutions is considered the state of the art in modeling unsteady nonlinear flow physics and offers an early and improved understanding of vehicle aerodynamics. In addition, these predictions can improve the accuracy of the structural analysis, performance predictions, and flight control design. This translates into reduced project risk and enhanced analysis of system performance prior to prototyping and first flight. The use of high fidelity aerodynamic simulations reduces the number of physical models and wind tunnel tests required until the design process converges to a design that optimizes an objective function, satisfies all mission requirements, and meets airworthiness standards. Specifically, the aircraft design should ensure the structural integrity in the presence of wind gusts of considerable intensity.

Calm atmospheric conditions rarely exist because of the continuous presence of random fluctuations in wind speed and direction. Wind gusts, in general, have continuous and random distributions and can occur in different directions. These gust profiles are described with the power spectral density technique. Sometimes, gust distributions can be represented as a discrete single function such as "one minus cosine". The impacts of these gusts (continuous and discrete) on the aerodynamics and structure of airplanes should be understood in order to improve the safety and functionality of designs economically. In particular, the aerodynamic responses to wind gusts are important for low-altitude and high-speed flight conditions or for large-size flexible aircraft with small natural frequencies [1]. FAR 25 (Federal Aviation Regulations, Part 25) requires that transport airplane structures should withstand the presence of static loads due to discrete gusts of 1-cosine with a length of 12.5 wing chord and a prescribed velocity at different flight envelope conditions [2]. In addition, the airplane should be certified when dynamic gust loads described by the power spectral density technique are encountered. A large factor of safety will make the structure heavy and less economical. Thus, the need for accurate prediction of aerodynamic gust responses is motivated by both safety and economic concerns.

Limited analytical solutions are available for gust predictions of two-dimensional test cases. Küssner [3] was the first to calculate the indicial lift response of a flat plate to a vertical sharp-edge gust in incompressible flow; his solution is known as the Küssner function. The Küssner function can be approximated by an exponential series as reported by Jones [4] for incompressible flow or by Mazelsky and Drischle [5] for compressible flow. As another example, Von Karman & Sears [6] derived the frequency response to a sinusoidal gust. Note that these theories are limited only to potential flow and a thin airfoil traversing gusts of low intensity. Techniques for gust generation in wind tunnels are complicated; initial efforts were made to generate an oscillating gust in the wind tunnel using a two-dimensional plunging airfoil mounted upstream of the test section [7]. Bennett and Gilman [1] described a gust generation method tested in the NASA Langley Transonic Dynamics Tunnel that used deflecting vanes. Compte-Bellot and Corrsin [8] used a hot wire technique to generate isotropic homogeneous turbulence in a wind tunnel. All these techniques are limited on the length scale of generated turbulence [9]. In addition, wind tunnel and flight testing are quite expensive and typically available late in the aircraft design stage. An alternative is the simulation of wind gust using computational methods with even further cost savings noticed by using a reduced order model.

The common industrial practice in computing the impacts of wind gusts on the aircraft structure is to use the double lattice method or strip theory [10]. However, these methods are not accurate in modeling nonlinear aerodynamics, e.g., the transonic regime, or when the viscosity effects are important. There are limited studies of wind gusts using solutions of the unsteady Reynolds-averaged Navier–Stokes (URANS) equations. This is because wind gust modeling is not typically available in most commercial CFD codes. A few methods of time-accurate simulations of wind gust are described in [11–18]. One approach is to add an artificial local gust velocity to each cell flow velocity [18,19]. The method can model any gust shape and the gust simulations can begin from the time that gust hits the most forward point of the vehicle. However, the technique is not available in commercial codes. Similar to analytical solutions, the method does not consider any effect of the vehicle on the gust profile. Another common approach to model gust in CFD is to impose the gust velocities at the inflow boundary of the computational domain. The main drawbacks include (1) the simulation time and cost needed to transport the gust from the inflow boundary to the vehicle and (2) the need of fine grid cells between the body and the upstream boundary. The second drawback can be mitigated by lowering the numerical dissipation in the code. Another method of modeling gust in CFD has been reported by Jirásek [13], who tried to mimic the experimental gust generators by using a source-based method in CFD to simulate a gust function of 1-cosine. The imposed gust velocity at the inflow boundary is the method used in this article. This method is simple to apply and capable of modeling body and gust interactions.

In more detail, the unstructured flow solver of COBALT is tested for simulating and modeling wind gusts. This code has been used at the U.S. Air Force Academy (USAFA) for flow field simulation of a variety of aerospace vehicles and for modeling nonlinear unsteady aerodynamics of maneuvering aircraft [20–25]. The code development began in February 1990 and has proven to be very robust and accurate since then. COBALT uses an arbitrary Lagrangian–Eulerian formulation and hence allows all translational and rotational degrees of freedom. This feature has been used to simulate aerodynamic behavior of a maneuvering aircraft [26] and to calculate the vehicle responses to a step change in the angle of attack and pitch rate for creating indicial response aerodynamic models [27–29]. Additionally, COBALT uses an overset grid method that allows the independent translation and rotation of each grid around a fixed or moving hinge line. This feature allows simulation of control surface deflections and calculation of aerodynamic indicial responses to a step change in the control surface deflection angle [30]. Once these indicial responses are calculated, a linear reduced order model (ROM) can be created to predict aerodynamic responses to arbitrary changes in the angle of attack, pitch rate, and control surface deflections through the superposition of indicial responses using Duhamel's integral. The model predictions are on the order of a few seconds and eliminate the need to run CFD for each new maneuver. Likewise, ROMs can be created using the vehicle responses to a sharp-edge gust in order to predict the aerodynamic responses to new gust profiles. This work is the first effort to demonstrate the COBALT capability in simulating and modeling aerodynamic responses of arbitrary configurations encountered wind gusts.

Test cases considered include a flat plate, a two-dimensional NACA0012 airfoil, and the high Reynolds number aero-structural dynamics (HIRENASD) [31,32] configuration, which resembles a typical large passenger transport aircraft. Test cases are assumed to be rigid and only longitudinal (upward) gust profiles are considered, though the developed codes can model any flow field and gust angled to a body. Gust profiles considered include a sharp edge, 1-cosine, a ramp up, and a 1-cosine using tabulated data consisting of gust intensity values at discrete time instants. Gust responses of two-dimensional test cases are calculated at Mach numbers in the range of 0.1–0.7. The lift responses to sharp-edge gusts are then compared with an analytical solution of Küssner. The effects of the Mach number on the gust responses are investigated for the two-dimensional cases as well. For the HIRENASD configuration, flow-field simulations are first validated with measurements corresponding to Mach 0.7 and Reynolds per length of 7 million. The sharp edge and 1-cosine gust responses of this vehicle are then calculated at Mach numbers of 0.1 and 0.7.

Reduced order aerodynamic models are created using the convolution integral and simulated vehicle indicial responses to a step change in the gust velocity (sharp-edge gust). These models are then used to predict responses to arbitrary gust distributions, e.g., 1-cosine. The ROM predictions are compared with time-accurate simulations of gust response (full-order models) to assess the accuracy of models. Finally, the computational costs of creating models are compared with the costs of time-accurate CFD solutions of gust responses. This article is organized as follows: First, gust modeling techniques are presented. Test cases and the flow solver are described next. The simulation and modeling results of gusts are then given followed by a discussion of the computational cost and concluding remarks.

2. Gust Modeling Methods

A wind gust is formed by "random fluctuations in the wind speed and direction caused by a swirling or eddy motion of the air" [33]. The random character of gust loads causes passenger inconvenience and discomfort. Gust-induced loads can significantly impact the aircraft stability and control (e.g., uncontrollable rolling moment) and structural integrity as well. Accurate predictions of gust loads is needed to assess the structural integrity and to estimate the aircraft stability and control characteristics. Additionally, the development of a gust alleviation system will benefit from accurate prediction of wind gusts [34].

Wind gusts can be characterized as vertical and horizontal. The horizontal gust can be divided into lateral and head-on gust types. Vertical gusts increases stress on the wing, fuselage, and horizontal tail structure. Horizontal lateral gusts change the loads acting on the fuselage, vertical tail, and pylon structure; finally, the horizontal head-on gust impacts the loads acting on the flap structure. For a transport aircraft, gusts are one of the largest source of structure fatigue. For a fighter aircraft, structure parts such as thin-outer wing or pylons should take into account gust loads in their designs.

Jones [35] have reviewed the history of gust modeling approaches. Typically, wind gust distributions have two trends: those which can be singled out as discrete gust profiles and those that occur in random pattern (turbulence). The random behaving gusts are described using Fourier's transform and power spectral density method. Earlier gust studies were limited to fixed and simple discrete gust distributions such as ramp and 1-cosine; these gust profiles are shown in Figure 1. These gusts are characterized by the gust gradient distance (H) and the gust maximum intensity ($w_{g,max}$). According to Jones [35], gust gradient distance is generally assumed to be 100 feet (in the United Kingdom) or 12.5 wing chords (in the United States). A 1-cosine gust profile is described with the following equation:

$$w_g = \frac{w_{g,max}}{2}(1 - \cos(\frac{\pi X}{H}))$$ (1)

where $w_{g,max}$ is the maximum gust velocity, x is the gust penetration distance, and H denotes the gust gradient distance. The gust penetration distance is time-dependent and depends on the gust-front speed (which equals aircraft speed in this work). For the purposes of this study, $\pi X/H$ is assumed to have a frequency of 1 Hz multiplied by the simulation time.

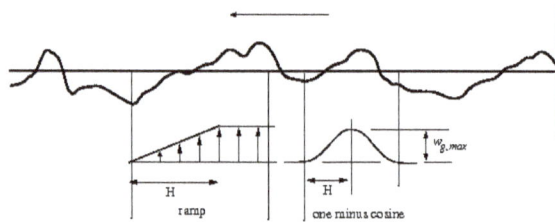

Figure 1. A continuous gust being singled out with a ramp and a one-minus-cosine gust profile. This figure was adapted from [36].

Though a step gust (sharp edge) is not a realistic gust and is very difficult (if not impossible) to being generated in a wind tunnel, the simulation of this gust gives valuable insight into the general characteristics of the aircraft response to arbitrary gust distributions. Assuming an airplane with quasi-steady aerodynamics and zero vertical motion encounters a sharp edge gust of intensity w_g, the incremental lift coefficient due to the gust is given by Fuller [37] as

$$\Delta C_L = C_{L\alpha}(w_g/V)$$ (2)

where $C_{L\alpha}$ is the lift curve slope, and w_g/V is the angle of attack in radians due to gust. The vertical acceleration increment, in units of g, due to this gust is

$$\Delta n = \rho C_{L\alpha} V w_g/(2W/S)$$ (3)

where W/S is the wing loading. According to this equation, the gust loads become important for aircraft with small wing loading and high flying speeds. This led to the first gust load regulations in

1934, which reduced the maneuver load factors in all passenger aircraft to the range of 2.5–4.0 g in order to take into account the incremental load factor when the airplane encounters a gust. Later, a factor of K was added to the sharp-edge gust equation to model the unsteady aerodynamic effects due to gust [37].

The exact unsteady lift response of a flat plate in incompressible flow to a unit step gust was calculated by Hans Georg Küssner [3]. His solution is named Küssner function. The Küssner function can be approximated by an exponential series in the form of [4]:

$$\Psi(s) = 1 - 0.5exp\left(-0.15s\right) - 0.5exp\left(-s\right) \tag{4}$$

where $\Psi(s)$ is the lift response of the flat plate encountering a step gust, and $s = 2Vt/c$ is the normalized time. The sharp-edge gust hits the plate leading edge at time zero. The lift-gust-curve slope is defined as follows:

$$C_{L_{wg}} = \frac{2\pi}{V}\Psi(s). \tag{5}$$

In this work, a convolution model is considered to predict linear aerodynamic responses to an arbitrary gust distribution. The model for predicting incremental lift coefficient responses to a gust takes the form of

$$\Delta C_L = \int_0^s C_{L_{wg}}(s - \sigma)\dot{w}_g(\sigma)d\sigma \tag{6}$$

where w_g is the forcing gust function (i.e., the 1-cosine) as a function of non-dimensionalized time and $C_{L_{wg}}(s)$ is the time-dependent lift-gust curve slope due to a sharp-edged gust excitation; $C_{L_{wg}}(s)$ can either being replaced with the Küssner approximation function or CFD data from the sharp edge wind gust simulation.

The following steps are to be taken to calculate sharp edge gust response in COBALT. Steady-state CFD data are first calculated for the calm case (zero gust velocities) at the desired Mach and Reynolds numbers. The gust response is then calculated by imposing a step change in the upward gust velocity at the inflow boundary. The gust will then travel with user-specified gust-front speed from a user-specified initial position outside the gridded domain. The gust travels over the airfoil/aircraft until a solution converges to its new steady conditions. The response function, $\Psi(s)$, is then computed by taking the differences between time-varying responses occurring after encountering the gust and the steady-state solution at calm conditions, and dividing it by the step magnitude of the gust velocity.

3. Flow Solver

The flow solver used for this study is the COBALT code [38] that solves the unsteady, three-dimensional, and compressible Navier–Stokes equations in an inertial reference frame. In COBALT, the Navier–Stokes equations are discretized on arbitrary grid topologies using a cell-centered finite volume method. Second-order accuracy in space is achieved using the exact Riemann solver of Gottlieb and Groth [39], and least squares gradient calculations are achieved using QR factorization. To accelerate the solution of the discretized system, a point-implicit method using analytic first-order inviscid and viscous Jacobians is used. A Newtonian sub-iteration method is used to improve the time accuracy of the point-implicit method. Tomaro et al. [40] converted the code from explicit to implicit, enabling Courant–Friedrichs–Lewy (CFL) numbers as high as 10^6. Some available turbulence models are the Spalart–Allmaras (SA) model [41], the Spalart–Allmaras with rotation/curvature correction (SARC) [41], Wilcox's k-ω model [42], and Mentor's SST (Shear Stress Transport) model [43].

To model an arbitrary wind gust, the "user-defined" boundary condition capability of COBALT is used. Note that this capability allows gust simulation for any configuration of interest (two- or three-dimensional cases). The user should then provide a subroutine (written in Fortran 90) to treat any boundary condition of the computational grid with customized functions. As an example, in normal (no gust) simulation of the flow around an airfoil, boundary conditions of far-field and no-slip wall available in COBALT are selected. The far-field conditions of COBALT use Riemann invariants which enforce the specified flow values at an inflow boundary, but no flow value is enforced at the outflow boundary. For the gust simulation purposes, the far-field boundary condition is replaced with a user-defined one. A subroutine should therefore be provided that defines inflow conditions. The subroutine should have free-stream conditions (Mach, pressure, temperature, etc.) for both the calm and gust conditions. In the calm case, the flow velocity components, pressure, and temperature are specified and are invariant in time. The gust velocities are set to zero as well. These data correspond to the desired Mach and Reynolds numbers. In order to verify the accuracy of scripts, the calm conditions can be compared with normal COBALT simulations using Riemann invariants at the free-stream. For the similar Mach and Reynolds numbers and flow angles, the user-defined solutions (e.g., integrated forces and moments) should exactly match those calculated using far-field boundary condition.

In the case of a gust, the gust conditions are specified in a time-varying manner appropriate for the type of gust to be modeled, i.e., step-function, cosine, or any arbitrary form. For a step-function gust simulation, the vertical gust component is specified that corresponds to the magnitude of the gust of interest. The gust will travel with user-specified gust-front speed from a user-specified initial position outside the gridded domain. Currently, the gust must start outside of the flow domain to ensure proper gust front formulation. The gust script was originally written by William Strang of Cobalt Solutions, LLC., Springfield, OH, USA, but has been built upon to simulate various gust profiles. Arbitrary gust profiles can be modeled using tabulated data in which gust intensity at different time/positions are listed. In order to assess the scripts, 1-cosine gust profile is considered and is modeled using Equation (1) and tables. The cosine gust is defined by using the number of gust cycles, the frequency of the gust, and the phase. All values in the script are non-dimensionalized using the speed of sound before applying Salas' farfield boundary condition [44] as well. The position of the gust front during the simulation depends on time and velocity, both of which depend upon where the gust begins and propagates to. Finally, the gust responses using Equation (1) and tabulated data are then compared.

4. Test Cases

4.1. Two-Dimensional NACA0012 Airfoil

This airfoil is selected because of the available static and dynamic experimental data. The solution of a sharp-edge gust traveling over this airfoil is compared with the Küssner approximation function and might also be compared with a flat plate response. The computational grid of this airfoil was generated using POINTWISE. The grid has 72,852 cells with a circular free-stream region with a radius of 50*c*. The grid cells around the airfoil surface can be seen in Figure 2. Boundary conditions are user-defined for the free-stream, and no-slip wall for the airfoil. The chord length of the NACA0012 airfoil is 1 m. The pitch axis and moment reference point is 0.25 m past the leading edge.

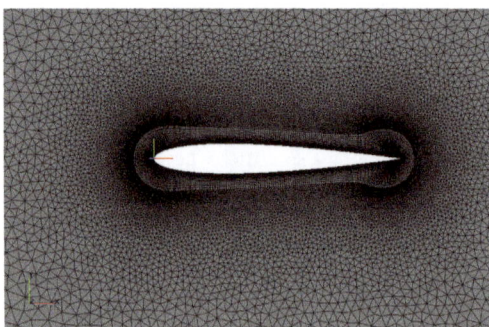

Figure 2. NACA0012 grid. Grid consists of 72,852 cells. Grid has pismatic sublayers around airfoil and tetrahedral cells in the outer region.

The computational grid had been previously validated with static and dynamic experimental data in [27]. The Spalart–Allamaras turbulence model is used for all two-dimensional simulations.

4.2. The HIRENASD Model

After the two-dimensional testing were completed, a three-dimensional test case was chosen to be accomplished using the HIRENASD configuration, a configuration funded by the German Aerospace Center (DLR) [32]. The HIRENASD configuration resembles a typical large passenger transport aircraft wing with a leading sweep angle of 34° with a supercritical wing profile of BAC 3-11. The wind tunnel model has a half span of 1.28571 m, a mean aerodynamic chord of 0.3445 m, and a wing reference area of 0.3925 m^2. The computational grid used can be seen in Figure 3 and has approximately 4 million cells. The Spalart–Allmaras with the rotational correction (SARC) turbulence model was selected for the three-dimensional computational case tested.

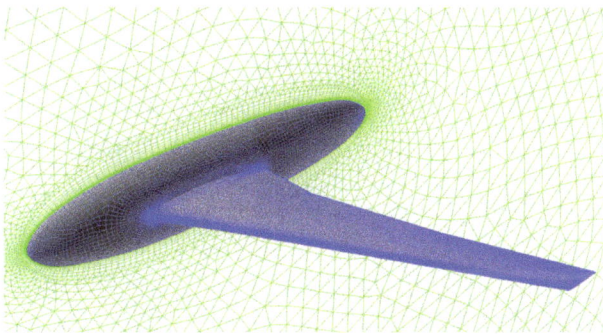

Figure 3. HIRENASD surface grid. Symmetry plane is shown in green.

Experimental data of the HIRENASD configuration is available from the European Transonic Wind tunnel (ETW), which is a cryogenic facility with a closed circuit with nitrogen gas as a fluid. The measurements include a pressure coefficient data at different span wise positions given in [32]. These measurements correspond to Mach 0.7 and a Reynolds per length of 7 million and is considered in this work as validation for the computational results.

A viscous grid is generated around this configuration. The grid has three boundary conditions: far-field, symmetry, and no-slip wall. The grid is shown in Figure 3 has prism layers around the walls and tetrahedra cells elsewhere. This grid has around 4 million cells. Figure 4 shows the grid calculated $y+$ values at the upper and lower surfaces at Mach 0.7. Figure 5 compares COBALT predictions of pressure coefficient data with experimental data measured at tap positions. The simulations use the SARC turbulence model and run for Mach 0.7, a Reynolds per length of 7 million, and an angle of attack of 1.5°. Figure 5 shows that the CFD data agree well with the experimental data at most locations, in particular at the lower surfaces. The biggest discrepancy is seen at the upper surface, at positions near the wing tip, and at the wing's leading edge, where CFD predictions over-estimate experimental pressure data. Note that, in simulations, the body is rigid, while this is considered a flexible model in the wind tunnel. Overall, these results confirm the validity and accuracy of computational methods used in this work for the HIRENASD configuration and any further results using this configuration.

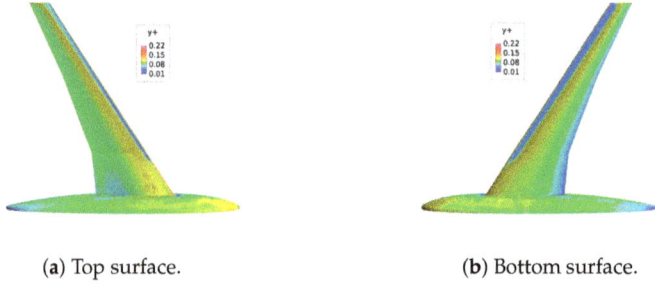

(a) Top surface.   (b) Bottom surface.

Figure 4. HIRENASD $y+$ distribution at Mach 0.7.

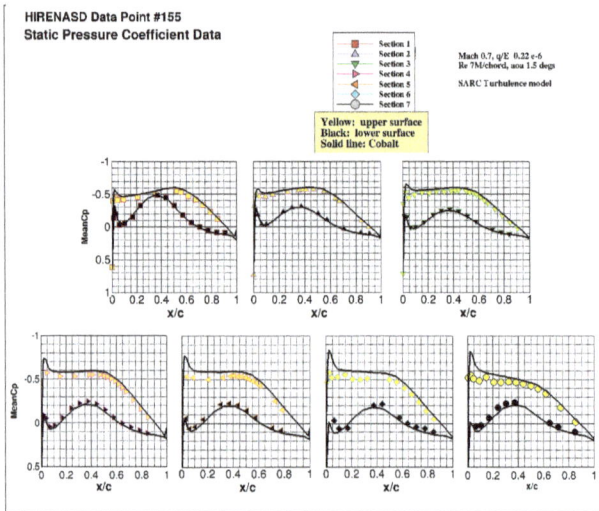

Figure 5. Validation of HIRENASD CFD models with experimental data. For section position, refer to [32]. In the figures, solid lines show COBALT predictions using SARC turbulence models. The upper surfaces are shown with yellow-filled markers. The black-filled markers show the bottom surface. Simulations and experimental data correspond to Mach 0.7, a Reynolds per length of 7 millions, and an angle of attack of 1.5°.

5. Results and Discussions

Computational resources were provided by the DoD's High Performance Computing Modernization Program. All airfoil simulations are run on the U.S. Air Force Research Laboratory (AFRL) 'Lightning' Cray XC30 system (2370 computing nodes with 24 cores per node running two Intel Xeon E5-2697v2 processors at a base core speed of 2.7 GHz with 63 GBytes of RAM available per node). HIRENASD simulations were run on the Engineering Research Development Center's (ERDC's) 'Topaz' SGI ICE X System (3456 computing nodes with 36 cores per node running two Intel Xeon E5-2699v3 processors at a base core speed of 2.3 GHz with 117 GBytes of RAM available per node). All gust simulations resumed from a calm condition solved with 2500 time steps to allow the flow around the vehicle to become steady. Second-order accuracy in time and three Newton sub-iterations are used as well. NACA0012 airfoil simulations used SA turbulence model and a time step of 1×10^{-4} s, while the HIRENASD configuration used the SARC turbulence model and a time step of 1×10^{-4} s.

5.1. Validation and Verification of User-Defined Codes

The first set of results presents the verification plots for the codes developed by the authors. In COBALT, the free-stream region is modeled with Riemann invariants that are integrated into the solver. In this work, however, the free-stream boundary of the computational grid is represented by a user-defined code. This code was scripted for modeling calm and gust simulations over any arbitrary configuration. In order to verify the scripts, Figure 6a shows the lift coefficient of NACA0012 airfoil calculated from (1) the far-field boundary condition (2) and the user-defined codes run for calm conditions. In the first method, the free-stream angle of attack, Mach number, pressure, and temperature are defined in the COBALT main input file. Simulations shown in Figure 6a correspond to Mach number of 0.1 and the static pressure and temperature values in order to have a Reynolds number of 5.93×10^6. The angle of attack changes from zero to $20°$. In the second method, the free-stream data of the main input file are discarded; instead, they are given in the user-developed script. Flow velocity components are specified to have a Mach number of 0.1 and an angle of attack in the range of zero to $20°$. Static pressure and temperature are defined to have a Reynolds number of 5.93×10^6. For the calm conditions, gust velocities are set to zero as well. Figure 6a shows that the calculated lift coefficient values from the used-defined code perfectly match with those obtained from simulations using the far-field boundary condition.

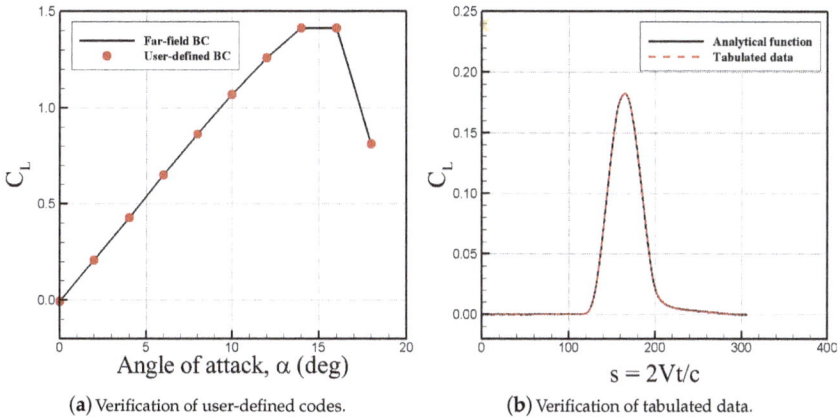

(a) Verification of user-defined codes.   **(b)** Verification of tabulated data.

Figure 6. Verification of user-defined codes. In (**a**), lift coefficient values are plotted vs. the angle of attack using the far-field boundary condition of COBALT and the user-defined scripts of calm conditions. In (**b**), a 1-cosine vertical gust profile is simulated using the analytical function and tabulated data. The 1-cosine function has an amplitude of 1 m/s, one cycle, and a gust gradient distance of about 17c.

For modeling an arbitrary gust distribution, the script written by the authors defines the gust data in tabular format. The table consists of gust velocity values at specific time instants. Figure 6b shows NACA0012 airfoil lift coefficient responses against a 1-cosine gust profile. The responses are from (1) an analytical function scripted and (2) tabulated data of the 1-cosine function. The gust profile has one cycle, amplitude of 1 m/s, and a gradient distance (H) of about 17c. Figure 6b shows that solutions from both methods of defining the gust perfectly match, confirming the capability of the code to model any gust profile. Notice that, in Figure 6b, gust should travel 60 m with a speed of about 34 m/s before it hits the airfoil leading edge. One cost associated with modeling wind gusts with imposing the gust velocities at the inflow boundary is the computational costs needed to propagate the wind across the flow domain until it has interacted with the object.

Finally, the sharp-edge predictions from COBALT at inflow Mach numbers of 0.1 and 0.5 are compared with analytical solutions of a thin airfoil at zero Mach number and Mach 0.5 in Figure 7.

In both runs, the gust velocity is 1 m/s, and the gust travels from $x = -30m$. Analytical approximation data correspond to Jones [4] for zero Mach number and Mazelsky and Drischle [5] for $M = 0.5$. In order to have a fair comparison, a flat plate of 1 m long was modeled in the solver. The computational domain has a circular free-stream around the plate with a radius of 25 m as shown in Figure 7a.

Overall, the calculated CFD trends and values (normalized by gust intensity and free-stream velocity) match very well with analytical solutions. However, CFD simulations at Mach 0.1 show oscillating behavior as the gust approaches the plate and crosses over it. This is possibly due to the strong interactions between the plate and the gust at small Mach numbers or the large ratios of gust to free-stream velocity. Another interesting observation is that the analytical solutions show zero lift increment until the time that gust hits the most forward point of the plate. In contrast, CFD lift data begin to rise before the gust hits the plate. Figure 7 shows that the sharp edge responses reach different steady-state values depending on the Mach number; this value increases by increasing the Mach number due to compressibility effects. In addition, the $M = 0.5$ response trend shows a longer transient behavior to reach its steady-state value than $M = 0.1$.

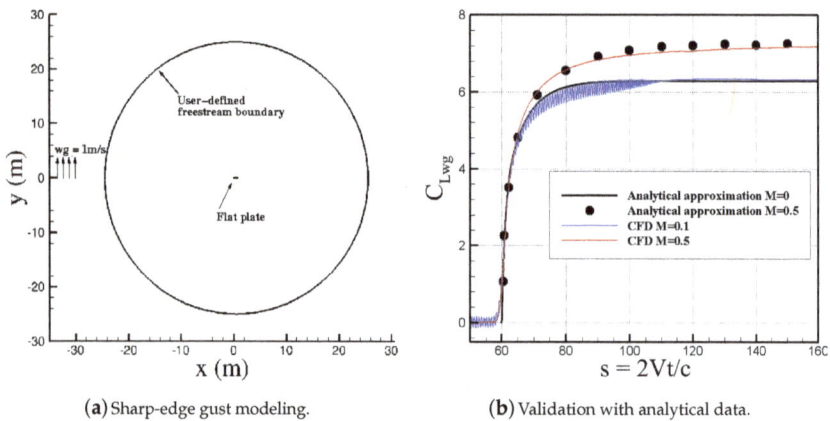

(**a**) Sharp-edge gust modeling. (**b**) Validation with analytical data.

Figure 7. CFD and analytical data of the lift-gust curve slope due to a unit sharp-edge gust at $M = 0.1$ and 0.5. The test case is a flat plate 1 m long. C_{Lwg}, the lift-gust curve slope, is defined as $\dfrac{2\pi}{V.wg}[C_L(wg \neq 0, t) - C_L(wg = 0, t = 0)]$ and has units of per radian. $M = 0$ analytical data are from Jones approximation function [4]; analytical data at $M = 0.5$ are from Mazelsky and Drischle [5].

5.2. Simulation Results of NACA0012 Airfoil

Firstly, the NACA0012 airfoil was used to simulate unit sharp edge gust responses across the flow domain for Mach numbers 0.1, 0.3, and 0.5. The calculated C_{Lwg} (lift-gust curve slope) values at these Mach numbers can be seen in Figure 8 and are shown with the analytical approximation function at a zero Mach number. Figure 8 shows that the airfoil response at $M = 0.1$ does not match with the analytical solution as it is valid only for thin airfoils. The responses reach larger steady-state values, as the Mach number increases due to compressibility effects. In addition, Figure 8 shows that the gust effects on airfoil could be seen even before the gust hits the airfoil leading edge. Oscillating behavior could be seen for the airfoil response at $M = 0.1$ as well.

Figure 8b–e show the flow solutions of a unit sharp edge gust traveling over the airfoil at a Mach number of 0.1 and 0.5. In these figures, gust travels from left to right. Inspecting velocity data shows that the sharp edge gust is not a very exact representative of the mathematical model because there is no visible gust front where, at its right side, vertical velocity is zero and then becomes 1 m/s. Instead, the vertical velocity changes from zero to 1 m/s over a small distance. Note that modeling an exact sharp edge in CFD is not possible because of discontinuity in the flow field. This might explain why airfoil responds sooner to gusts than do analytical solutions. Additionally, Figure 8b–e show that the gust profile will be affected as it approaches and crosses over the airfoil. However, the interaction time between airfoil and gust is much shorter for larger Mach numbers. These interactions cause the oscillation behavior seen at small Mach number plots.

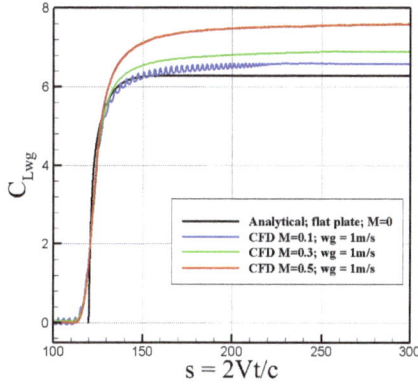

(**a**) NACA0012 airfoil sharp-edge gust responses.

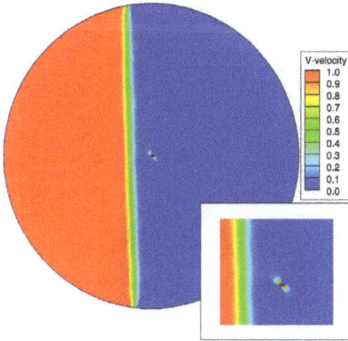

(**b**) $M = 0.1$ and $s = 105$.

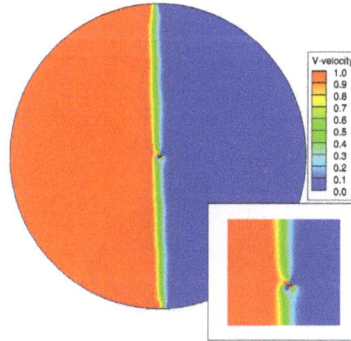

(**c**) $M = 0.1$ and $s = 119$.

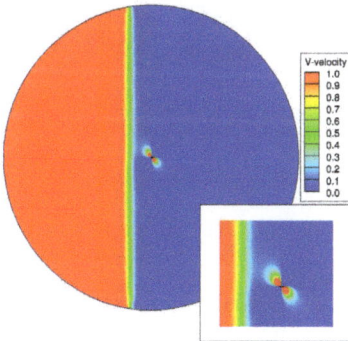

(**d**) $M = 0.5$ and $s = 105$.

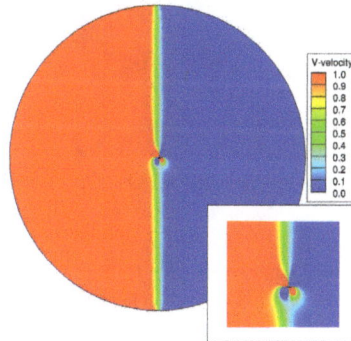

(**e**) $M = 0.5$ and $s = 119$.

Figure 8. NACA0012 airfoil responses to a unit sharp-edge gust at Mach numbers of 0.1, 0.3, and 0.5.

Figure 9 shows 1-cosine gust responses using an analytical equation. Figure 9a presents lift coefficient increments from calm conditions, ΔC_L, for Mach numbers 0.1, 0.3, and 0.5 at wind gust speeds of 1 m/s, 3 m/s, and 5 m/s, respectively. In all simulations, the gust front travels from $x = -60m$, which is outside the inflow boundary towards the airfoil with free-stream velocity, V, at a zero angle of attack. Though w_g/V is similar in these runs, but the increments are bigger for larger Mach number again due to compressibility effects. Notice that gust hits the airfoil at different times because of different gust front speeds, such that the larger the Mach number is, the sooner the gust impacts the airfoil. Figure 9b–c show the wind gust as it propagates over the flow domain until it reaches the airfoil. The vertical velocity data show the 1-cosine gust profile with a centered maximum velocity. Figure 9c shows that gust profile changes due to its interaction with the airfoil flow field.

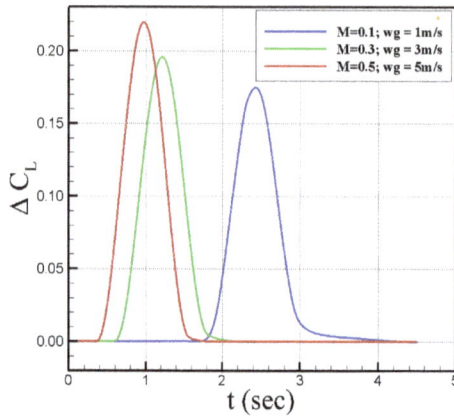

(a) NACA0012 airfoil 1-cosine gust responses.

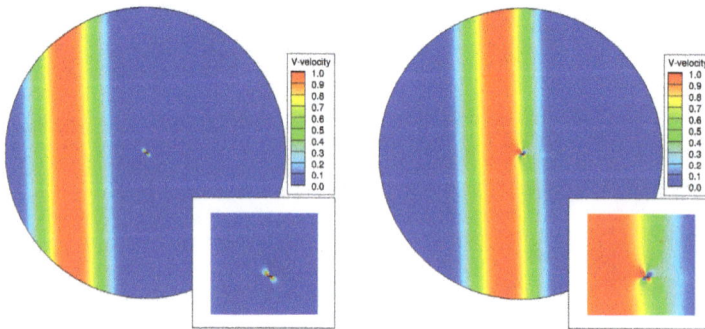

(b) $M = 0.1$ and $s = 105$.

(c) $M = 0.1$ and $s = 146$.

Figure 9. NACA0012 airfoil responses to a 1-cosine gust at Mach numbers of 0.1, 0.3, and 0.5. ΔC_L is defined as $C_L(wg \neq 0, t) - C_L(wg = 0, t = 0)$.

The ramp gusts can be seen in Figure 10, with Figure 10a displaying the ΔC_L (lift coefficient increments from calm conditions) for Mach numbers 0.1, 0.3, and 0.5 at wind gust speeds of 1 m/s, 3 m/s, and 5 m/s, respectively. In Figure 10b–c, the ramp wind gust is seen to propagate nicely over the flow domain until it reaches the airfoil.

Plotted in Figure 11 is the CFD simulation data, the quasi-steady model using Equation (2), and the convolution model (6) for both cosine and ramp gust shapes. The convolution model is able to use the sharp edge gust simulation results, and its predictions perfectly match with the CFD data obtained from COBALT, matching the peak magnitude, shape, phase (with respect to the wind gust), and transient effects towards the end. The quasi-steady model is able to approximate everything but is not nearly as accurate. It lacks phase correction, accurate magnitudes, and transient effects such that the smaller the Mach number, the more poorly ΔC_L matches the CFD or convolution model data.

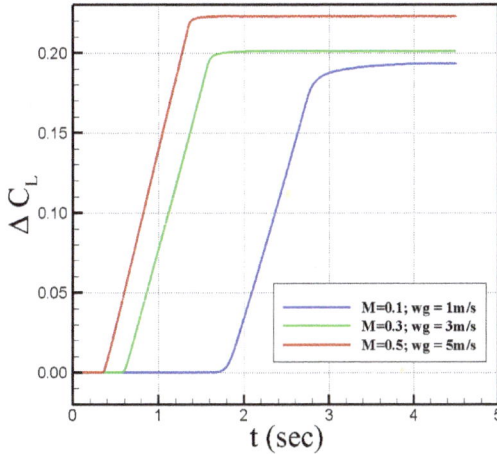

(a) NACA0012 airfoil ramp gust responses.

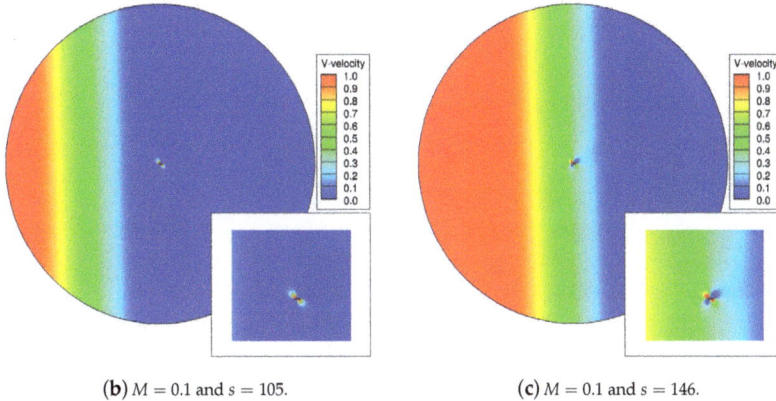

(b) $M = 0.1$ and $s = 105$.

(c) $M = 0.1$ and $s = 146$.

Figure 10. NACA0012 airfoil responses to a ramp gust at Mach numbers of 0.1, 0.3, and 0.5. ΔC_L is defined as $C_L(wg \neq 0, t) - C_L(wg = 0, t = 0)$.

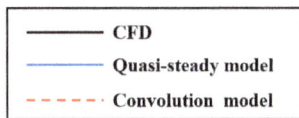

(a)

Figure 11. *Cont.*

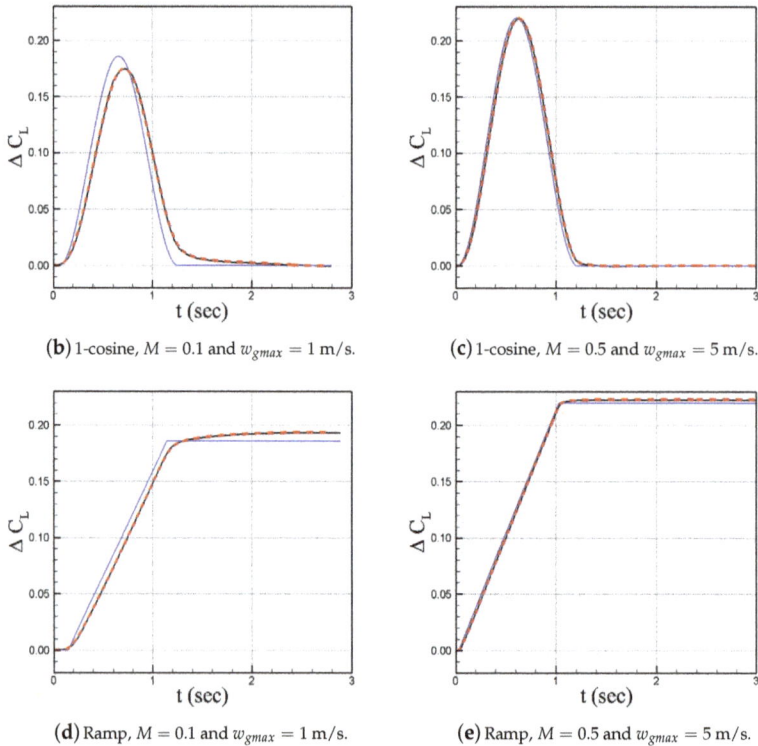

(**b**) 1-cosine, $M = 0.1$ and $w_{gmax} = 1$ m/s.

(**c**) 1-cosine, $M = 0.5$ and $w_{gmax} = 5$ m/s.

(**d**) Ramp, $M = 0.1$ and $w_{gmax} = 1$ m/s.

(**e**) Ramp, $M = 0.5$ and $w_{gmax} = 5$ m/s.

Figure 11. NACA0012 airfoil gust response modeling.

5.3. Simulation Results of HIRENASD Configuration

The HIRENASD configuration underwent test environments similar to those that of the NACA0012 simulations. At Mach numbers of 0.1 and 0.7 and a wind gust of 1 m/s, a unit sharp edge gust was tested, and the $C_{L_{wg}}$ (lift-gust curve slope) can be seen in Figure 12, where the result is shown with those of the NACA0012 simulations. The HIRENASD gust begins from $x = -40m$ and the geometry has a mean aerodynamic chord different from that of the airfoil. The HIRENASD plots in Figure 12 are therefore translated to match the starting point of the NACA0012 airfoil. Figure 12 shows that the HIRENASD indicial response trends are very different from those of the NACA0012 airfoil. Interestingly, it has a shorter build-up time to a steady-state solution. Note that the steady-state values are lower than those found for the airfoil due to wing tip losses and other effects. The only method to estimate 3D aircraft responses is CFD, since analytical solutions are not applicable and it is not feasible to test a sharp edge gust in a wind tunnel.

In more detail, Figure 13 depicts a sharp edge gust flowing across the flow domain on the HIRENASD configuration. The vertical velocity is depicted on the symmetry plane, and the pressure coefficient is presented on the HIRENASD geometry. In each successive subplot, the gust is seen propagating from right to left until it has almost made complete contact in Figure 13c,d and reaches a steady-state value by Figure 13f. The gust shape again changes as it interacts with the vehicle. The wing upper surface shows pressure changes as the gust crosses over it.

Additionally, 1-cosine and ramp wind gusts are tested at Mach 0.7 with upward wind gusts of 1 m/s and 7 m/s coming from the inflow boundary at $x = -40m$. The results of the CFD predictions are shown in Figure 14, where the trends of ΔC_L caused by the cosine and ramp simulations appear to reacting similarly to how the NACA0012 simulations did. The convolution model is also applied by using the data from the sharp edge gusts to predict the cosine and ramp gusts at Mach 0.7 with wind gust speeds of 1 m/s and 7 m/s. The results show that the model is able to accurately predict time-accurate simulation data, as seen in Figure 14b–e.

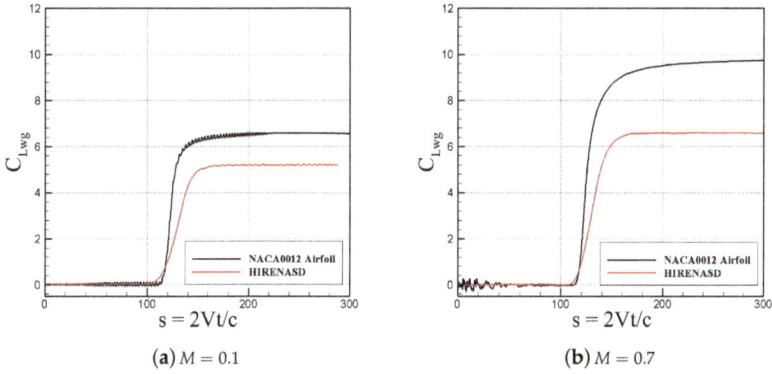

(**a**) $M = 0.1$ (**b**) $M = 0.7$

Figure 12. HIRENASD and NACA0012 airfoil responses to a unit sharp-edge gust at Mach numbers of 0.1 and 0.7.

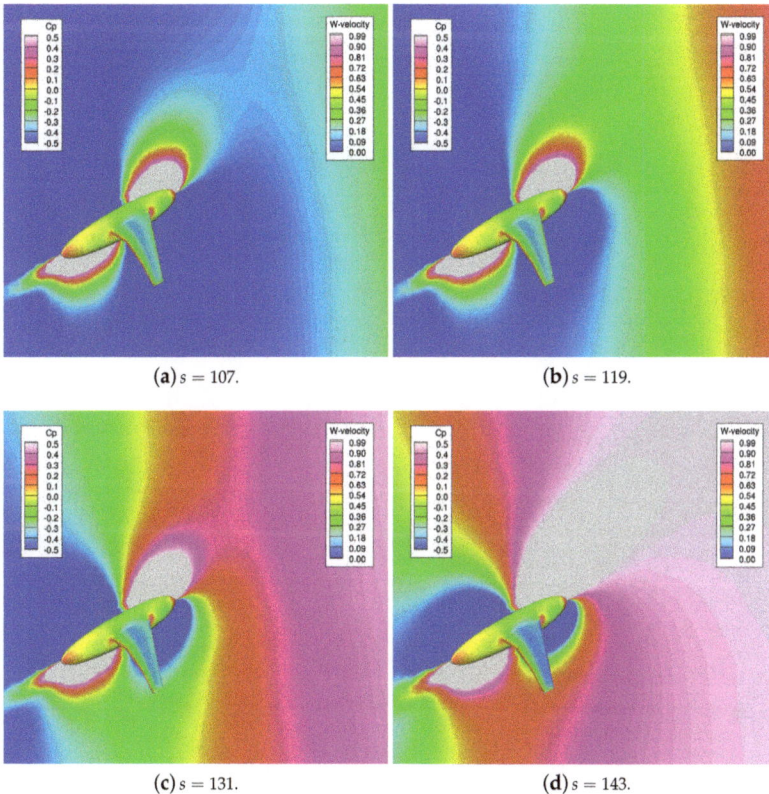

(**a**) $s = 107$.   (**b**) $s = 119$.

(**c**) $s = 131$.   (**d**) $s = 143$.

Figure 13. *Cont.*

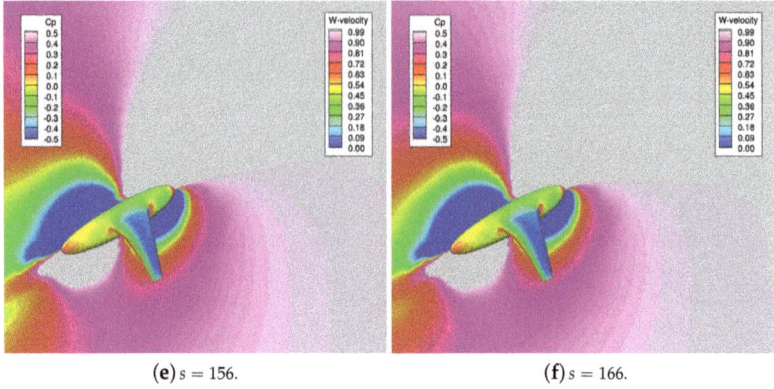

(**e**) $s = 156$.   (**f**) $s = 166$.

Figure 13. HIRENASD responses to a unit sharp-edge gust at Mach number of 0.1. Symmetry plane is colored by vertical velocity. Wing is colored by pressure coefficient.

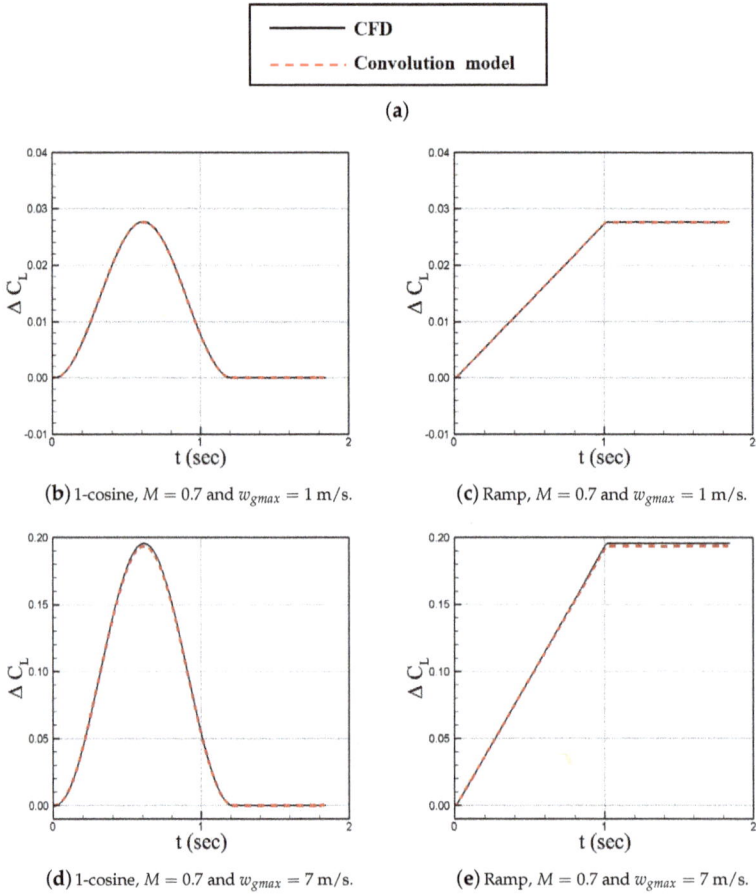

(**b**) 1-cosine, $M = 0.7$ and $w_{gmax} = 1$ m/s.   (**c**) Ramp, $M = 0.7$ and $w_{gmax} = 1$ m/s.

(**d**) 1-cosine, $M = 0.7$ and $w_{gmax} = 7$ m/s.   (**e**) Ramp, $M = 0.7$ and $w_{gmax} = 7$ m/s.

Figure 14. HIRENASD gust response modeling.

5.4. Computational Costs

Running a simulation where the wind gust has to start outside of the flow domain and propagate across until it reaches the vehicle can become computationally costly. For the flat plate example, the sharp edge gust propagates from outside its domain with $x = -30m$ at zero time. The gust front speed was set to free-stream speed. Figure 7 shows that CFD prediction matches with the analytical approximation data. There is a "time-to-hit-gust" in the CFD data; the data obtained during this time are not of any significant importance in this work.

For the NACA0012 airfoil at $M = 0.1$, a starting gust position $x = -60m$, and a time step of 1×10^{-4} s, it takes approximately 17,632 time steps in COBALT for the gust front to reach $x = 0$, which is the location of the leading edge of the airfoil. The computational cost is inversely proportional to both the size of the domain and the Mach number of the flow field being investigated. The computational cost increases as the size of the object increases and the size of the gust increases in order to obtain a complete representation of flow features that develop after the gust has passed and reached steady-state values.

However, in order to remain consistent in making scripts and handling data, 2500 startup (zero gust speed) iterations followed by 47,500 simulation time steps were chosen for the two-dimensional NACA0012 cases. This design choice led to simulations using approximately 1.5 CPU hours (a few minutes on 24 cores) for the startup iterations and 50–72 CPU hours (2.5 h of real time on 24 cores) for the wind gust simulations. The total CPU hour usage for the 2D portion of this study to obtain sharp edge gust and cosine simulation data was approximately 2952 CPU hours on the "Lightning" system, but this did not include the CPU time used in diagnosing errors or ramp simulations. The cost of creating ROM was only 246 CPU hours.

The 3D HIRENASD simulation took significantly more processing power to run a single simulation due to the increased number of elements. In obtaining the results, 2500 startup iterations used approximately 1500 CPU hours, and the remaining 15,000 iterations to obtain the wind gust simulation data used approximately 8300 CPU hours (4 hours of wall clock time on 2016 cores) for a total of 19,600 CPU hours used on the 'Topaz' system to calculate the sharp edge and cosine gust simulations, and this does not include the CPU time used in diagnosing errors. This particular case could net a total of 50% savings or more, i.e., 9800+ CPU hours, if the sharp edge gust only were calculated and if the convolution model were applied to obtain the cosine gust.

6. Conclusions

A user-defined script written for COBALT was developed and used to model wind gusts by forming them at the far-field boundary and allowing them to propagate across the flow domain. The script was proven to offer no difference in answers calculated by traditional methods of running COBALT simulations. Sharp edge, ramp, and cosine gusts can now be modeled using COBALT, offering enough precision to match calculations from other works and analytical solutions.

By using a sharp edge gust and the convolution model, the coefficient of lift was estimated and matched to a cosine gust simulated in COBALT. The convolution model was found to be as accurate as the simulations themselves and would finish calculations on the order of a minute or two compared to the many CPU hours needed in finding a solution with COBALT.

Acknowledgments: This article was approved for Public Release with Distribution A. This material is based in part on research sponsored by the US Air Force Academy under agreement number FA7000-17-2-0007. The U.S. Government is authorized to reproduce and distribute reprints for governmental purposes notwithstanding any copyright notation thereon. The views and conclusions contained herein are those of the authors and should not be interpreted as necessarily representing the official policies or endorsements, either expressed or implied, of the organizations involved with this research or the U.S. Government. The authors would like to thank William Strang of Cobalt Solutions, LLC., in Springfield, Ohio, for giving us a script to model wind gusts in the Cobalt flow solver. The numerical simulations have been performed on the HPCMP's Topaz cluster located at the ERDC DoD Supercomputing Resource Center (DSRC) and the Air Force Research Laboratory (AFRL) Lightning cluster.

Author Contributions: Mehdi Ghoreyshi and Ivan Greisz run simulations, generated all plots, and prepared this article. Adam Jirasek generated the computational grids and provided a script to model gust profiles using tabulated data. Matthew Satchell is the program manager of this research at USAFA.

Conflicts of Interest: The authors declare no conflict of interest.

Nomenclature

a	acoustic speed, $\mathrm{m\,s^{-1}}$
c	chord length
CFD	computational fluid dynamics
CPU	central processing unit
C_L	lift coefficient
$C_{L\alpha}$	lift curve slope
$C_{L_{wg}}$	lift-gust curve slope
Cp	pressure coefficient, $(p - p_\infty)/q_\infty$
CREATE	Computational Research and Engineering Acquisition Tools and Environments
DDES	Delayed Detached Eddy Simulation
DoD	Department of Defense
ETW	European Transonic Windtunnel
H	gust gradient distance, m
HIRENASD	high Reynolds number aero-structural dynamics
HPC	High Performance Computing
ΔL	incremental lift
M	Mach number, V/a
Δn	vertical acceleration increment
p	static pressure, $\mathrm{N/m^2}$
p_∞	free-stream pressure, $\mathrm{N/m^2}$
$\Psi(s)$	Küssner exponential series approximation or sharp edge gust data
q_∞	free-stream dynamic pressure, $\mathrm{N/m^2}$
s	normalized time
S	wing area
SA	Spalart–Allmaras
SARC	Spalart–Allmaras with rotational and curvature correction
RANS	Reynolds Averaged Navier Stokes
t	time, s
USAFA	United States Air Force Academy
V_∞	free-stream velocity, $\mathrm{m\,s^{-1}}$
u,v,w	velocity components, $\mathrm{m\,s^{-1}}$
w_g	gust vertical velocity, $\mathrm{m\,s^{-1}}$
$w_{g,max}$	maximum gust vertical velocity, $\mathrm{m\,s^{-1}}$
X	gust penetration distance, m
x,y,z	grid coordinates, m
$y+$	non-dimensional wall normal distance

References

1. Bennett, R.M.; Gilman, J., Jr. A Wind-Tunnel Technique for Measuring Frequency-Response Functions for Gust Load Analyses. *J. Aircr.* **1966**, *3*, 535–540. [CrossRef]
2. Noback, R. Comparison of Discrete and Continuous Gust Methods for Airplane Design Loads Determination. *J. Aircr.* **1986**, *23*, 226–231. [CrossRef]
3. Küssner, H.G. Zusammenfassender Bericht über den instationären Auftrieb von Flügeln. *Luftfahrtforschung* **1936**, *13*, 410–424.
4. Jones, R.T. *The Unsteady Lift of a Wing of Finite Aspect Ratio*; Technical Report; National Advisory Committee for Aeronautics: Langley Field, VA, USA, 1940.

5. Mazelsky, B.; Drischler, J.A. *Numerical Determination of Indicial Lift and Moment Functions for a Two-Dimensional Sinking and Pitching Airfoil at Mach Numbers 0.5 and 0.6*; Technical Report; National Advisory Committee for Aeronautics: Langley Field, VA, USA, 1952.
6. Von Karman, T.; Sears, W.R. Airfoil Theory for Non-Uniform Motion. *J. Aeronaut. Sci.* **1938**, *5*, 379–390. [CrossRef]
7. Hakkinen, R.J.; Richardson, A.S., Jr. *Theoretical and Experimental Investigation of Random Gust Loads Part I: Aerodynamic Transfer Function of a Simple Wing Configuration in Incompressible Flow*; Technical Report; National Advisory Committee for Aeronautics Report; National Advisory Committee for Aeronautics: Washington, DC, USA, 1957.
8. Comte-Bellot, G.; Corrsin, S. The Use of a Contraction to Improve the Isotropy of Grid-Generated Turbulence. *J. Fluid Mech.* **1966**, *25*, 657–682. [CrossRef]
9. Roadman, J.; Mohseni, K. *Gust Characterization and Generation for Wind Tunnel Testing of Micro Aerial Vehicles*; AIAA Paper 2009-1290; AIAA: Reston, VA, USA, 2009.
10. Valente, C.; Lemmens, Y.; Wales, C.; Jones, D.; Gaitonde, A.; Cooper, J.E. *A Doublet-Lattice Method Correction Approach for High Fidelity Gust Loads Analysis*; AIAA Paper 2017-0632; AIAA: Reston, VA, USA, 2017.
11. Weishäupl, C.; Laschka, B. Euler Solutions for Airfoils in Inhomogeneous Atmospheric Flows. *J. Aircr.* **2001**, *38*, 257–265.
12. Heinrich, R.; Reimer, L. Comparison of Different Approaches for Gust Modeling in the CFD Code TAU. In Proceedings of the International Forum on Aeroelasticity and Structural Dynamics, Bristiol, UK, 24–26 June 2013.
13. Jirasek, A. CFD Analysis of Gust Using Two Different Gust Models. In Proceedings of the RTO AVT-189 Specialists' Meeting on Assessment of Stability and Control Prediction Methods for Air and Sea Vehicles, Portsmouth, UK, 12–14 October 2011.
14. Raveh, D.E.; Zaide, A. Numerical Simulation and Reduced-Order Modeling of Airfoil Gust Response. *AIAA J.* **2006**, *44*, 1826–1834. [CrossRef]
15. Wales, C.; Jones, D.; Gaitonde, A. Prescribed Velocity Method for Simulation of Aerofoil Gust Responses. *J. Aircr.* **2014**, *52*, 64–76. [CrossRef]
16. Da Ronch, A.; Tantaroudas, N.D.; Badcock, K.J. *Reduction of Nonlinear Models for Control Applications*; AIAA Paper 2013-1491; AIAA: Reston, VA, USA, 2013.
17. Singh, R.; Baeder, J.D. Direct Calculation of Three-Dimensional Indicial Lift Response Using Computational Fluid Dynamics. *J. Aircr.* **1997**, *34*, 465–471. [CrossRef]
18. Förster, M.; Breitsamter, C. Aeroelastic Prediction of Discrete Gust Loads Using Nonlinear and Time-Linearized CFD-Methods. *J. Aeroelast. Struct. Dyn.* **2015**, *3*, 19–38.
19. Bartels, R.E. *Development, Verification and Use of Gust Modeling in the NASA Computational Fluid Dynamics Code FUN3D*; AIAA Paper 2013-3044; AIAA: Reston, VA, USA, June 2013.
20. Siegel, S.; Cohen, K.; McLaughlin, T. *Feedback Control of a Circular Cylinder Wake in Experiment and Simulation*; AIAA Paper 2003-3569; AIAA: Reston, VA, USA, 2003.
21. Bergeron, K.; Cassez, J.F.; Bury, Y. *Computational Investigation of the Upsweep Flow Field for a Simplified C-130 Shape*; AIAA Paper 2009-0090; AIAA: Reston, VA, USA, 2009.
22. Ghoreyshi, M.; Jirasek, A.; Cummings, R.M. *CFD Modeling for Trajectory Predictions of a Generic Fighter Configuration*; AIAA Paper 2011-6523; AIAA: Reston, VA, USA, 2011.
23. Ghoreyshi, M.; Jirasek, A.; Cummings, R.M. *Closed Loop Flow Control of a Tangent Ogive at a High Angle of Attack*; AIAA Paper 2013-0395; AIAA: Reston, VA, USA, 2013.
24. Ghoreyshi, M.; Bergeron, K.; Seidel, J.; Jirasek, A.; Lofthouse, A.J.; Cummings, R.M. Prediction of Aerodynamic Characteristics of Ram-Air Parachutes. *J. Aircr.* **2016**, *53*, 1802–1820. [CrossRef]
25. Ghoreyshi, M.; Darragh, R.; Harrison, S.; Lofthouse, A.J.; Hamlington, P.E. Canard–wing interference effects on the flight characteristics of a transonic passenger aircraft. *Aerosp. Sci. Technol.* **2017**, *69*, 342–356. [CrossRef]
26. Ghoreyshi, M.; Cummings, R.M. Unsteady Aerodynamics Modeling for Aircraft Maneuvers: A New Approach Using Time-Dependent Surrogate Modeling. *Aerosp. Sci. Technol.* **2014**, *39*, 222–242. [CrossRef]
27. Ghoreyshi, M.; Jirasek, A.; Cummings, R.M. Computational Investigation into the Use of Response Functions for Aerodynamic-Load Modeling. *AIAA J.* **2012**, *50*, 1314–1327. [CrossRef]

28. Ghoreyshi, M.; Jirasek, A.; Cummings, R.M. Reduced Order Unsteady Aerodynamic Modeling for Stability and Control Analysis Using Computational Fluid Dynamics. *Prog. Aeros. Sci.* **2014**, *71*, 167–217. [CrossRef]
29. Ghoreyshi, M.; Cummings, R.M.; DaRonch, A.; Badcock, K.J. Transonic Aerodynamic Load Modeling of X-31 Aircraft Pitching Motions. *AIAA J.* **2013**, *51*, 2447–2464. [CrossRef]
30. Ghoreyshi, M.; Cummings, R.M. Unsteady Aerodynamic Modeling of Aircraft Control Surfaces by Indicial Response Methods. *AIAA J.* **2014**, *52*, 2683–2700. [CrossRef]
31. Ballmann, J.; Braun, C.; Dafnis, A.; Korsch, H.; Reimerdes, H.G.; Brakhage, K.H.; Olivier, H. *Numerically Predicted and Expected Experimental Results of the HIRENASD Project*; Annual Meeting of the German Aerospace Association (DGLR); Paper DGLR-2006-117; DGLR: Braunschweig, Germany, 2006.
32. Ballmann, J.; Dafnis, A.; Korsch, H.; Buxel, C.; Reimerdes, H.G.; Brakhage, K.H.; Olivier, H.; Braun, C.; Baars, A.; Boucke, A. *Experimental Analysis of High Reynolds Number Aero-Structural Dynamics in ETW*; AIAA Paper 2008-0841; AIAA: Reston, VA, USA, 2008.
33. Campbell, G.S.; Norman, J.M. *An Introduction to Environmental Biophysics*; Springer: New York, NY, USA, 2000.
34. Guo, S.; Jing, Z.W.; Li, H.; Lei, W.T.; He, Y.Y. Gust Response and Body Freedom Flutter of a Flying-Wing Aircraft with a Passive Gust Alleviation Device. *Aerosp. Sci. Technol.* **2017**, *70*, 277–285. [CrossRef]
35. Jones, J.G. *Documentation of the Linear Statistical Discrete Gust Method*; Technical Report; DOT/FAA/AR-04/20; Office of Aviation Research: Washington, DC, USA, 2004.
36. Hoblit, F.M. *Gust Loads on Aircraft: Concepts and Applications*; American Institute of Aeronautics & Astronautics: Reston, VA, USA, 1988.
37. Fuller, J.R. Evolution of Airplane Gust Loads Design Requirements. *J. Aircr.* **1995**, *32*, 235–246. [CrossRef]
38. Strang, W.; Tomaro, R.; Grismer, M. *The Defining Methods of Cobalt-60-A Parallel, Implicit, Unstructured Euler/Navier-Stokes Flow Solver*; AIAA Paper 1999-0786; AIAA: Reston, VA, USA, 1999.
39. Gottlieb, J.J.; Groth, C.P.T. Assessment of Riemann Solvers for Unsteady One-dimensional Inviscid Flows of Perfect Gasses. *J. Fluids Struct.* **1998**, *78*, 437–458.
40. Tomaro, R.F.; Strang, W.Z.; Sankar, L.N. *An Implicit Algorithm For Solving Time Dependent Flows on Unstructured Grids*; AIAA Paper 1997-0333; AIAA: Reston, VA, USA, 1997.
41. Spalart, P.R.; Allmaras, S.R. *A One Equation Turbulence Model for Aerodynamic Flows*; AIAA Paper 1992-0439; AIAA: Reston, VA, USA, 1992.
42. Wilcox, D.C. Reassesment of the Scale Determining Equation for Advanced Turbulence Models. *AIAA J.* **1988**, *26*, 1299–1310. [CrossRef]
43. Menter, F. Eddy Viscosity Transport Equations and Their Relation to the k-ε Model. *ASME J. Fluids Eng.* **1997**, *119*, 876–884. [CrossRef]
44. Thomas, J.L.; Salas, M.D. Far-field Boundary Conditions for Transonic Lifting Solutions to the Euler Equations. *AIAA J.* **1986**, *24*, 1074–1080. [CrossRef]

aerospace

MDPI

Article

Numerical Simulation of Heat Transfer and Chemistry in the Wake behind a Hypersonic Slender Body at Angle of Attack

Matthew J. Satchell [1,*], Jeffrey M. Layng [2] and Robert B. Greendyke [3]

[1] High Performance Computing Research Center, Department of Aeronautics, USAF Academy, El Paso County, CO 80840, USA

[2] Department of Aeronautics, USAF Academy, El Paso County, CO 80840, USA; Jeffrey.Layng@usafa.edu

[3] Air Force Institute of Technology, AFIT 2950 Hobson Way, WPAFB, Greene County, OH 45433, USA; Robert.Greendyke@afit.edu

* Correspondence: matthew.satchell.1@us.af.mil

Received: 13 December 2017; Accepted: 24 February 2018; Published: 11 March 2018

Abstract: The effect of thermal and chemical boundary conditions on the structure and chemical composition of the wake behind a 3D Mach 7 sphere-cone at an angle of attack of 5 degrees and an altitude of roughly 30,000 m is explored. A special emphasis is placed on determining the number density of chemical species which might lead to detection via the electromagnetic spectrum. The use of non-ablating cold-wall, adiabatic, and radiative equilibrium wall boundary conditions are used to simulate extremes in potential thermal protection system designs. Non-ablating, as well as an ablating boundary condition using the "steady-state ablation" assumption to compute a surface energy balance on the wall are used in order to determine the impacts of ablation on wake composition. On-body thermal boundary conditions downstream of an ablating nose are found to significantly affect wake temperature and composition, while the role of catalysis is found to change the composition only marginally except at very high temperatures on the cone's surface for the flow regime considered. Ablation is found to drive the extensive production of detectable species otherwise unrelated to ablation, whereas if ablation is not present at all, air-species which would otherwise produce detectable spectra are minimal. Studies of afterbody cooling techniques, as well as shape, are recommended for further analysis.

Keywords: hypersonic; wake; chemistry; slender-body; angle of attack; detection; after-body

1. Introduction

1.1. Introduction and Theory

As hypersonic vehicles grow in number and capability, a keen interest is growing in the observability and detection of these vehicles by states seeking to defend against threats made possible by rapid advances in hypersonic technology. Gradual evolutions in the maneuverability of reentry vehicles and the development of scramjet powered vehicles are now yielding unprecedented range and speed of strike for hypersonic weapons while potentially enabling evasion of detection [1]. Figure 1 shows the trajectory of a traditional ballistic missile and also that of a suppressed "boost-glide" trajectory, wherein the flight vehicle separates from its boost stage and burns back towards the earth before turning into a glide at significantly lower altitude than a traditional original ballistic trajectory. This study will focus on a single point within the glide trajectory.

Owing to their unconventional trajectory, hypersonic weapons may decrease the odds of detection, and so methods of detection are of increasing interest to nations hoping to defend against such threats. Since many forms of detection rely upon active and passive use of the electromagnetic spectrum, the emissions of hypersonic vehicles from across the EM spectrum are of interest. Analysis of the emissions from a given system across the entire spectrum being quite a far reaching task, this paper will focus on assessment of the thermal and chemical environment, which can then be extrapolated for a particular case of interest in order to determine the emissions characteristics.

Hypersonic flow differs from regular supersonic flow primarily in that temperatures regularly reach sufficient levels to change the chemical nature of the gas. As shock layer and stagnation temperatures reach upwards of 800 K, vibrational modes are excited in the air molecules, leading to a change in the specific heats as a function of temperature [2]. As the temperature increases to 2000 K, the vibrations become sufficiently violent as to lead to the dissociation of the molecular O_2, and by 4000 K most of the oxygen has been dissociated into atomic oxygen [2]. The dissociated oxygen is free to chemically react with the other species, resulting in formation of various combinations of nitrogen and oxygen, to be discussed later in this section. As temperatures increase, ionization begins to occur in significant quantities, pumping free electrons into the flow and eventually forming a plasma. As temperature continues to increase, the shock and boundary layers can begin to radiate energy into the body and surrounding flowfield, cooling the shock and boundary layers but heating the body. Between these and other nonlinearities in the shift of flow properties in air at high temperatures, actual gas temperature is dramatically different than that which predictions for an ideal gas would yield [2].

The flow in the vicinity of leading edges, or the nose in the case of a cone, tends to be in chemical nonequilibrium downstream of the subsonic stagnation region [2]. Nonequilibrium occurs when the relaxation time for a given chemical reaction is on the same order as the local speed of the flow [2]. However, as molecules collide, relax and recombine in irreversible processes, they form species different from those in the original freestream composition of the air and indeed different from those species formed by a gas in purely chemical equilibrium [2]. As the hot gases generated in the stagnation region then travel downstream through the boundary layer, they tend towards an equilibrium state [2]. Because of high entropy introduced by the severe gradients and chemical reactions around the nose, the downstream boundary layer is thickened considerably as compared to a non-reacting flow. Finally, in many circumstances, the proximity of the shock to the body near the leading edges causes the viscous boundary layer to interact with the shock layer, further complicating the physics and chemistry of the flow [2].

The current effort is focused on a 3.5 m long, 7 degree half-angle cone traveling at Mach 7 at 30 km altitude (yielding a Reynolds number of 4.1 M), with an angle of attack of 5 degrees, summarized in Table 1.

Table 1. Summary of Flight Conditions.

Flight Condition	Value
Mach	7
Altitude (Km)	30
Velocity (m/s)	2115
Angle of Attack (deg)	5
Density (kg/m^3)	0.01841
Temperature (K)	226.5
Reynolds Number	9,237,700

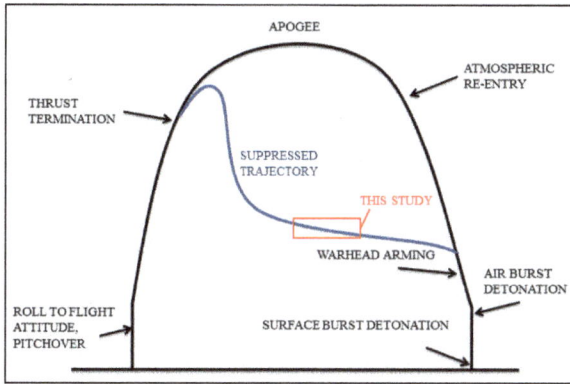

Figure 1. A ballistic and a suppressed flight trajectory.

The goals of this investigation are:

1. Determine a bounding series of predictions into which the subject vehicle's wake compositions should fall
2. Determine the role of body chemistry on wake composition
3. Determine role of body's thermal history in the wake
4. Explore the role of chemical reactions in wake composition

At Mach 7, temperatures behind the shock are not sufficiently high as to merit concern for strong ionization or radiation of the flowfield [2,3]. Temperatures for a Mach 7 vehicle at 30 km can be estimated to be just over 2000 K behind the shock for a chemically reacting flow [2]. Indeed the temperatures at the Mach number and altitude of interest only surpass the 2000 K threshold for oxygen dissociation near the nose for most thermal boundary conditions, as "in general, the surface of a hypervelocity entry vehicle is much cooler than the flow temperature directly behind the shock" [4]. Additionally, substantial nonequilibrium effects tend to persist along the body on the order of unity in millimeters for the more extreme case of a wedge at 6638 m/s, or roughly three times the speed in this study at similar altitude conditions [2]. Thus, for a slender vehicle where much of the body is far from the shock, the majority of flow around the vehicle will be in chemical equilibrium at the present flight conditions, with the composition of the flow remaining roughly constant until the end of the body is reached [2]. Because this study is interested in the chemical composition of the wake, which is significantly influenced by the chemical reactions occurring at the nose, accurate assessment of the chemistry at the nose and subsequent relaxation must be performed.

A number of chemistry models are available to the numerical investigator of hypersonic flows. Figure 2 demonstrates the various chemistry models that are most commonly used for a given flight regime. Depending upon the flight altitude, velocity and thermal protection system (TPS), a broad array of potential combinations of atomic and molecular species may be present in the flow. For flight speeds ranging up to around 5 km/s, a 5-species model consisting only of O, O_2, N, N_2, and NO is generally sufficient to accurately capture the effects of chemistry within a flow [3]. Above these speeds, a higher number of species must be used in order to capture the effects of ion formation via the formation of NO^+ and the associated free electrons, and still more species must be considered at even higher speeds. Figure 2 also shows the limits of the ranges for chemical and thermal equilibrium as well as non-equilibrium conditions. It is clear that this study falls squarely within the chemical equilibrium, 5-species flight regime. However, as shall be discussed later, the 7-species model must be used in order to adequately capture the effects in the wake.

Of particular significance to the present effort are the electron density and that of nitric oxide (NO). Elevated electron densities are detectable via microwave reflection from the wake, while NO eventually recombines into the formation of NO_2, which itself produces optically detectable spectral emissions [5]. Additionally, the nitrosonium ion NO^+ forms by the collision of N and O molecules, releasing an electron to form NO^+ and thereby introducing electrons into a flow incapable of producing ionization by virtue of temperature alone [5]. NO^+ can also be formed by the collision of NO with either N_2 or O_2 [3]. The free electrons may then be observed via microwave beam reflection by a casual observer and so NO^+ is also tracked in order to determine the degree to which electrons are produced by this reaction [5]. Carbon dioxide, which is formed by a carbon ablator, is also investigated. High temperature CO_2 is known to produce emissions in the infrared spectrum [6]. The cyanogen molecule, CN, is also produced from an ablating carbon surface in air, and is a strong emitter at high temperatures. Finally, the density of molecular carbon in the wake is also investigated, as ablated carbon molecules in the flowfield may emit as black-bodies which remain at elevated temperatures far into the wake, thus providing still further fodder for detection.

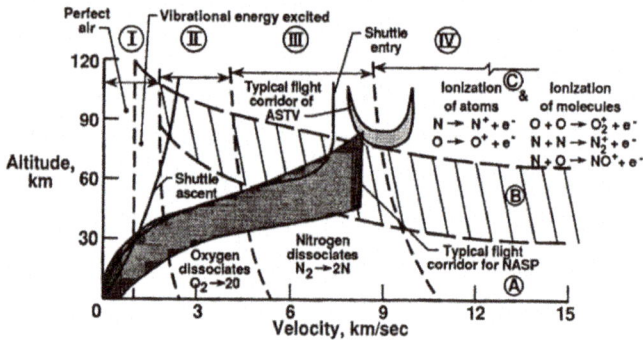

Figure 2. Review of chemical makeup vs flight regime for a 30.5 cm radius sphere. Reproduced from NASA RP1232 [3].

A further consideration is how chemical species are resolved at the wall. The catalysis model, which determines the gradient of chemical compositions at the wall, can significantly influence the chemical composition in the boundary layer, the heat transfer to the body and ultimately the wake [4]. Species which have dissociated or chemically reacted through the shock and outer portions of the boundary layer tend to relax towards their equilibrium values as they approach the wall through the boundary layer as they collide with other atoms or molecules to redistribute their elevated energy states [2,4]. However, in the case of a nonequilibrium flow at the surface wherein not all species have fully relaxed before hitting the wall, when the molecules collide with the generally cooler wall they are able to impart some of their energy into the wall and are thereby catalyzed to relax substantially [7]. This chemical reaction on the wall can also significantly heat the body, and strongly depends upon the composition and behavior of the TPS [7]. For example, Gnoffo et al., 2010 states that "metallic surfaces are generally considered to be strongly catalytic but the potential formation of an oxide coating introduces uncertainty in the proper formulation of this boundary condition" [7]. Furthermore, according to Barnhardt et al. 2009. "It has been conjectured that surface catalysis may be enhanced by the presence of larger concentration gradients" [8]. Thus the setting of any given catalysis model for the wall is assuming behaviors not just for the body's response to the flow, but for the flow itself, presenting a challenging conundrum in understanding the chemical composition of the boundary layer for an unspecified object.

For the purposes of bounding the problem, Gnoffo et al., 2010 used a super-catalytic and also a non-catalytic wall condition [7] in order to predict conditions on the FIRE II capsule. The super-catalytic condition sets species values at the wall equal to freestream values, while the non-catalytic condition sets species gradients equal to zero at the wall. Another species boundary condition is the fully-catalytic wall, which assumes that all dissociated or ionized species catalyze to molecular species. A final general catalysis model is the equilibrium-catalytic model, which assumes that all species are catalyzed to equilibrium values for the temperature and pressure at the wall, and specifies a mass-fraction gradient for each species equal to zero at the wall [9]. In some CFD codes, material responses have been characterized in such a way as to provide material-specific catalysis models for some common ablators, adding to the arsenal of available material response options [9].

Figure 3, which highlights RMS of density gradients from a simulation in the present effort, closely matches in its features the structures reported by theoretical, CFD and DSMC computational efforts found in the literature[1,10–13]. Sharpness or bluntness of the body merely adjusts the length scales and strength of each structure in terms of chemical and gas property gradients [12,13].

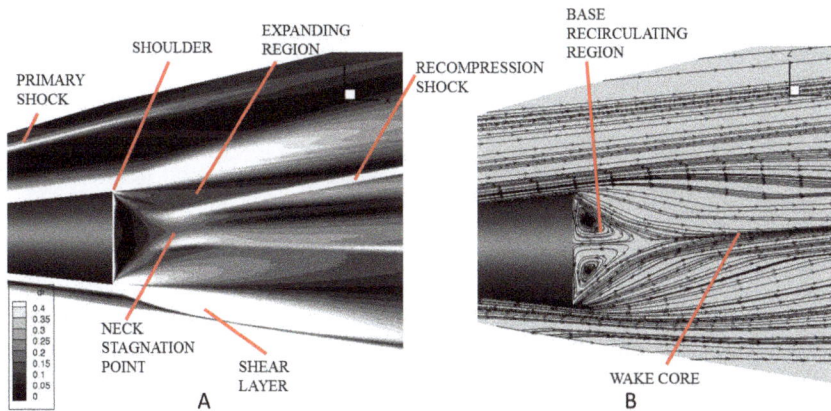

Figure 3. Density gradients within a hypersonic wake (**A**) and Streamlines of a hypersonic wake (**B**).

Figure 3 shows several key features of a hypersonic wake, which were thoroughly explored in the 1960's [10] and the extents and variations of which have since been greatly studied [13–15]. First are the boundary layer streamlines which are pulled in towards the vacuum of the passing body and compressed by their meeting at the neck stagnation point, at which location the wake temperature is the greatest and tends to closely mirror the temperatures reached on the body at the nose [1,16]. The size of this region is quite sensitive to the freestream number density, growing in length with decreased density as the molecules gain more freedom to move [12]. It is additionally sensitive to wall temperatures, and so may vary substantially in size based on thermal boundary conditions, and sensitive also to angle of attack and Reynolds number [13–15]. A recompression shockwave is formed by the converging boundary flows, leaving a central core of expanding flow filled with the products of the boundary layer. Flow outside the recompression shock is characterized by an expanding region as the flow turns over the shoulder, which is in turn surrounded by a free shear layer as the remainder of the boundary layer mixes with the expanding flow below and the near-freestream conditions above it.

The effects of angle of attack on the flow both over the body and the wake are significant [14,17]. The flowfield becomes asymmetric, with the windward side shock sitting quite close to the body, while the leeward side shock is drawn away from the body according to the angle of attack. Combined with a vortical cross-flow formed by the strong acceleration of the fluid around the nose from the windward side to the leeward side, the windward side and leeward side of the flow differ dramatically both in terms of chemical composition and boundary layer properties. Additionally, the boundary layer on the leeward side is much thicker than the windward side [17]. These very different layers then pour into the vacuum behind the vehicle, causing an asymmetric wake to form, in which the hotter gases from the windward side produce a higher concentration of chemically active species on the windward side of the wake and the stronger recombination shock located there [16]. Overall, the effect of non-zero angle of attack on the properties of the flowfield around and behind a cone is well-documented by authors from the 1960's to present [14,17], but little data has been published, numerically or experimentally, to explore the composition of the wake of a slender vehicle under the present flight conditions. This study will endeavor to examine the thermochemical properties of the flow as driven by these various features.

2. Methods

The Langley Aerothermodynamic Upwind Relaxation Algorithm (LAURA) v5.5-74986 was used for all simulations. Reviews of the numerics and physics models are provided, amongst other places, in [1,7,18,19] and will not be covered in detail here. LAURA is a multiblock, structured solver which has both a point-implicit and a line-implicit solution method available. For the purposes of this study the point-implicit method was utilized on all wake computations due to the three-dimensional, cross-block nature of the wake flow [8,14].

2.1. Grid Generation and Alignment

An unaligned 3D grid of 24 blocks 33 × 33 × 72 in dimension arrayed in a butterfly topology around a 25th nose block of 32 × 32 × 72 was developed using Pointwise, with the x-direction oriented along the body's centerline, y-direction facing the symmetry, and z completing the right-handed coordinate system pointing across the body's diameter. The grid modeled half the body, an axisymmetric case being impossible because of angle of attack effects, and a 2D case likewise not being able to capture the important 3D phenomenon associated with a cone at angle of attack. A butterfly topology for a structured mesh sets the block boundaries such that a singularity at the stagnation region on the nose and along the wake core in the rear is avoided, which has been shown to produce non-physical results in some regions [1]. The topology is shown in Figure 4. The 25 block grid was then coarsened to 8 × 8 × 16 and the flowfield initialized to Mach 7 at 5 degrees angle of attack with laminar flow with a line-implicit relaxation method. After 1000 iterations, the turbulence model, to be discussed in detail later, was switched on. The very coarse grid was run until an L_2 Norm of 10^{-6} was achieved. L_2 Norm is given as:

$$L_2 = \Sigma_{i=1}^{N} \left\{ \frac{\| R_i \|}{\rho_i} \right\}^2 \tag{1}$$

Equation (1) shows that the L_2 Norm is a measure of global convergence of the continuity, momentum and energy equations, summing the residuals R_i at each point from all equations and scaling by the local density.

LAURA's adaptive mesh refinement routine, align_shock was run every 5000 iterations. align_shock shifted cells in the body-normal k-direction such that tight clustering was achieved in the shock and boundary layers, with a variable growth rate at the wall which was set to achieve a non-dimensional distance y^+ no greater than 1 at the wall and retain 75% of cells in the boundary layer, with tight clustering around the shock. Grid resolution studies conducted in the past with the LAURA code have been performed across many conditions, ultimately resulting in some general guidelines for the appropriate wall-normal grid spacing for flows with nonequilibrium regions. Of particular interest in the literature is the concept of Cell Reynolds Number, which is given by the equation:

$$Re_{cell} = \frac{\rho a_T \Delta z}{\mu} \qquad (2)$$

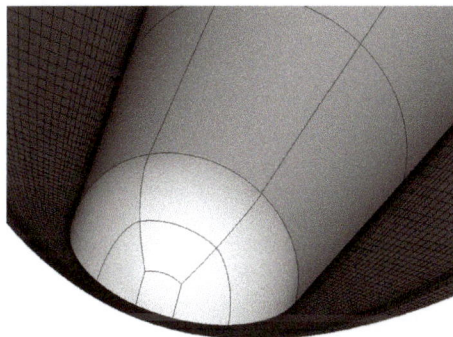

Figure 4. Butterfly topology on the nose of the vehicle.

Here, a_T is a local speed of sound to account for thermal velocity, and Δz is the height of the first cell off the wall. Values of Re_{cell} on the order of unity are able to reliably simulate non-equilibrium chemical reactions and heat transfer to the wall [18–22]. This ultimately places the height of the cells on the wall significantly lower than the height required to adequately capture turbulent flow structures with most Reynolds Averaged Navier-Stokes (RANS) solvers, which mandate a y^+ value of less than one [23]. y^+ is given by Equation (3):

$$y^+ = \frac{y\rho u_*}{\mu} \qquad (3)$$

In Equation (3), u_* is the friction velocity, μ is viscosity and ρ is the density. Growth rates were 4% close to the wall, and ranged up to 100% in the inviscid region between the boundary layer and the shock. The end result was a maximum $y^+ = 1$ near the rear of the vehicle, with a minimum of $y^+ = 0.0061$ at the nose as seen in Figure 5a. The y^+ did change with the various boundary conditions utilized in this study, but remained on the same order for all studies. Re_{cell} was 0.3 at the nose and reaching a maximum of 10 near the rear of the vehicle, well within established bounds for accurate surface heading prediction [18,20]. Clustering near the shock was set by Equation (4). In Equation (4), ϵ was set to 5 to ensure high resolution around the shock, and f_{sh} was set to 0.8 to hold the shock at 80% of the wall-normal distance from the wall, the results of which can be seen in Figure 5b.

$$k_i = \epsilon_0 k_2^2 (1 - K_i)(K_i + f_{sh}) + k_i \qquad (4)$$

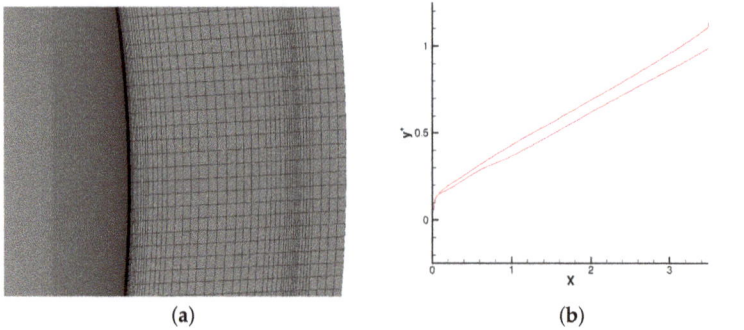

Figure 5. (a) Mesh at the stagnation region and (a) y^+ on the wall of the vehicle along the symmetry.

The coarse grid was then refined to its original resolution using LAURA's prolongate routine, which refines the mesh back to the original mesh dimensions, but leaves in place the body-normal refinements made in the coarse grid. Finally, the solution was run to an L_2 Norm of 10^{-6}, with surface temperature and heat flux monitored to confirm a steady solution. During this computation, body-normal refinements were again run every 5000 iterations initially and then every 10,000 as sensitivity in the L_2 Norm diminished, with the temperature and heat flux eventually showing no change with further refinement. The resulting grid is seen in Figures 5 and 6.

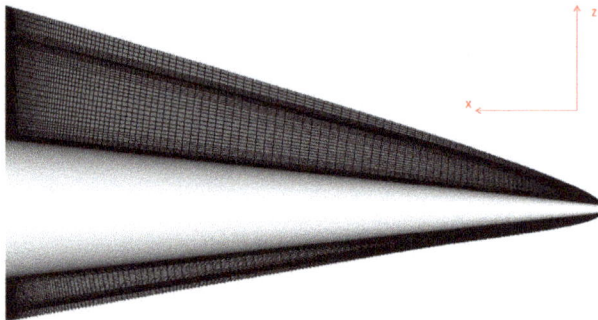

Figure 6. Body flowfield mesh.

The flow-aligned grid was then modified using Pointwise with an additional 70 blocks in the wake, each of dimensions $32 \times 32 \times 72$. Body-normal spacing on the rear surface of the cone leading into the wake was set to 0.001 m, with a growth rate of 1%. This value was selected as a "safe value" based on findings in Barnhardt et al., 2009 and 2012, which found for a much higher speed flow that a spacing of 0.0025 m off the base wall was sufficient to capture wake structures as accurately as possible with the Spalart-Allmaras (SA) turbulence model used in a DES formulation [8,14]. Given the heavy computational demands required to obtain a converged mesh of the body for a given spacing in the cross-body and flow-wise directions, and the subsequently time-consuming and expensive process of developing and running the wake mesh to match the flow-aligned grid, a grid-resolution study of the wake was deemed not to be feasible. However, given the discussion above, some confidence may be nevertheless be placed in the grid resolution required for the flow phenomenon at hand. The full 95-block grid is pictured in Figure 7. Figure 7 also has inlays showing increasing zoom towards the rear of the vehicle in order to show resolution at the corner.

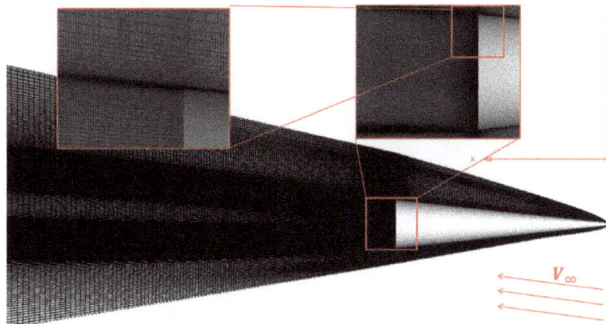

Figure 7. Full Volume Grid along the symmetry.

2.2. Numerics and Models

A LAURA flow solution was resolved using the two-temperature model for both 7-species and 13-species cases which were each tracked according to its own set of conservation equations in the chemically reacting flows. The 7-species model was used for the non-ablating cases, and consists of the molecular nitrogen and oxygen, their respective atomic counterparts, NO, NO^+ and electron density. The 13-species model used the same species along with elements from a pure carbon ablator, allowing for the formation of CN in addition to the most common molecular carbon combinations. Each set of species is reviewed in Table 2.

Table 2. The 7-Species Model (**a**) and 13-Species Model (**b**).

7-Species Model (a)	
Species	Concentration
N	6.22×10^{-20}
O	7.76×10^{-09}
N_2	0.737795
O_2	0.262205
NO	1.00×10^{-09}
NO^+	4.57×10^{-24}
e^-	8.35×10^{-29}

13-Species Model (b)	
Species	Concentration
N	6.22×10^{-20}
O	7.76×10^{-09}
N_2	0.737795
O_2	0.262205
NO	1.00×10^{-09}
NO^+	4.57×10^{-24}
e^-	8.35×10^{-29}
C	1.00×10^{-25}
C_2	1.00×10^{-25}
C_3	1.00×10^{-25}
CO	1.00×10^{-25}
CO_2	1.00×10^{-25}
CN	1.00×10^{-25}

Mach 7 is sufficiently low in the spectrum of hypersonic flows that many of the chemical kinetics that affect heat transfer to the body in higher speed flows are relatively insignificant to the heating of the subject body. As such a 5-species solution is traditionally used in this flow regime. In particular, at Mach 7, ionization does not occur at rates high enough to significantly affect the flow structure or heat transfer [2]. However, a small degree of ionization is present even at Mach 7 owing to the formation of nitrosonium, which does form at the present flight conditions, and so the 7-species model was required for the non-ablating cases, and the 13-species model for the ablating cases. LAURA computes variable Prandtl and Schmidt numbers, computing conductivities and diffusivities directly from collision cross-sections [9].

In addition to the chemical composition of the wake flowfield, this study endeavors to determine some reasonable bounds within which the significant variation of possible TPS materials can affect the subject species concentrations. As the body encounters the extreme energies of a hypersonic flow, the TPS can be designed to respond in a number of possible ways. Furthermore, different portions of the TPS and body structure can be produced with different thermal management systems, thus producing further variation in surface thermal conditions. The following is a brief discussion of some of the design options available and their associated models in the present effort.

The TPS may act as a "hot structure", wherein the structure absorbs enormous quantities of heat, eventually leading in the extreme case to a condition of zero heat transfer to the body at the surface. Such a case yields a state similar to the adiabatic thermal boundary condition. Although a system design which yielded an adiabatic wall condition would certainly be destroyed by the resultant temperatures in the hypersonic environment, this condition provides a helpful upper bound to the temperatures and therefore chemical compositions that might be encountered in the wake.

The opposite of the adiabatic condition is the cold-wall condition, wherein some active cooling mechanism in the flight vehicle keeps the wall cooled to a specified temperature, or when the flight time is relatively short. For example, the HIFIRE-1 flight vehicle was shown to maintain a cold-wall condition even towards the end of its trajectory [24]. This condition is also commonly used to simulate wind-tunnel experiments wherein the body is not exposed to the flow for sufficient timescales as to heat up significantly. Two extremes are examined herein; that in which the entire body is kept at a constant temperature of 500 K, and also the case where the nose is allowed to heat significantly and also ablate, but the rest of the body is kept to the temperature of 500 K. While the former case is unlikely, it provides a lower bound within which the chemical composition of the wake might be estimated.

A third alternative is also explored, wherein the wall is treated as adiabatic except for the degree to which the wall is able to radiate away heat imparted by the flow; the so-called 'radiative equilibrium' condition. The flow of radiation away from the wall is not coupled with the flow solver in this study, although the wall is allowed to benefit from the radiative cooling. Previous studies have shown that under many circumstances this condition can provide a reasonable approximation for surface temperature as compared to flight data [7]. The radiative equilibrium condition states that heat transfer into the wall is equal to:

$$q_{wall} = \epsilon \sigma T_{wall}^4 \tag{5}$$

This boundary condition effectively states that the wall will reject all heat from the flow except that which it is able to radiate away. Taken together, these thermal boundary conditions were selected to give a high, low, and reasonable mid-spectrum estimation of thermal conditions on the surface of the flight vehicle. Ultimately, the surface conditions on the flight vehicle determine the temperature and chemical composition of the wake, and so determining these bounds in turn establishes reasonable bounds for the thermal environment within the wake.

LAURA uses two relaxation factors, which are respectively multiplied against the viscous and inviscid Jacobian computations in the point-implicit matrix with each iteration. These factors help to stabilize the code, and may be set to quite high values in order to ease the code through tough transients [18,19] without affecting the solution, so long as they are ultimately returned to a value on the order of 1 and 3, respectively [1]. In this study, the factors were both set initially to 40 for initialization of the case. The inviscid and viscous relaxation factors were gradually lowered to 3 and 1.5, respectively as the solution converged. Setting the relaxation factors to their final values did significantly impact the solution as compared to a converged solution arrived at with the factors set to 10 and 6, respectively. A steady simulation was selected based on previous findings that indicated variational frequencies for flow properties at the specified conditions are sufficiently high as to allow a time-averaged solution to produce the observable wake phenomenon of interest [1]. A Courant-Friedrichs-Lewy number (CFL) of 5 was initially selected and was allowed to grow as high as 50 as convergence was reached on the fine grid.

The Spalart-Allmaras one-equation turbulence model with the Catris compressibility correction was selected to model all turbulence. Barnhardt et al. (2009) found that "RANS greatly overpredicts heating near the base center and the distribution does not correlate well to the experiment. Detached Eddy Simulation (DES) captures the essential distribution within measurement uncertainty" as compared to the NASA Reentry-F test flight in 1968, which studied a reentering 5-degree half-angle cone [8,14]. Because DES was able to predict the base temperature within acceptable error as compared with flight test data, these results suggest that the error lay in the RANS modeling of the wake structures. Thus, given the large degrees of uncertainty in the RANS models' predictions for the recirculation zone [14], the unavailability of DES in LAURA, and the parametric nature of this study, the choice of SA-Catris as the turbulence model was driven by a desire for stability and simplicity. Furthermore, the SA model was shown in Barnhardt et al., 2009 to produce reasonable estimations of heat flux on the surface of a cone at much higher speeds [14]. Transition was not modeled in this study, with the entire domain initialized as turbulent in keeping with the original study by Kania et al., 2015 [1], and for additional reasons discussed next. This likely produced slightly higher surface temperatures over the front portions of the vehicle as compared to the potentially laminar flow which some studies have shown is likely to exist there [14,25]. For comparison, two laminar cases were also run for a non-ablating case, the results of which are discussed later.

With the assumption of turbulent flow throughout the flowfield comes the assumption of a turbulent wake. Many studies have focused on laminar flow at either much lower freestream Reynolds numbers, zero angle of attack, blunt bodies with or without sharp corners on the rear surfaces such as those encountered on a traditional cone, or been in two dimensions [13–15,26,27]. Thorough reviews of the many wake studies to date can be found in the works just cited, which together support the presence of laminar near-wakes for many flow conditions, but indicate that turbulence may be also be present in a broad variety of flows. These studies together conclude that: transition to turbulence in a near-wake is poorly understood [15], transition in the near-wake is highly sensitive to angle of attack, geometry, freestream Mach number and Reynolds number [14,27], and very little current experimental data exists to validate numerical predictions of near-wake flows for the cases of interest to this study [14,26]. All told, the authors could find no basis upon which to conclusively determine whether the flow could be adequately captured by a laminar simulation or must necessarily be turbulent, in which case the wake flow could be studied only qualitatively regardless of the accuracy of any preceding assumptions owing to the poor reliability of RANS simulation for base flows [14]. Based on the findings that boundary layer transition is highly dependent upon Mach number, angle of attack, and Reynolds number [14,17,24], the observation that the boundary layer on the leeward side is almost certainly separated quite early on [24,28] and the emphasis of this study on the poorly characterized near-wake as opposed to the body, a turbulent scheme was selected for the entirety of the geometry for most cases. However, for completeness, a laminar simulation was also run on the non-ablating and ablating radiative equilibrium cases with no catalysis.

In light of the above discussion on handling of turbulence in this study, and the potential impacts upon surface temperature of a fully turbulent flow, some further analysis of the validity of the results is warranted. Referring again to Figure 2, we see that this study should necessarily bound on the lower edge of chemically reacting flow. Furthermore, the Counter-Rocket, Artillery and Mortar (RAM-C) flight test, which featured a considerably blunted 10° cone which traversed a Mach number and altitude quite close to that of the present study, showed the significant presence of free electrons owing to mechanisms discussed previously for the present effort under those flight conditions [5,29]. In the present study, the peak temperature for the ablating cases was 2411 K on the surface, a temperature only just sufficient to produce non-equilibrium effects at the stagnation point which would subsequently produce potentially observable phenomenon in the wake. It can thus be concluded that, if the above handling of turbulence overpredicted temperature, it was not to such great extent as to push the results beyond the realm of reasonable limitations. Furthermore, the non-ablating cases run in this study provide a lower bound in the case of a substantially weaker set of chemical reactions.

The present study is concerned with trying to bound the ranges within which anticipated chemistry of an unknown incoming hypersonic object might be found. Thus no assumption is made respecting the nature of the TPS except that it be made purely of a carbon ablator subject to the steady-state ablation assumption, which "specifies that the pyrolysis ablation rate is proportional to the char ablation rate and the in-depth conduction is proportional to the enthalpy at the surface" [9]. The steady-state ablation model solves the surface energy balance by according to the following equation:

$$- q_c - \alpha q_{rad} + \sigma \epsilon T_w^4 + (\dot{m}_c + \dot{m}_g) h_w = 0 \tag{6}$$

In Equation (6), α_{rad} is the heat radiated into the surface from the flowfield (assumed to be zero in this study), $\sigma \epsilon T_w^4$ is heat radiated away from the wall with σ as the Stefan-Boltzmann constant and ϵ the emissivity of the surface, and $(\dot{m}_c + \dot{m}_g) h_w$ is the heat transfer caused by the massflow yielded pyrolysis $\dot{m}_g$ and charring $\dot{m}_c$, and h_w is the enthalpy at the wall. q_c is the heat convective heat transfer at the wall, which is assumed proportional to the enthalpy at the wall. The steady-state ablation assumption then relates the ablation massflows as:

$$\dot{m}_g = \left(\frac{\rho_v}{\rho_c} - 1\right) \dot{m}_c \tag{7}$$

In Equation (7), ρ_v and ρ_c are the virgin and charred material densities, respectively. The char material density is assumed to be 256.29536 kg/m^3, which is that of the heritage AVCOAT material, and the virgin density is also that of the virgin AVCOAT, $\rho_v = 544.627742$ kg/m^3. These are the default values in the LAURA code, which were selected in keeping with the recommended best practices for an unspecified ablative procedure in the LAURA manual [9]. The parametric nature of this study is again emphasized here; the subject geometry, TPS, and flight conditions do not lend themselves well to reproducing a specific flight test and are not intended to attempt to do so. However, the role of ablation in the production of observable species at the subject conditions is explored, and so the above set of simple and reliably stable representations of the TPS were selected with that aim in mind. A full treatment of the ablation in LAURA is given in the literature [7,30].

These assumptions were applied to the first four blocks, extending to $x = 0.00259$ m along the nose for cases where radiative-equilibrium or a cold-wall was assumed for non-ablating sections downstream of the nose, and the first seven blocks extending to 0.021 m for those cases assuming an adiabatic body downstream of the nose. The temperature downstream of the ablating portions substantially impacts the conditions within the ablating zones. The subsonic flow within the boundary layer allows heating or cooling from the non-ablating portions of the flow to propagate upstream to the ablating portions of the flow via molecular diffusion, changing ablation rates and chemical composition of the boundary layer in both ablating and non-ablating zones. Thus the behavior of the TPS downstream of the ablating nose becomes even more significant in accurate prediction of the composition of the wake.

From the above discussion of the ablation assumptions utilized in this effort, several key implications upon the results thus produced can be drawn:

1. Note that conduction into the wall is treated by the ablation model available in LAURA as proportional to the enthalpy at the surface. Depending on the actual properties of the TPS and thermal history of the vehicle, the actual surface temperature, and thus chemistry both along the body and in the wake may vary significantly.

2. Addition of further elements such as Hydrogen, Silicon and Nitrogen, and also changes in the response of the TPS material to high temperatures would certainly impact the chemistry and heating conditions on the wall, driving changes to both the wake temperature and wake chemistry. However, the effect on wake properties of any particular additional species, nonetheless combination thereof, would be quite difficult to qualitatively ascertain without performing additional simulations owing to the complexity and quantity of endothermic and exothermic reactions which would arise at differing levels depending upon temperature, pressure and quantity of the added species. Thus, although this study aims to produce qualitative discussion of the variation of detectable species with changes in the TPS under the given flight conditions, effects caused by the addition of further chemical species beyond carbon are entirely unexplored herein.

3. Use of different materials is unlikely to produce a hotter wall than the adiabatic case, or a cooler wall than the cold-wall case. The extent to which these temperatures bound the chemical reactions in the wake is explored in the results and discussion.

2.3. The Cases

The cases studied in this investigation are reviewed in Table 3. The radiative equilibrium case has been shown to produce results with reasonable agreement to experiment [1,7,19], and so is the most extensively explored because of its potential to most closely match an actual system. However, the adiabatic and cold-wall cases are also explored in order to assess the degree to which the structure and composition of the wake vary with thermal conditions on the body. Ablating and non-ablating cases for each condition are run in order to ascertain the extent to which ablation impacts thermochemistry in the wake. Finally, catalysis is varied for each condition in order to understand the role of surface catalysis in wake properties. All cases are performed for the same 3.5 m long, 7 degree sphere-cone with a 0.25 m sphere radius at 30,000 m altitude and Mach 7.

Table 3. The various cases run for this study. All were performed at M = 7, 30 km altitude, 5° angle of attack. '*' denotes that these cases were run but are not explored in detail.

Case	Wall Boundary Condition	Wall Catalycity	Ablation
1	Radiative Equilibrium *	Non-Catalytic	Off
2	Radiative Equilibrium	Fully-Catalytic	Off
3	Radiative Equilibrium	Non-Catalytic	On
4	Radiative Equilibrium	Fully-Catalytic	On
5	Radiative Equilibrium	Equilibrium Catalytic	On
6	Adiabatic *	Non-Catalytic	Off
7	Adiabatic	Non-Catalytic	On
8	Adiabatic	Fully-Catalytic	On
9	Cold	Non-Catalytic	Off
10	Cold	Non-Catalytic	On

2.4. Detection

A final note regarding the detection of the species studied herein is warranted before proceeding to discussion of the results of the simulations described above. Detection of radiative emissions is a complex task which depends upon a multitude of case-specific factors such as viewer optical path, emission absorption by the intermediate medium in the particular spectra of interest, and background radiation *as viewed by the observer*, amongst many others [5,6]. Furthermore, the intensity of emissions from a given species depends upon the temperature and density of that species as integrated across a particular viewing angle, with the prior considerations then applied [5,6]. For these reasons, discussion of actual radiative emissions and the detectability of them is entirely foregone in the present effort. However, for several points of reference are here discussed.

The Earth emits essentially as a blackbody as seen from space at a temperature of 255 K [31], producing a substantial background radiance against which an object must stand out. This radiance has a peak energy occurring at a wavelength of $500/cm^{-1}$ at $\mathcal{O}(10^{-7})w/cm^3$-ster [31], where CO_2 has been shown for a Mach 6 sphere at 24.4 km to produce an intensity nearly an order of magnitude greater than the background level at $2400/cm^{-1}$ [6]. Thus CO_2 can play a significant role in detection. CN emits strongly around $390/cm^{-1}$, which is much closer to the peak energy spectrum emitted by the Earth, but may nevertheless emit sufficiently strongly as to be easily detected. For example, at 8000 K, CN emits at $\mathcal{O}(10^{-4})w/cm^3$-ster at this wavelength, thus overpowering the background noise by several orders of magnitude. Also for reference, NO emits at this same order of magnitude at the given temperature, making it likewise a strong potential candidate for detection, depending on optical depth, receiver sensitivity and nominal wavelength, and many other factors. These numbers are provided more to motivate the results to follow than to provide helpful bounds within the results of this study might reasonably vary; given the number of potential complicating factors, such bounds are beyond the scope of this effort.

3. Results

3.1. Validation of Results

Quantification of error is notoriously difficult for hypersonic simulations owing to a general lack of comparable experimental data and also to the complexity of the phenomenon of interest. Although a massive collection of data is available for near-surface hypersonic phenomena, particularly in application to blunt-body or spherical geometries at zero angle of attack, little emphasis has been placed upon flow features away from the body, in particular the wake. The problem of validation is further complicated by the tendency for studies to be performed at zero angle of attack. Since flow structures, and thus viscous and thermochemical phenomenon vary substantially with small changes in flight condition [32]. Furthermore, data for chemistry within the wake is particularly scarce, given that most wake studies identified in the literature have focused on afterbody heating and pressure distributions as opposed to the thermochemistry away from the body.

In light of the absence of applicable validation data for chemistry within the wake, which is the focus of this study, a first-glance sanity check of the flow around the body using some experimental results found in the literature for the flow around the body is now explored. Some theoretical prediction methods are also available to give a first-order approximation for a check of the results. The difficulty of matching flight velocities and densities in ground-based test systems, and measuring the wake at all in an actual flight vehicle, makes it difficult to find relevant data against which temperature-driven phenomenon and chemical kinetics may be checked. As a result, the most common data available is for pressure and Mach distributions in various forms. The flowfield solutions are thus checked against such data, but no attempt is made to validate the chemical species except to observe trends within the bounds of the parametric analysis.

The first mechanism of validation will be a comparison of stagnation region chemical makeup against theory and experiment. Huber (1963) extensively cataloged the properties of air behind a normal shockwave at various altitudes and velocities with accuracies estimated to be within 0.2% for a given set of input properties [33]. Huber predicts a post-shock temperature of 2200 K for a Mach number of 7.18 and altitude of 100,000 feet, with the stagnation temperature estimated—with a high degree of uncertainty—to be 2439 K. The temperature behind the shock is found to be 2390 K with a stagnation temperature of 2410 K. Given the 10% higher ambient pressure in the present study, the use of a sea-level gas composition in the Huber study, the close agreement of the present study with the predicted stagnation temperature, and the general uncertainty surrounding the chemical rates used herein, this agreement is found to be sufficient.

Since the bulk of chemical reactions occur in the stagnation region, the chemical makeup of the flow outside a region immediately off the wall should remain roughly constant along the body for the [2]. Hence if the bulk flow behaviors are demonstrated to be accurate, those portions of the flow whose chemistry is determined entirely by stagnation region phenomenon should remain as accurate as the stagnation region calculations allow, with the species merely being transported by convective phenomenon and mixed by any turbulent phenomenon prior to reaching the wake. Also, surface pressures downstream of the stagnation region do not change significantly with temperature or catalysis, which is also consistent with theory [2], lending further weight to the importance of demonstrating accurate bulk flow behavior. Finally, as shall be demonstrated later, the chemical effects of catalysis trapped captured within the boundary layer (or beneath it, on the leeward side) and largely limited to the base and core regions of the wake.

The accuracy of the pressure distribution downstream of the stagnation region is a helpful test against which the volumetric flow properties may be assessed. Classical Newtonian flow prediction for a cone of half-angle θ at zero angle of attack gives a rapid first estimate. Newtonian flow predicts a roughly constant coefficient of pressure across a hypersonic body, and does not consider the stagnation pressure or its downstream propagation. A full derivation of the pressure coefficient approximation is provided in Anderson's text, "Hypersonics" [2], and it ultimately yields:

$$C_p = 2\sin^2(\theta) = 0.0297 \qquad (8)$$

Figure 8a shows the coefficient of pressure for the subject cone, which is at an angle of attack. As expected, the coefficients of pressure for the top and bottom surfaces are higher and lower, respectively, than the Newtonian flow prediction, but fall within good agreement of the theory. Thus the plots in Figure 8 are for the non-ablating radiative equilibrium case only. Figure 9 shows the 3D pressure distribution along the body.

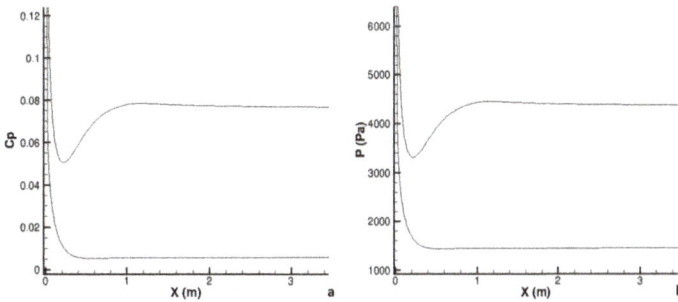

Figure 8. Coefficient of pressure (**a**) and Pressure (**b**) along the windward and leeward rays.

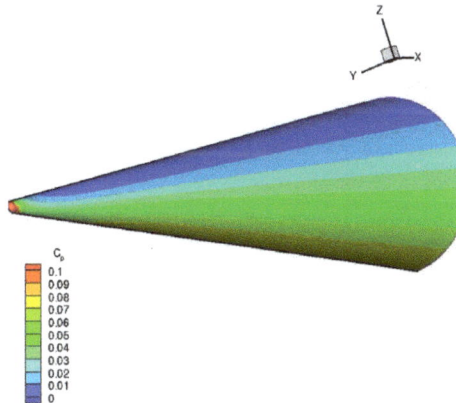

Figure 9. Coefficient of pressure for the non-ablating radiative equilibrium case.

Until relatively recently, thermochemistry in hypersonic wakes has not enjoyed a great deal of examination and indeed no experimental studies were identified which focused on chemical species within the wake near the relevant Mach number Reynolds number. Several studies of flow structures in laminar wakes behind blunt bodies were performed in the 1960's and backed by experimental data, and an excellent review of such literature until 1966 is provided by P.S. Lykoudis of the RAND Corporation [10]. Those studies were unfortunately limited by the diagnostic capabilities and also the detection technology of the time, which did not allow for substantial examination of the wake species or their utilization for tracking purposes. However, the studies gathered and summarized by Lykoudis did determine with confidence the behavior of the wake structure. Efforts at the time, and also the majority of available efforts since then have focused on the wakes behind blunt bodies in an effort to ascertain the environment encountered by interplanetary probes and manned spacecraft entering planetary atmospheres. Fewer studies have focused on the wakes behind sharp vehicles, and an accompanying void of published experimental data also exists for sharp flight vehicles, particularly at an angle of attack [13]. However, several recent studies do provide some basis for comparison of results from the present effort with an eye towards reproduction of general trends.

Examining data from the HIFiRE-1 test results [24], which flew a 7° half-angle cone with a 0.025 m spherical nosetip, at 481 s into its trajectory, the angle of attack was roughly 13°, with a freestream unit Reynolds number of roughly 2.5×10^6 per meter and an altitude in the vicinity of 30 km. These estimations are rough, as the uncertainty on any freestream value was approximately 25% [24]. Although the significantly higher angle of attack at this time would increase the windward surface pressure and temperature and decrease those of the leeward regions, this data provides a helpful glimpse into actual surface conditions around the cone in the flight regime of interest.

Figure 10 shows the pressure at four points along the surface of the HIFiRE-1 cone. The cone was determined to be rotating beneath flow-fixed features such that the transducers stationed radially around the vehicle were able to each obtain a read on the windward and leeward sides, yielding the oscillatory behavior seen above. From Kimmel, 2014: "Surface pressure traces from four different stations located at body-fixed coordinates (0.7013 m, 55°), (0.7013 m, 325°), (0.9263 m, 10°), and (1.0513 m, 10°) are shown in Figure 10. These coordinates correspond to the transducers PLBW10, PLBW14, PLBW4, and PLBW5" [24]. Given the significant differences in pressure readings from each transducer, the actual pressures can only be estimated, but range on the leeward side between 1 kPa and 3 kPa, reaching between 9 and 12 kPa on the windward side. Recalling that the angle of attack is close to 8 degrees higher than the present study, pressures on the two sides can be assumed to be significantly closer to one another and between the bounds seen above. Referring back to Figure 8b, the surface pressures fall squarely within such a range, between 1.5 kPa and 4.5 kPa.

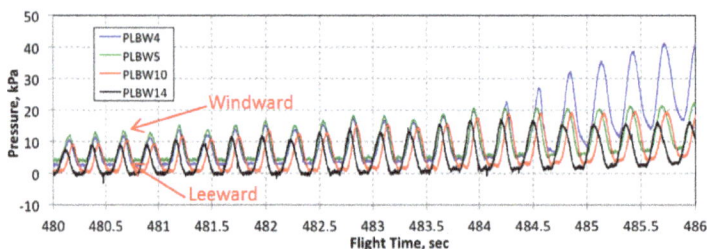

Figure 10. Pressure readings along the HIFiRE surface, adapted from the HIFiRE-1 and HIFiRE-5 Test Results report in 2014 [24].

A final note on HIFiRE-1 regarding its surface temperature is warranted here. HIFiRE-1 had thermocouples along its length and distributed radially in order to help locate the boundary layer transition front. The vehicle was designed to be non-ablating, and followed a ballistic trajectory. As a result, the total flight time was around 10 min, leading to essentially a cold-wall condition over the cone surface. Thus the surface thermocouples detected a temperature at the wall of 430 K at the end of the cone at 482 s. Thus the wall conditions for the cold-wall case here could be expected to produce similar results as the HIFiRE-1 flight [24]. Other flight vehicles with longer range and a lifting body intended for hypersonic glide would spend substantially more time in an environment of extreme heating, and would thereby be better represented by the hotter wall boundary conditions.

Several other experimental studies were identified that were conducted under varying circumstances which taken together provide some insight into the accuracy of the flowfields presented here. The first is the study conducted by Lin et al., 2006, "Hypersonic Reentry Vehicle Wake Flow Fields at Angle of Attack" [16]. Lin et al., 2006 compiled much data regarding the base pressure from multiple wind tunnels and methods of experimentation with Mach numbers ranging between 6 and 10, with Reynolds numbers ranging from 0.5 million to 4 million per foot. The experiments were conducted on cones with base ratios ranging from 0 (sharp) to 15%, with a 7 to 10 degree half-angle. Conic surface pressure, or the pressure on the surface of the cone at zero angle of attack, was used to non-dimensionalize the data, forming the pressure ratio $\frac{P}{P_{cone}}$. Ultimately, the experiments ranged in base pressure ratios from 0.03–0.1 depending on the method and conditions used. The cone of interest to this study has a bluntness ratio of 6% at a per meter Reynolds number of 2,900,000 or 883,920 per foot, and so it again falls nicely between the experiments' ranges. In this study, the conic pressure at zero angle of attack was estimated from Newtonian flow theory as:

$$P_{cone} = C_p q_\infty + P_\infty = 2443 \, \text{Pa} \tag{9}$$

Here, $C_p = 0.0297$ as estimated previously, and $q_\infty = 41{,}200$ and $P_\infty = 1220$ Pa are the freestream dynamic and static pressures, respectively. Examining Figure 11, the base pressure for this study ranges roughly from 0.03 to 0.05 depending on the surface conditions. This data matches very well with the tunnel data summarized in Lin et al., 2006, lending confidence to the present study's analysis of the near-wake region.

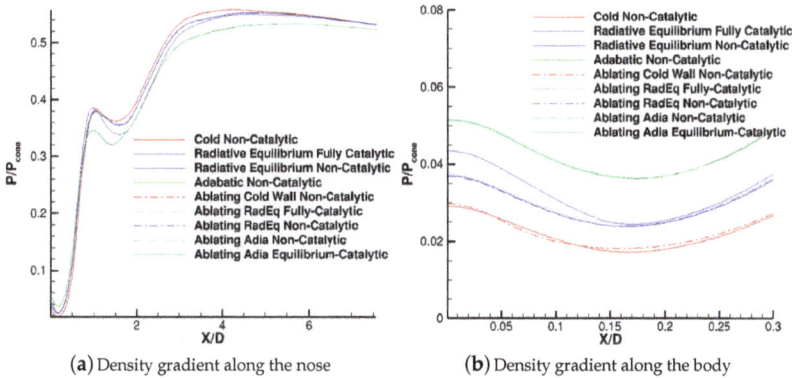

(**a**) Density gradient along the nose

(**b**) Density gradient along the body

Figure 11. Non-dimensional pressure ratio (**a**) along the wake centerline and (**b**) also those in the base region of the wake along the centerline. Normalized by $P_{cone} = 2443$ Pa.

Furthermore, Lin et al., 2006 plots the centerline Mach number from a number of wind-tunnel experiments, as well their own CFD for a Mach 6, 10° half-angle cone against a non-dimensional $\frac{X}{D}$. The trends from experiment and CFD in that study are very closely matched. Using several representative locations for example [16], Figure 12 shows excellent agreement between the present CFD and experiment. Although some variation from the experimental data is expected due to Mach and geometric differences, one of the key findings of Lin et al. was that variations in the profile seen in Figure 11 are slow with Mach and geometry changes.

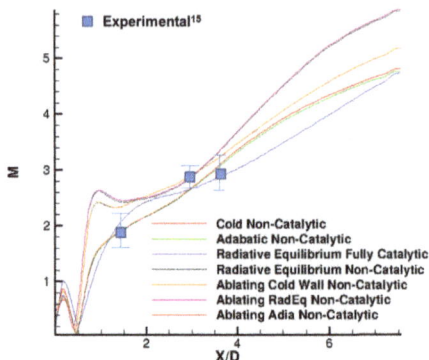

Figure 12. Wake Centerline Mach number. Bars on experiment show range of experimental results [16].

3.2. Flowfield Structure

Confident that the bulk qualitative behaviors of the flow have been accurately produced according to several methods of analysis, we can now examine the flow structure more closely. Figure 13 shows the total density gradients throughout the flowfield, calculated as the RMS of density gradient in each direction:

$$d\rho = \sqrt{\left(\frac{\partial \rho}{\partial x}\right)^2 + \left(\frac{\partial \rho}{\partial y}\right)^2 + \left(\frac{\partial \rho}{\partial z}\right)^2} \tag{10}$$

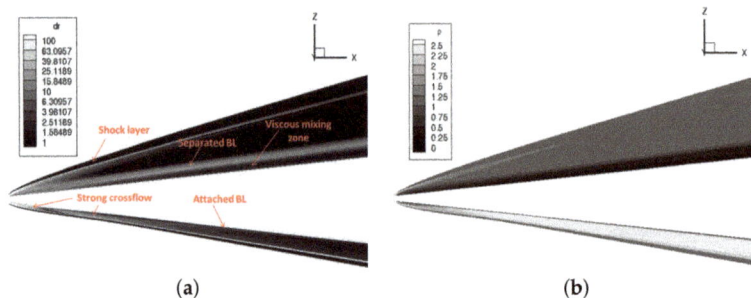

(a)   (b)

Figure 13. Cone flow features at Mach 7, 30 km altitude, and 5 degrees angle of attack. (**a**) Density gradients along the cone; (**b**) Densities along the cone.

The structure of the flow is strongly influenced by angle of attack, producing vastly different results on the windward and leeward rays. The windward side boundary layer remains attached along the length of the body, despite a powerful cross-flow as hot, pressurized flow from the windward side is pulled into the vacuum of the leeward side. Although its effect on the flow is most dramatic near the front of the body, the cross-flow continues along the length of the body, leading to complex

3-dimensional features in the wake base. Additionally, this cross-flow has been shown to cause significant boundary layer instability, further exacerbating the problem of identifying the location of transition to turbulence [24]. The curvature seen in the windward shock is also caused by this cross-flow, although the grid alignment routine pulled the outer boundary of the domain closer to the shock than the surrounding flow in that area, exaggerating via optical illusion the actual curvature of the shock. The flow over the leeward body can be divided into five distinct regions moving in from the outside:

1. Freestream
2. Shock layer
3. Inviscid zone
4. Separated Boundary Layer
5. Viscous Mixing Zone

Each of the latter three contribute significantly to the structure and composition of the wake, and their composition and behavior depend greatly on the surface conditions encountered. In the stagnation zone, a thin boundary layer forms immediately as the flow works its way around the spherical nose of the cone, separating as it goes due to the angle of attack. Powerful gradients in temperature and pressure through the boundary layer drive rapid thermochemical changes, which then interact with the viscous mixing region beneath the separated boundary layer. This mixing region draws species from the surface catalysis and is a primary source of observable species in the wake.

The boundary layer and mixing zone draw from generally different pools of chemical species, the former drawing greatly from the stagnation region while the latter is fed by catalysis along the length of the body. Downstream of the nose, because of mixing caused by shear forces between the layers, the species from the outer layer are able to diffuse to the surface, and vice-versa. Above these two layers, the inviscid region does not experience substantial gradients except at its boundary with the shock layer. This layer does not contain substantial quantities of chemically reactive products, as most of the products generated either by chemistry near the surface have been caught up in the viscous interaction close to the nose and remain close to the body. The oblique shock is not sufficiently strong in this study to produce significant chemical reactions. However, some species do diffuse out of the lower two layers. In particular, carbon-species ablated near the nose, which are chemically quite stable, slowly diffuse out of the separated boundary layer.

As the layers travel downstream along the body, each experiences relaxation and recombination of its species at rates unique to each zone. When the end of the body is reached, several distinct phenomenon can be observed as the zones are forced to meet by the wake structures. Showing x-oriented density gradients $\frac{\partial \rho}{\partial x}$, and pairing it with a chart of streamlines, Figure 14 shows that much of the mixing zone beneath the boundary layer is pulled directly into the recirculation zone, heating up through a shock bounding the subsonic recirculation zone and expanding supersonic zone just outside of it. The rest of the viscous zone is pulled along by the boundary layer, expanding in all directions and mixing in the complex three-dimensional flow between the neck and the viscous mixing layer. The viscous mixing layer itself is turned upwards first by the expansion fan at the corner, and still further by the recompression shocks formed at the neck. Importantly, the 3D cross-flow swirling flow from bottom to top adds to the complexity of the base flow, bringing from out of plane the same phenomenon described above with the addition of a strong vortex.

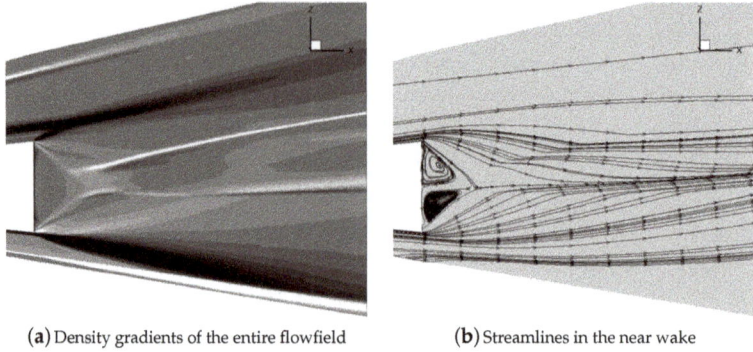

(**a**) Density gradients of the entire flowfield (**b**) Streamlines in the near wake

Figure 14. (**a**) Plots of density gradients $\frac{\partial \rho}{\partial x}$ in the near-wake and (**b**) Streamlines to match it for the ablating adiabatic, non-catalytic case.

The complex interactions of these various layers in the wake, combined with the chemical reactions which took place in each layer as flow traveled along the body, create the conditions for still more complex chemistry. Even for flows which were considered to be in chemical equilibrium along the body, the wake is frequently not in chemical equilibrium but is rather chemically frozen [5], making accurate characterization of the chemistry along the body, which forms the boundary condition for the chemically reacting wake, even more critical.

The chemical composition of the wake is seen to vary dramatically depending on body temperature, which is a function of and strongly affects the two innermost layers. Since the nose of a slender vehicle is frequently of a different internal structural material and TPS than the body (see HIFiRE [24] or the RAM-C experiments [29] for examples), the various thermal boundary conditions discussed previously come into play as we try to predict the chemical behaviors exhibited by the flow. Figure 15 shows the powerful role that surface temperature plays within the wake. Figure 15a shows the surface and wake temperatures with each thermal boundary condition at the non-catalytic setting. Each condition has two lines, one representing the top surface and one representing the bottom. At no point in any simulation is the top surface the hotter, so the top line consistently represents the bottom surface, and the bottom line represents the top surface.

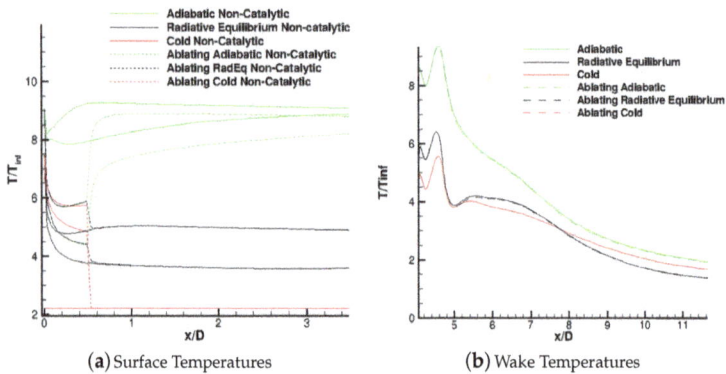

(**a**) Surface Temperatures (**b**) Wake Temperatures

Figure 15. (**a**) Temperatures along the top and bottom surfaces of the body for each thermal boundary condition (**b**) and along the wake centerline.

Clearly, the boundary conditions downstream of the ablating nose play a significant role in conditions in the wake. Interestingly, however, each ablating case was run with the same surface energy balance on the nose; the temperature profile for the ablating portion of the body is identical over the first 0.45 m for the ablating cases. After $x = 0.45$ m the various surface temperature conditions being applied downstream of that section. The surface temperature at the stagnation region reached 2050 K for all ablating cases. However, as Figure 15b shows, the response of the body downstream of the nose ultimately determined the temperature within the wake independent of conditions at the nose. Figure 15a shows the surface thermal boundary conditions which led to the observed wake temperatures.

3.3. Cone Flow Chemical Composition

3.3.1. Stagnation Region

Although the wake temperature was found to be primarily a function of conditions downstream of the nose, the species generated in the stagnation region, especially ablating species, are found to play a very strong role in determining wake species. The next series of figures shows the chemical composition of the flow based on surface temperature in the form of either a number density, N of each species or a number density divided by a normalized local density according to the following equation:

$$n_S^* = n_S \frac{\rho_\infty}{\rho} \tag{11}$$

Here, S is a given species, the "$*$" denoting a normalized number density. This normalization shows production or depletion of a given species independent of flow density, an increase in which would otherwise show an increase in number density regardless of whether a species was being chemically produced or not. However, for an observer attempting to detect the flight vehicle as it passes by, actual number density is more helpful. Thus for discussion of wake species directly related to detection—carbon dioxide, nitrosonium, nitric oxide and free electrons—number densities are used, $n[S]$, but for discussion of species production or depletion along the body, number density is normalized by the local flow density. Figure 16 compares the trends for number density vs. normalized number density for nitric oxide in the stagnation streamline; although the number density of NO increases nearly an order of magnitude across the shock in (a), when normalized by density the actual production of NO, which one would not expect to observe in significant levels until higher temperatures, is seen to be negligible through the shock.

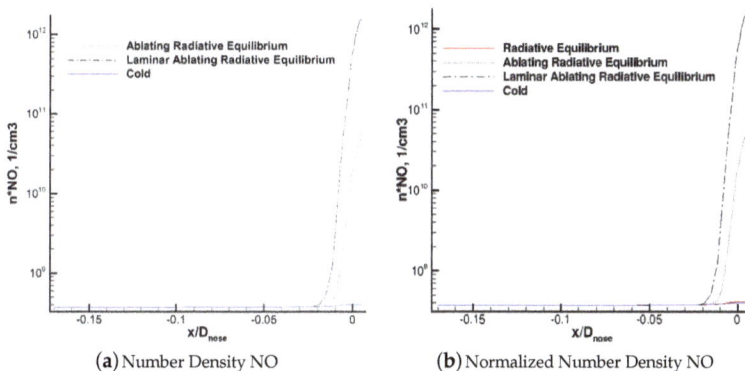

(a) Number Density NO (b) Normalized Number Density NO

Figure 16. (a) Nitric oxide NO number density and (b) Normalized number density of NO along the stagnation streamline.

Again referring to Figure 16, the first species of interest is nitric oxide NO, because of its chemiluminescent properties in the far wake as it recombines with an atomic oxygen molecule into an electrically excited NO_2, which in turn emits a photon to relax [5]. Of note, NO exists in trace levels in the atmosphere owing to pollution and, to a lesser extent, its production by lightning. The higher the density of NO, the more chemiluminescence that will be observed in the far wake [5]. NO is formed by the reaction $N_2+O==>NO+N$. Since oxygen is the first species to dissociate, nitric oxide can occur in significant quantities under conditions in which it neither recombines into NO_2 nor reacts with free oxygen atoms in the reaction $NO+O==>O_2+N$ [5]. Thus the levels of NO may remain high if atomic oxygen is being consumed at high rates via some other reaction (from, say, reactions with ablative materials). Here, neither the cold nor the non-ablating radiative equilibrium case at first appear to produce significant levels of NO. However, examining the production of N in Figure 17, we find that at the temperature of interest the rate of NO breakdown into O_2 and N outpaces that of production of NO [5], and so no sign of NO appears until closer to the nose, whereas levels of N increase substantially.

It is emphasized here that atomic nitrogen is not being produced via dissociation, given that the temperature is far too low for that reaction. Rather, the mild dissociation of oxygen produces a chain of chemical reactions which ultimately results in the production of atomic nitrogen, while leaving molecular oxygen unchanged. Low level production of atomic nitrogen at Mach 7 is supported by Bussing et al., 1989 [34].

In the ablating cases, the presence of nitric oxide increases by 3 and 4 orders of magnitude, respectively. The reason for this is that the ablating cases introduce carbon species to the mixture, which consume large quantities of O_2 and O, thereby preventing the reaction $NO+O==>O_2+N$. The higher rates of NO then lead to increased levels of NO^+ due to the collision $NO+M==>NO^+ + e^- + M$, where M can be either O_2 or N_2. The increased production of nitrosonium due to increased NO levels is seen in Figure 18a for the ablating cases. The turbulent case introduces more carbon species into the fluid because of its higher mixing rate, which in turn consumes more atomic oxygen. Figure 19 shows the plummeting levels of oxygen species close to the surface where carbon species including C, CN and CO_2 appear at substantial levels. C_2 and C_3 are not present at significant levels, but at higher temperatures can be present in substantial quantities. Although the turbulent case produces more pure carbon species, the laminar case produces higher levels of CO_2 owing to its higher levels of atomic oxygen through the reaction $O+O+C==>CO_2$.

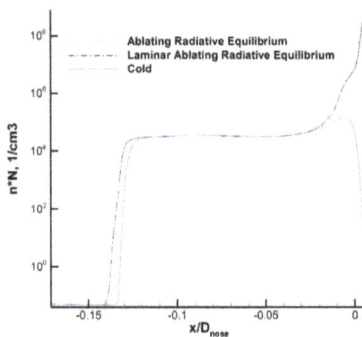

Figure 17. Stagnation region atomic Nitrogen production due to $NO+O==>O_2+N$.

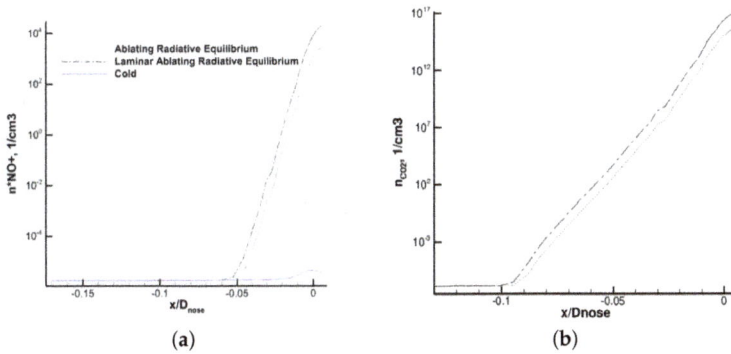

Figure 18. Normalized number densities of nitrosonium and carbon dioxide in the stagnation region. (a) Nitrosonium due to $NO+M==>NO^+ + e^- + M$; (b) Carbon Dioxide.

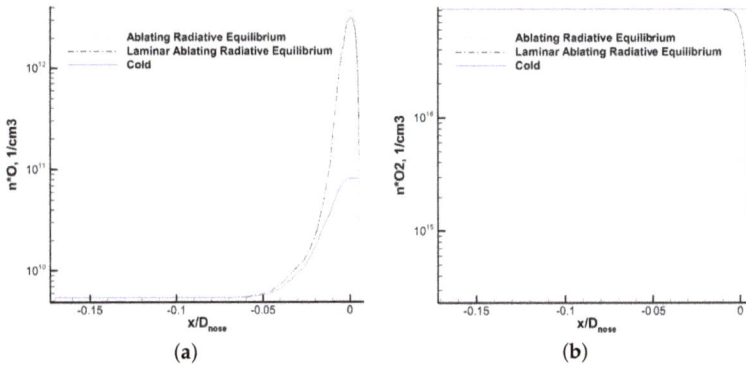

Figure 19. Stagnation region oxygen species by normalized number density. (a) Normalized number Density O; (b) Normalized number Density O_2.

3.3.2. Characterization of the Wake

With observable species in the stagnation region identified and their origins understood, the significance of thermal boundary conditions downstream of the ablating nose can now be investigated. Although only the cold-wall case is represented for the non-ablating possibilities, a radiative-equilibrium, non-ablating case was also solved. All comments applying to the cold-wall case similarly apply to the radiative equilibrium case, with insignificantly small deviations in observed species values. Figure 20 shows kinetic and vibrational temperatures of the flowfield for the ablating radiative equilibrium case. While the values shift significantly between the various surface boundary conditions, the overall structure is generally consistent across cases. The largest structural change is a small shift in neck location depending on the temperature at the back of the body, and also a substantially different density distribution within the core. A plot of density through the wake core is provided in Figure 21 for reference.

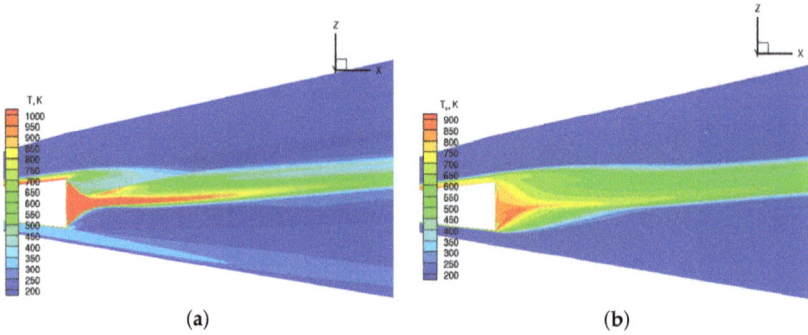

Figure 20. Temperature profiles for the ablating radiative equilibrium case. Values shift significantly for other surface temperature cases, but the structure remains quite similar. (**a**) Kinetic Temperature; (**b**) Vibrational Temperature.

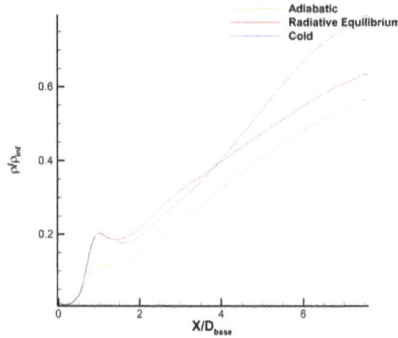

Figure 21. Wake Centerline density for the ablating cases normalized by its freestream value.

Another interesting perspective on the thermal environment of the recirculation zone and the formation of the neck is provided by examining the gradients of temperature in that zone. Figure 22 examines the temperature gradient with respect to x for the adiabatic non-catalytic abalting case, $\frac{\partial T}{\partial X}$. This data is presented in greyscale in order to clearly show zones of cooling (black) vs. heating (white).

As the flow expands around the corner, it cools down, but the area of cooling is visibly separated by the shock which contains the subsonic recirculation zone, which heats the flow again. Within the recirculation zone, the flow continues to cool as it continues expanding around the corner. However, as the flow is compressed towards the center with its rearward motion, it heats up again. The recompression shocks from this meeting of top and bottom flows are clearly visible, although the bottom recompression shock is significantly weakened and moved upwards by the bottom expansion fan. The dark spot of cooling moving from the top cone corner into the neck is formed by a powerful swirl that encompasses the recirculation zone pulling the cooling, expanding flow from the cooler edge of the cone in the third dimension into the neck. Although not shown here, substantial heating in the y and z directions is visible in that area.

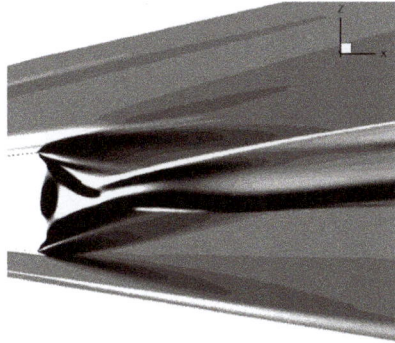

Figure 22. Gradients of temperature in the near-wake for the adiabatic non-catalytic ablating case.

With the above discussion of conditions which feed the wake in mind, this section will examine wake species. Figure 23 shows the distribution of cyanogen and electron number density for the turbulent ablating radiative equilibrium case. For all cases, whether ablating or not, the distribution of any observable species is nearly identical. Although some ablative materials such as CO_2 and CO, surround the core in a broad halo, having diffused from the edges of the stagnation zone and separated boundary layer, the temperatures outside the core and viscous mixing zone are sufficiently cool as to merit these products' contribution to radiation insignificant.

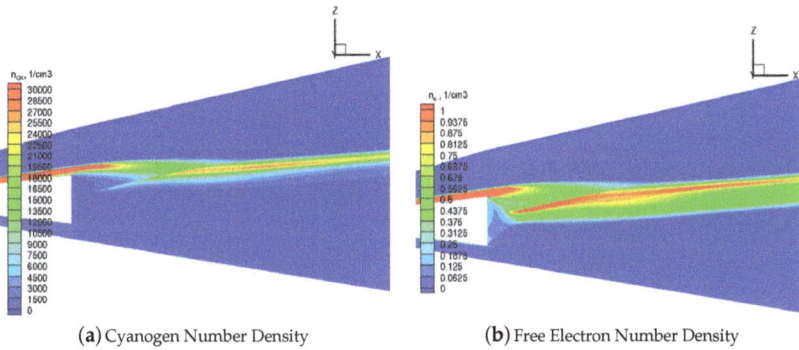

(a) Cyanogen Number Density **(b)** Free Electron Number Density

Figure 23. CN (**a**) and e⁻ (**b**) number densities in the turbulent ablating radiative equilibrium case.

The wake is seen to be non-symmetric because of the angle of attack. Thus species are concentrated more highly above the core centerline than below it, and indeed the upper recompression shock forces much of the flow from the viscous mixing zone to remain above the wake core, and seen by the divergence of the mixing and boundary layers' streamlines in Figure 14 downstream of the neck. Note also that the number density of electrons is identical to that of the nitrosonium density since nitrosonium is the only source of free elctrons. Thus, the two shall be considered synonymous for much of the rest of this investigation.

Species will now be discussed in terms of a cross-sectional slice seen in Figure 24, which is placed just upstream of the neck region. Because of the varying thermochemical conditions which feed the recirculation zone, the location of the neck shifts slightly with each case, allowing a consistent analysis to be performed across the cases. The placement of this slice allows a clear observation of the various wake structures as produced by the flow structures over the body. Figure 24b shows the temperatures in the cross-sectional wake.

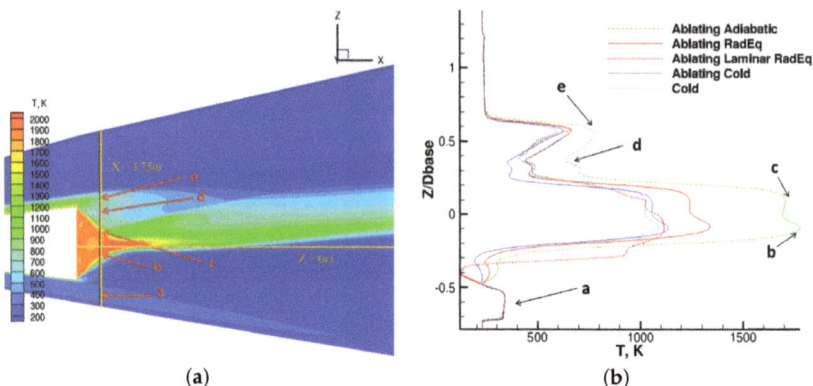

Figure 24. Location of and temperatures in the wake cross-section. (**a**) Adiabatic wall condition with wake cross-sectional slice; (**b**) Wake cross-sectional temperatures.

The locations highlighted in Figure 24 are created as the flow rounds the corner behind the cone. These locations are:

(a) Lower oblique shock
(b) Base zone bounding lower shock
(c) Base zone bounding upper shock
(d) Upper Expansion zone
(e) Expanding mixing and boundary layers

In the following figures, these features are evident in many of the plots, in particular the laminar ablating case. The dependence of the observable species upon conditions at the nose shall now be investigated using the same density-normalized notation as was used in the stagnation region. Figure 25 shows the distribution of NO and NO^+ in the wake cross section. The NO levels at the nose were of the order $1 \times 10^{11}/cm^3$ in the stagnation region for the turbulent cases. In the wake, they remain close to that order for the hotter-walled cases, at $8 \times 10^{10}/cm^3$. However, for the ablating cold-wall case, the order has dropped to $5 \times 10^9/cm^3$. For the non-ablating cold-wall case, where very little NO was created in the first place, no NO remains above free-stream levels. Importantly, none was created in the recirculation zone, which remained at the relatively low temperature of 1100 K due to the cold wall. Also, the level of NO increases dramatically for the warmer-walled cases just above the upper bounding shock, point A in Figure 25a, indicating that the bulk of NO in the wake is dumped in from the separated boundary layer and viscous mixing zone. In light of these observations, it is concluded that NO produced at the nose is the primary source of that species, but its level is dramatically effected by the degree of ablation and wall temperature downstream of the nose.

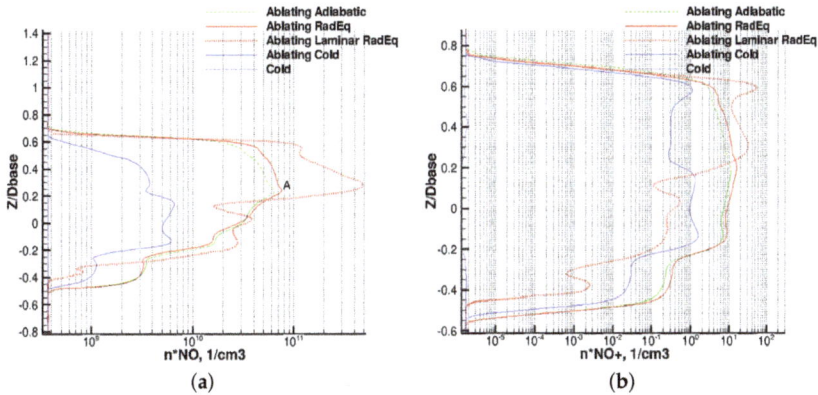

Figure 25. Normalized NO and NO$^+$ distribution in the wake cross section normalized by non-dimensional local density. (**a**) n^* Nitric Oxide; (**b**) n^* Nitrosonium.

Nitrosonium, following similar trends but with a depletion from $1 \times 10^4/cm^3$ in the stagnation region to the order of $1 \times 10^1/cm^3$ in the wake. The cold-wall ablating case fared better here, reaching the order of unity but still remaining quite low. Carbon dioxide and cyanogen, pictured in Figure 26, follow very similar trends. CO_2 levels peak at $1 \times 10^{15}/cm^3$ as in the stagnation region, and CN peak at $1 \times 10^5/cm^3$ for all cases, a good deal lower than the $1 \times 10^7/cm^3$ levels reached in the stagnation region.

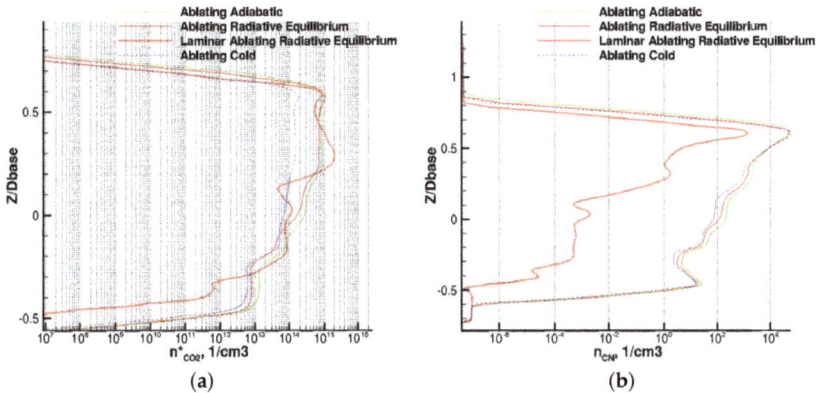

Figure 26. Normalized CO_2 and CN distribution in the wake cross section, normalized by non-dimensional local density. (**a**) N^* Carbon Dioxide; (**b**) N^* Cyanogen.

With the sources of observable species accounted for, discussion shall hereafter focuses less on the chemistry and more on the products of the chemistry; the non-normalized number densities of those chemical species which lead to the possibility of detection. Figures 27 and 28 show the non-normalized number densities of the observable species in the wake cross-sectional slice. With the normalization removed, most species show a decrease in number density. CO_2 is seen to reach $1 \times 10^{15}/cm^3$ in the wake, while CN reaches $1 \times 10^4/cm^3$. Levels of NO actually drop below freestream values in the recirculation zone, but otherwise reach one order of magnitude higher than freestream levels. NO$^+$ reaches 5 orders of magnitude higher than the freestream values. The non-ablating cold-wall case failed to produce any detectable species, emphasizing the critical role that the ablative species play in production of an electromagnetically active wake. Given the substantially lower levels of air-species NO and NO^+ in the cold-walled case than its warmer-walled counterparts, it is concluded that, carbon-based species from ablation aside, cooler walls—regardless of nose conditions—produce significantly less visible wakes.

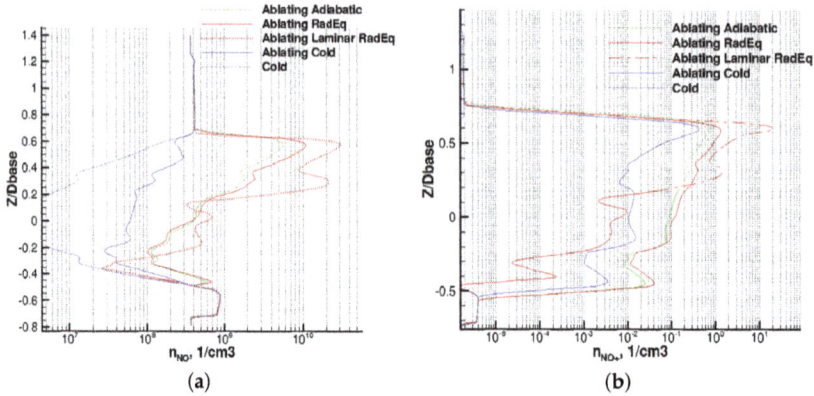

Figure 27. NO and NO^+ number density in the wake cross section. (**a**) N Nitric Oxide; (**b**) N Nitrosonium.

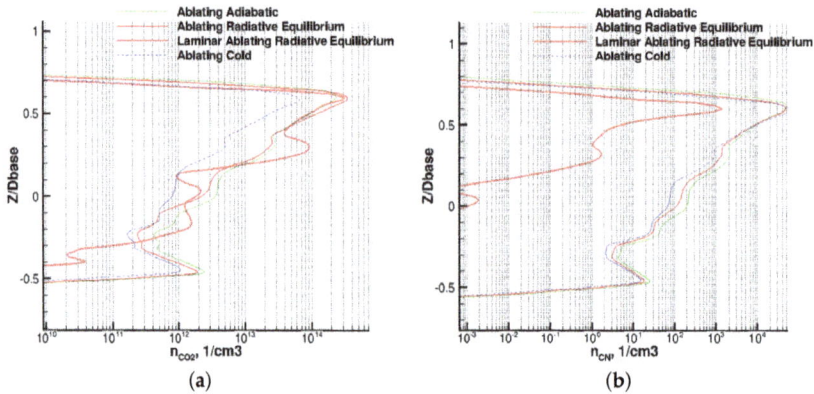

Figure 28. CO_2 and CN number density in the wake cross section. (**a**) N Carbon Dioxide; (**b**) N Cyanogen.

Given the results discussed above, the non-ablating cases are not explored further. The laminar case is also explored no further, because downstream of the neck, to the extent that laminar flow may exist at the present angle of attack over the body of the vehicle, it is the opinion of the authors that flow downstream of the neck is certainly turbulent. As a result, the laminar case produces non-physical results downstream of the neck which, regardless of relaxation settings, CFL or other numerical settings the authors could not remove. Given the highly supersonic nature of the wake flow downstream and outside the neck, the downstream instabilities are not anticipated to have influenced the upstream results significantly.

Visible species in the wake core will now be investigated. Figure 29 repeats Figure 15 for convenience, including now the number density of CO_2 in the wake core, revealing several interesting features. First is that, as the wake rapidly cools beyond the neck, CO_2 and other species from the impinging boundary layer and mixing region pour into the core, increasing the number densities of a every species significantly. All species increase the number density by a factor of 4 owing to the recovery to freestream densities from vacuum of the base. Despite this increase, given the very cool temperatures downstream of the neck, it is unlikely that the carbon-based species will be radiating at levels above the ambient noise. Furthermore, the species then begin to dissipate as they mix with the outer layers of the wake and the freestream beyond.

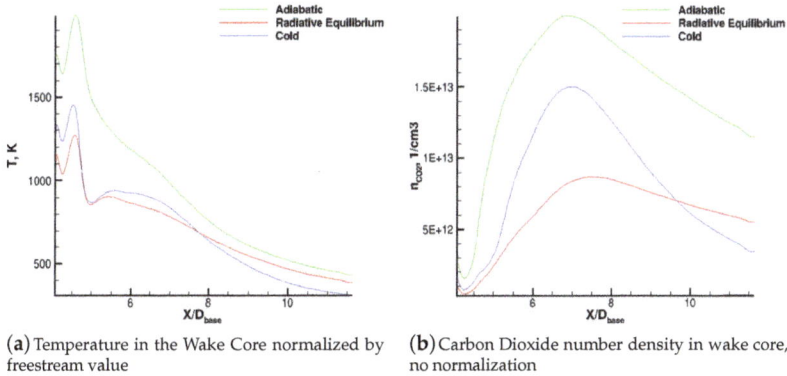

(**a**) Temperature in the Wake Core normalized by freestream value

(**b**) Carbon Dioxide number density in wake core, no normalization

Figure 29. (**a**) Temperature and (**b**) carbon dioxide number density in the wake core.

A second observation is that the recirculation zone produces a very hot environment rivaling the nose stagnation temperature for the adiabatic case but is otherwise significantly cooler than the recompression zone. Number densities of the species of interest climb for all cases as they approach the throat, and diminish from there. This may seem counterintuitive; if the recirculation zone is quite hot and is furthermore fed by the boundary layer products, one may intuit that it should in fact have a higher chemical composition. In fact, the chemical concentrations of the chemically reactive species are quite high in the recirculation zone as seen previously, but because the density is so low in that zone, the neck and wake actually contain higher number densities.

3.4. Catalysis

The cases discussed thus far have been uniformly non-catalytic. However, catalysis can also produce substantial differences in heat transfer to the body and the resultant chemistry around the body and in the wake [4]. As flow temperatures increase, the influence of surface catalysis also increases. The reason for this is that with higher temperatures, increased levels of atomic species are recombined on the surface, and the increased energy required to hold the atomic state, released by the catalysis, is dispatched into the wall with the result of higher heat transfer at the wall [4]. The preceding discussion is not repeated in full to include the effects of catalysis, but the results are shown at the wake cross section in Figure 30 for the ablating radiative equilibrium case. Catalysis was set to "Fully Catalytic", meaning that all atomic species are catalyzed into molecular species, making the ratio of specific heats of each species, $\gamma_S = 1$ on the surface where γ_S is the ratio of specific heats of each species. This is in contrast to the setting by the non-catalytic condition of $\gamma_S = 0$.

Figures 30–32 reveal that, although the wake temperature is not necessarily strongly affected by the catalysis, the chemistry is substantially shifted by its presence. Catalysis results in a roughly two-thirds reduction in observable species across the board, with the exception of CN. The reason for this reduction is the lower levels of atomic species which are generally responsible for production of observable species. In the catalytic condition, these species become more scarce near the surface, which is the only zone where they were generally to be found, as seen in Figure 32b.

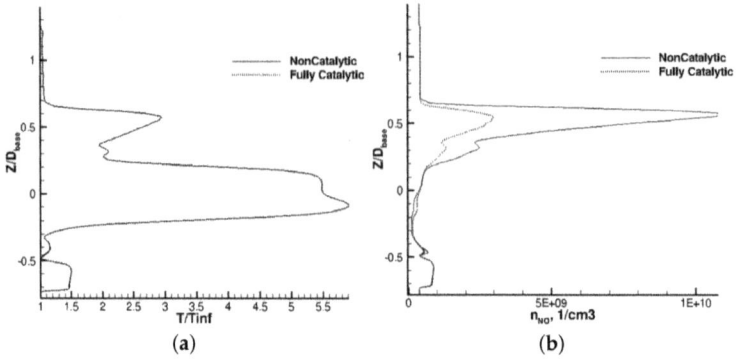

Figure 30. Effects of catalysis on the radiative equilibrium ablating case. (**a**) Temperature; (**b**) Nitric Oxide.

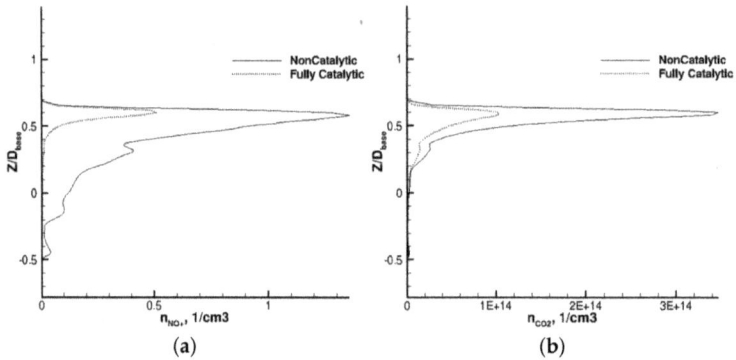

Figure 31. Effects of catalysis on the radiative equilibrium ablating case, continued. (**a**) Nitrosonium; (**b**) Carbon Dioxide.

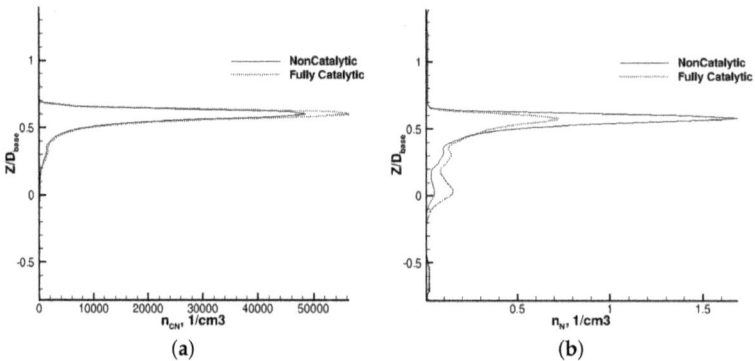

Figure 32. Effects of catalysis on the radiative equilibrium ablating case, continued. (**a**) Cyanogen; (**b**) Atomic Nitrogen.

3.5. Reduction in Ablative Surfaces

The final area of analysis is the effect of increasing or decreasing the amount of ablating material in the flowfield. While from the above discussion it may seem reasonable to conclude that less ablation might lead to a smaller degree of observable species, the following demonstrates that the relationship between ablating products and visible products in the wake is non-linear. Figure 33 shows the comparative sizes of the regions of ablation. The cases discussed previously were run with the "Big" ablation scenario, wherein a portion of length 0.45 m was allowed to ablate, although the blowing rates show that the amount of ablation downstream of immediate nosetip was only 20% that of the stagnation zone. However, despite the small rate, the extra length added an order of magnitude more CO_2 into the flow, as seen in Figure 34b. Figure 34 is taken at the wake cross-sectional slice and also shows the temperature distribution through the wake, which is unaffected by the temperatures at the nose. The decrease in CO_2 allowed more atomic oxygen produced in the stagnation zone to persist, producing counterintuitive results depending on wall temperature downstream of the ablating portion.

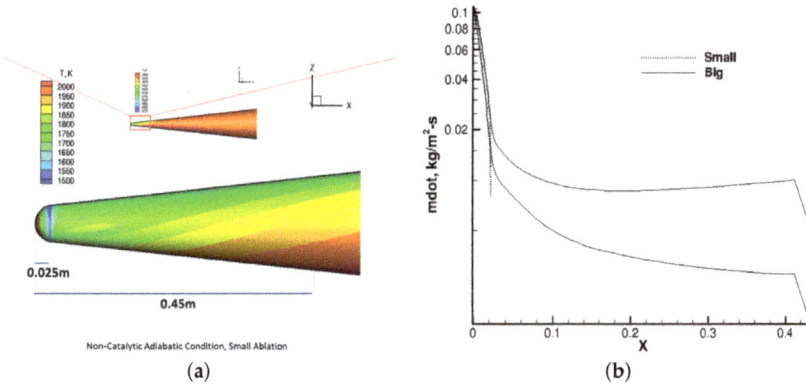

Figure 33. Size of ablating portions of cone for the non-catalytic adiabatic case. (**a**) Zones of ablation; (**b**) Ablation Blowing Rates.

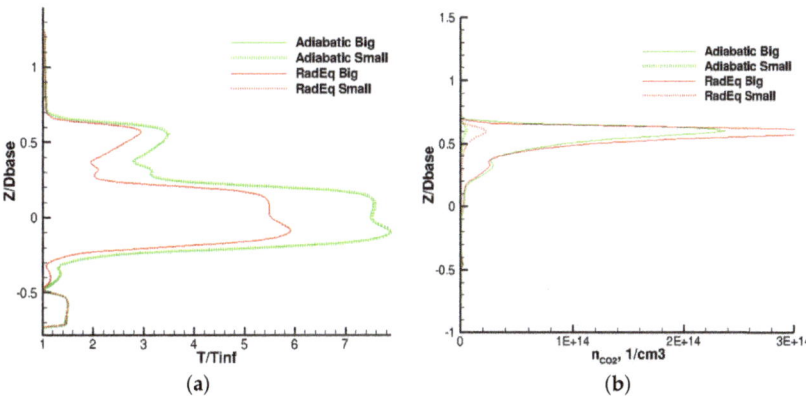

Figure 34. Wake slice temperature and carbon dioxide distribution for the large and small ablating cases. (**a**) Normalized Temperature; (**b**) Number Density of CO_2.

Figure 35 shows the NO and electron number densities. NO is seen to increase by a factor of two for the smaller radiative equilibrium case, whereas it decreases by two thirds for the adiabatic case. The reason for this becomes apparent in Figure 36, where the adiabatic case produces twice as much CN due to its very high temperatures, consuming the freed atomic nitrogen. In the radiative equilibrium case, the atomic nitrogen is instead consumed by the formation of N_2 and NO, as is the freed atomic oxygen. Finally, referring again to Figure 35, electron density for NO jumps an order of magnitude for the smaller radiative equilibrium case as the extra NO collides with molecular oxygen and nitrogen to form nitrosonium. Electron density also jumps by a factor of five for the smaller adiabatic case by the collision of the freed atomic nitrogen and oxygen to form nitrosonium and its associated ions.

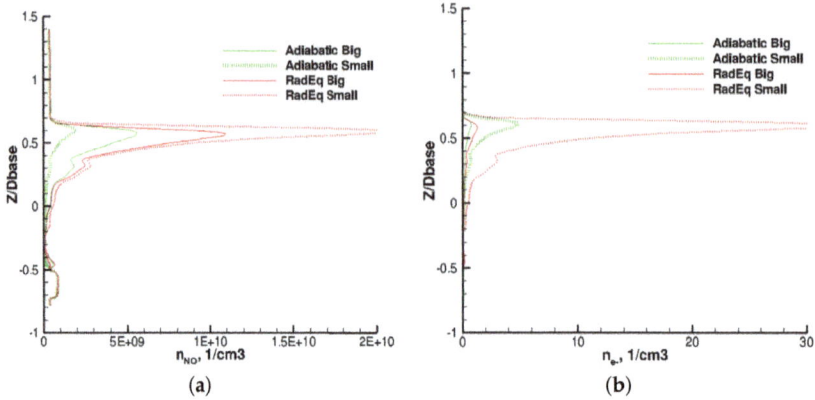

Figure 35. Wake slice nitric oxide and electron distribution for the large and small ablating cases. (a) Number Density of NO; (b) Electron Number Density.

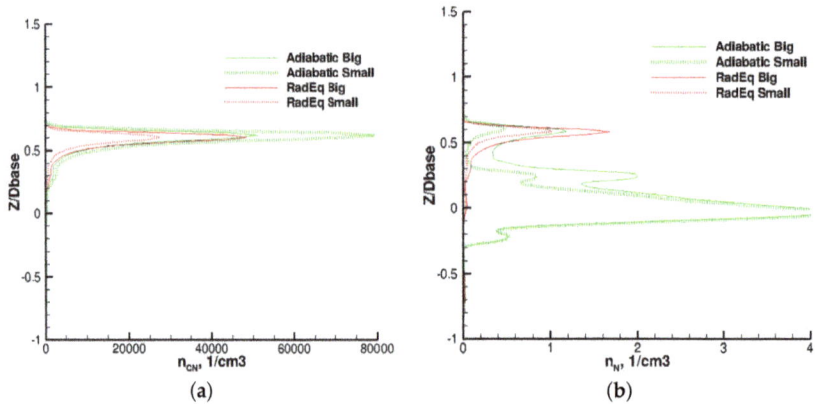

Figure 36. Wake slice cyanogen and atomic nitrogen for the large and small ablating cases. (a) Number Density of CN; (b) Number Density of N.

4. Conclusions and Recommendations

The purpose of this study was to understand the fundamental flow phenomenon which affect the structures and chemistry within the wake of a generic sphere-cone traveling at moderate angle of attack. A special emphasis was placed upon the chemistry which leads to possible avenues of detection through the electromagnetic spectrum. The carbon dioxide, nitric oxide, cyanogen and electron density were used to assess the overall visibility of a vehicle. Because of the parametric nature of this study, little information was assumed about the structural materials used or the thermal protection system implemented to protect the internal structure except that ablating material was made purely of carbon. Given the broad range of potential designs of hypersonic reentry systems, in addition to the possibility for a broad range of changes to the form factor which would keep the geometry close to that of the present shape while allowing some degree of maneuverability, those features of the vehicle which might significantly affect the structure and chemistry of the wake were examined. In particular, the effects of ablation, catalysis and heat transfer into the body were analyzed in order to provide some reasonable bounds within which such a system might be expected to operate. The impact on wake behavior of using a nose of a different material from the rear surfaces was also assessed.

Heat transfer to the body was modeled in the form of various boundary conditions. A cold-wall boundary condition was used to simulate a maximally efficient, actively-cooled wall. A wall which was able to radiate heat away, or otherwise absorb into the internal structures energy at a rate which was commensurate to that rate at which a carbon black body would emit the same energy, was also studied. Finally, a system which was able to absorb the enormous heat fluxes of hypersonic flight and survive while operating at extremely high temperatures was also studied. The effects of surface catalysis on each type of system were investigated. Key findings related to detectability:

1. For the flight regime explored herein, temperature downstream of the ablating nose plays a dominant role in determining the levels of detectable species present in the wake.
2. The production of NO and NO^+ depends in very large part upon the presence of ablating species to catalyze production of the nitric species.
3. Reducing surface temperature and catalysis behind an ablating nose reduces the carbon dioxide produced by as much as an order of magnitude. A very cold wall further reduces CO_2 and all other emissions. The opposite is also true.
4. Minimizing the size of the ablating portion of the nose is not helpful to reduction of detectable species if the body heats up as a result. However, having no ablation at all is enormously helpful in reduction of all detectable species.
5. Neither catalysis, ablation, nor nose conditions had any affect on temperature on the cone base. The cone base flow temperature was found to be a strong function of main body surface temperature.
6. Based on the above, a study of cooling methodologies on the rear portion of the main body surface for (a) minimization of communications interference due to free electrons, (b) radiative species and (c) for protection of the base is recommended.

The structures on the body which affect the wake were also investigated. Key findings here include:

1. Flow structures formed at the nose and shoulder were found to survive and persist until the end of the body, leading directly to distinct structures and associated layers of species within the wake.
2. Gradients caused by nose bluntness and acceleration around the shoulder were found to meaningfully affect the components of the wake, and so an analysis of wake composition changes based on nose geometry is recommended.
3. The recompression shock formed by the neck causes the highest gradients within the wake, and its effects could be diminished by disruption of the boundary layer which feeds it. A study of afterbody shape factor is therefore recommended.
4. The location of the neck was seen to shift only minutely with wake and body temperature.
5. Neither temperature, ablation nor catalysis were found to significantly affect surface pressure distribution.

Author Contributions: Robert Greendyke conceived of this effort, having sponsored a Master's Thesis topic on the material in the past (Kania 2015), which formed the basis for the effort. Greendyke asked Matthew Satchell to investigate further. Matthew performed all simulations used herein and was sole author of the paper. Matthew tasked Jeffery Layng with some early exploration whose results were not ultimately presented but were tremendously informative in ultimate identification of the scope of the effort. Jeffery was also tremendously helpful in identifying improvements to the paper.

Conflicts of Interest: The authors declare no conflict of interest.

References

1. Kania, M.A. Analysis of Hypersonic Vehicle Wakes. Master's Thesis, Air Force Institute of Technology, Greene County, OH, USA, 2015.
2. Anderson, J.D. *Hypersonic and High Temperature Gas Dynamics*; AIAA Education Series; American Institute of Aeronautics and Astronautics: Reston, VA, USA, 2006.
3. Gupta, R.N. *A Review of Reaction Rates and Thermodynamic and Transport Properties for an 11-Species Air Model for Chemical and Thermal Nonequilibrium Calculations to 30,000 K*; NASA Reference Publication 1232; NASA Langley Research Center: Langley, VA, USA, 1990.
4. Gnoffo, P.A.; James, W.K.; Harris Hamilton, H., II; Olynicky, D.R.; Venkatapathyz, E. *Computational Aerothermodynamic Design Issues for Hypersonic Vehicles*; AIAA: Reston, VA, USA, 1999.
5. Park, C. *Nonequilibrium Hypersonic Aerothermodynamics*; John Wiley & Sons: Hoboken, NJ, USA, 1990.
6. Tropf, W.J.; Thoams, M.E.; Harris, T.J.; Lutz, S.A. *Performance of Optical Sensors in Hypersonic Flight*; Johns Hopkins APL Technical Digest V08; The Johns Hopkins University Applied Physics Laboratory: Laurel, Maryland, 1987.
7. Gnoffo, P.A.; Johnston, C.O.; Thompson, R.A. Implementation of Radiation, Ablation, and Free Energy Minimization in Hypersonic Simulations. *J. Spacecr. Rocket.* **1971**, *47*, 251–257.
8. Barnhardt, M.D.; Candler, G.V. Modeling and Simulation of High-Speed Wake Flows. Ph.D. Thesis, University of Minnesota, Minneapolis, MN, USA, 2009.
9. Mazaheri, A.; Gnoffo, P.A.; Johnston, C.O.; Kleb, B. *LAURA Users Manual: 5.5-65135*; NASA TM-2013-217800; NASA: Langley, VA, USA, 2017.
10. Lykoudis, P.S. A review of hypersonic wake studies. *AIAA J.* **1966**, *4*, 577–590.
11. Muramoto, K.K. Model for Predicting Hypersonic Laminar Near-Wake Flowfields. *J. Spacecr. Rocket.* **2015**, *33*, 305–307.
12. Zhong, J.; Ozawa, T.; Levin, D.A. Comparison of High-Altitude Hypersonic Wake Flows of Slender and Blunt Bodies. *AIAA J.* **2008**, *46*, 251–262.
13. Park, G.; Gai, S.L.; Neely, A.J. Base Flow of Circular Cylinder at Hypersonic Speeds. *AIAA J.* **2016**, *54*, 458–468.
14. Barnhardt, M.; Candler, J.V. Detached-Eddy Simulation of the Reentry-F Flight Experiment. *J. Spacecr. Rocket.* **2012**, *49*, 691–699.
15. Hinman, W.S.; Johansen, C.T. Reynolds and Mach Number Dependence of Hypersonic Blunt Body Laminar Near Wakes. *AIAA J.* **2017**, *55*, 500–508.
16. Lin, T.C.; Sproul, L.K.; Kim, M.; Olmos, M.; Feiz, H. Hypersonic Reentry Vehicle Wake Flow Fields at Angle of Attack. In Proceedings of the 44th AIAA Aerospace Sciences Meeting and Exhibit, Reno, NV, USA, 9–12 January 2006.
17. Stainback, P.C. *Effect of Unit Reynolds Number, Nose Bluntness, Angle of Attack and Roughness on Transition on a 5° Half-Angle Cone at Mach 8*; NASA TN-D-4961; NASA Langley Research Center: Hampton, VA, USA, 1969.
18. Gnoffo, P.A. *An Upwind-Biased, Point-Implicit Relaxation Algorithm for Viscous, Compressible Perfect-Gas Flows*; NASA TP 2953; NASA Langley Research Center: Langley, VA, USA, 2003.
19. Gnoffo, P.A. Computational Aerothermodynamics in Aeroassist Applications. *J. Spacecr. Rocket.* **2003**, *40*, 305–312.
20. Gnoffo, P.A. CFD Validation Studies for Hypersonic Flow Prediction; In Proceedings of the 39th Aerospace Sciences Conference, Reno, NV, USA, 8–11 January 2001.
21. Gnoffo, P.A.; White, J.A. Computational Aerothermodynamic Simulation Issues on Unstructured Grids. In Proceedings of the 37th AIAA Thermophysics Conference, Portland, OR, USA, 28 June–1 July 2004.
22. Gnoffo, P.A. A Perspective on Computational Aerothermodynamics at NASA. In Proceedings of the Fluid Mechanics and Thermodynamics Conference, Gold Coast, Australia, 3–7 December 2007.
23. Wilcox, D.C. *Turbulence Modeling for CFD*, 2nd ed.; DCW Industries: La Cañada Flintridge, CA, USA, 1994.
24. Kimmel, R.L.; Adamczak, D.W. HIFiRE-1 Preliminary Aerothermodynamic Measurements. In Proceedings of the 41st AIAA Fluid Dynamics Conference and Exhibit, Honolulu, HI, USA, 27–30 June 2011.
25. Schneider, S.P. Hypersonic Boundary-Layer Transition with Ablation and Blowing. *J. Spacecr. Rocket.* **2010**, *47*, 225–237.

26. Hruschka, R.; O'Byrne, S.; Kleine, H. Comparison of velocity and temperature measurements with simulations in a hypersonic wake flow. *Exp. Fluids* **2011**, *51*, 407–421.

27. Grasso, F.; Pettinelli, C. Analysis of laminar near-wake hypersonic flows. *J. Spacecr. Rocket.* **1995**, *32*, 970–980.

28. Balakumar, P.; Owens, L.R. Stability of Hypersonic Boundary Layers on a Cone at an Angle of Attack. In Proceedings of the 40th Fluid Dynamics Conference and Exhibit, Chicago, IL, USA, 28 June–1 July 2010.

29. Jones, W.L., Jr.; Cross, A.E. *Electrostatic-Probe Measurements of Plasma Parameters for Two Reentry Flight Experiments at 25,000 Feet Per Second*; NASA TN D-6617; NASA Langley Research Center: Hampton, VA, USA, 1972.

30. Johnston, C.O.; Gnoffo, P.A.; Mazaheri, A. Influence of Coupled Radiation and Ablation on the Aerothermodynamic Environment of Planetary Entry Vehicles. In Proceedings of the Radiation and Gas-Surface Interaction Phenomena in High Speed Re-Entry Meeting: Rhode-St-Genèse, Belgium, 6–8 May 2013.

31. Harries, J.; Carli, B.; Rizzi, R.; Serio, C.; Mlynczak, M.; Palchetti, L.; Maestri, T.; Brindley, H.; Masiello, G. The Far Infrared Earth. *Rev. Geophys.* **2008**, *46*, RG4004, doi:10.1029/2007RG000233.

32. Cummings, R.M.; Bertin, J.J. Critical Hypersonic Aerothermodynamic Phenomena. *Annu. Rev. Fluid Mech.* **2006**, *38*, 129–157.

33. Huber, P.W. *Hypersonic Shock-Heated Flow Parameters for Velocities to 46,000 Feet per Second and Altitudes to 323,000 Feet*; NASA TR R-163; National Aeronautics and Space Administration: Washington, DC, USA, 1963.

34. Bussing, T.R.A.; Eberhardt, S. Chemistry associated with hypersonic vehicles. *J. Thermophys. Heat Transf.* **1989**, *3*, 245–253.

aerospace

MDPI

Article

Computational Study of Propeller–Wing Aerodynamic Interaction

Pooneh Aref [1], Mehdi Ghoreyshi [1,*], Adam Jirasek [1], Matthew J. Satchell [1] and Keith Bergeron [2]

[1] High Performance Computing Research Center, U.S. Air Force Academy, Air Force Academy,
    CO 80840, USA; Pooneh.Aref@usafa.edu (P.A.); Adam.Jirasek@usafa.edu (A.J.);
    Matthew.Satchell@usafa.edu (M.J.S.)

[2] U.S. Army Natick Soldier Research, Development & Engineering Center, Natick, MA 01760, USA;
    keith.bergeron2.civ@mail.mil

* Correspondence: Mehdi.Ghoreyshi@usafa.edu

Received: 12 April 2018; Accepted: 18 July 2018; Published: 25 July 2018

Abstract: Kestrel simulation tools are used to investigate the mutual interference between the propeller and wing of C130J aircraft. Only the wing, nacelles, and propeller geometries are considered. The propulsion system modelled is a Dowty six-bladed R391 propeller mounted at inboard or outboard wing sections in single and dual propeller configurations. The results show that installed propeller configurations have asymmetric blade loadings such that downward-moving blades produce more thrust force than those moving upward. In addition, the influence of installed propeller flow-fields on the wing aerodynamic (pressure coefficient and local lift distribution) are investigated. The installed propeller configuration data are compared with the non-installed case, and the results show that propeller effects will improve the wing's lift distribution. The increase in lift behind the propeller is different at the left and right sides of the propeller. In addition, the propeller helps to delay the wing flow separation behind it for tested conditions of this work. Finally, the results show the capability of Kestrel simulation tools for modeling and design of propellers and investigates their effects over aircraft during conceptual design in which no experimental or flight test data are available yet. This will lead to reducing the number of tests required later.

Keywords: wing–propeller aerodynamic interaction; p-factor; installed propeller; overset grid approach

1. Introduction

For low speed operations, propeller-driven aircraft are more effective than jet engines. The propellers of large size aircraft are usually placed on and in the front of the wing which can drastically alter the aerodynamics of the wing and other parts of the aircraft that are immersed in the propeller slipstream. Propellers of these aircraft typically operate at a constant (desired) rotational speed. The propeller blade angle is then adjusted according to the flight speed in order to achieve the maximum efficiency. The propellers can rotate in the same or opposite directions as well. Understanding the effects of these propellers on the aerodynamic performance, aircraft stability and control, vibration, and noise is a challenging task and expensive using wind tunnel or flight testing. There are significant deficiencies when using simple analytical methods such as momentum theory of Froude [1] and Rankine [2]. An alternative is to use computational methods that allow rapid and accurate prediction of the mutual interference between the propeller and wing. Additionally, there is a growing interest in the use of propellers in new and novel design concepts such as flying taxis, or in the unmanned aerial vehicles or drones for the reconnaissance and payload carrying missions. No historical data exist for these concepts and thus the design of these vehicles would be helped by the early availability of high quality computational models to allow control laws to be defined.

Advances in computational modeling of propellers are reported in literature [3]. In a simple manner, propellers may be physically replaced with thin actuator disks using Froude–Rankine momentum theory. This approach assumes an infinite number of thin propeller blades and inviscid flow through the disk. The model then should ensure the mass flow continuity between front and rear faces of disk. Depending on the input thrust and rotational speed, the rear face will have a jump in total pressure, total temperature, and velocity. Advanced computational methods of sliding interfaces, *Chimera* or *overset* grids have been used for propeller flow simulations as well [4–7]. Results of such simulations have compared well with available wind tunnel data. Periodic slipstream unsteadiness has been captured in wing lift and drag, and increased suction peaks at the wing leading edge have also been documented for wing mounted engines. In addition to propeller slipstream interaction with the wing, other components of the aircraft may also be affected by the local unsteadiness depending on relative position of the propeller and the aircraft component. It is well known for traditional single engine aircraft, the wake–fuselage and wake–tail interactions are significant at high power and low airspeed configurations, such as during takeoff. For these conditions, the aircraft experiences a yaw to the left if no control input is made to counter the resultant force. In addition, at high angles of attack, asymmetric blade effects lead to an asymmetric relocation of the propeller's center of thrust, P-factor. For propeller driven aircraft with multiple engines mounted along the wing, the P-factor effect can be mitigated by using counter-rotating propellers on either side of the aircraft. Note that the C-130H/J propellers rotate in the same direction (clockwise when viewed from the rear) while the P-38 propellers are mounted to rotate in opposite directions depending on the whether the engine is on the port or starboard side of the aircraft. The propellers' slipstream characteristics are not only a design consideration for traditional aircraft performance metrics, but they may also contribute to constraints and limitations on the aircraft's use. The focus of this work is to investigate the spinning propeller effects on C130-J wing aerodynamics.

The aerodynamic modeling of C130 aircraft in air drop configuration has been the subject of recent studies at the U.S. Air Force Academy and U.S. Army Natick Soldier Research, Development & Engineering Center Center [8–10]. Propellers have been modeled with a very thin actuator disk in References [8,9]. These studies investigated the wake and flow in the vicinity of the cargo ramp and open troop doors. In a subsequent study [10], the C130H/J test cases were simulated with fully resolved blade geometries and using an overset grid approach. The simulation results were compared with previous studies that assumed the propellers as thin actuator disks, and they showed that propeller effects increase the averaged velocities around the open door and in most locations behind the open cargo ramp at the measured positions. The propeller effects on the wing aerodynamics are briefly described in Reference [10] as well. The current work extends these studies and investigates the mutual interference between C130J propeller(s) and its wing. Only wing, nacelle, and propellers components of the aircraft are considered. The propulsion system modelled is a Dowty six-bladed R391 propeller mounted at inboard and/or outboard wing sections. The installed and non-installed performances of this propeller are investigated. The installed performance includes a single propeller mounted at the inboard or outboard nacelle and two propellers mounted at the inboard and outboard nacelles. Propellers can spin clockwise (CW) or counterclockwise (CCW) at different blade angles. Finally, the stall behavior of the wing with and without propellers are presented.

This work uses the High Performance Computing Modernization Program (HPCMP) Computational Research and Engineering Acquisition Tools and Environments (CREATE)^{TM}-Air Vehicles (AV) Kestrel simulation tools (version 8.0) to investigate the propeller wing aerodynamic interaction of the C130J aircraft. The article is organized as follows: first, the Computational Fluid Dynamics (CFD) solver and test cases are described. The propeller performance and propeller/wing aerodynamic interaction are then briefly presented. Next, the article concludes with the a presentation of the results of the C130J wing and propeller aerodynamic interaction.

2. CFD Solver

The flow solver used in this work is the fixed wing computational tool of CREATE™-AV program, i.e., Kestrel. The Department of Defense (DoD)-developed solver is funded by the DoD HPCMP. The CREATE™ focuses on addressing the complexity of applying computationally based engineering to improve DoD acquisition processes [11], and it consists of three computationally based engineering tool sets for design of air vehicles, ships, and radio-frequency antennae. The fixed wing analysis code, Kestrel, is part of the Air Vehicles Project (CREATE™-AV) and is a modularized, multidisciplinary, virtual aircraft simulation tool incorporating aerodynamics, jet propulsion integration, structural dynamics, kinematics, and kinetics [11]. The code has a Python-based infrastructure that integrates Python, C, C++, or Fortran-written components [12]. New modules can easily integrated into the code.

Kestrel version 8.0 is used in this work. The flow solver of the code discretizes Reynolds-Averaged Navier Stokes (RANS) equations into a cell-centered finite-volume form. The code then solves unsteady, three-dimensional, compressible RANS equations on hybrid unstructured grids [13]. The code uses the Method of Lines (MOL) to separate temporal and spatial integration schemes from each other [14]. The spatial residual is computed via a Godunov type scheme [15]. Second-order spatial accuracy is obtained through a least squares reconstruction. The numerical fluxes at each element face are computed using various exact and approximate Riemann schemes with a default method based on HLLE++ scheme [16]. In addition, the code uses a subiterative, point-implicit scheme method (a typical Gauss–Seidel technique) to improve the temporal accuracy.

Kestrel receives an eXtensible Markup Language (XML) input file generated by Kestrel User Interface and stores the solution convergence and volume results in a common data structure for later use by the Output Manager component. Some of the turbulence models available within Kestrel include turbulence models of Spalart–Allmaras (SA) [17], Spalart–Allmaras with rotational/curvature correction (SARC) [18], Mentor's SST model [19], and Delayed Detached Eddy Simulation (DDES) with SARC [20].

Kestrel allows single and multi-body (overset) simulations. For the C130 example, the aircraft is defined as the body in Kestrel and propellers are defined as children of the main body. In this way, any motion applied to the aircraft will be applied to the propellers as well. Likewise, flaps should be defined as children of the aircraft's body in the code, but for a store separation problem, different bodies should be defined. Kestrel uses an overset grid approach that allows the independent translation and rotation of each body and its children. Overlapping grids are generated individually, without the need to force grid points aligned with neighboring components. However, some small gaps should be present between bodies to avoid body intersections in the code. In addition, Kestrel allows prescribed or six degrees of freedom motions of rigid aircraft [12]. Bodies and their children can have their own motions. For example, propellers of C-130 can spin around their rotation axis while the whole aircraft undergoes a turn maneuver.

The propeller blades can be fully resolved in Kestrel using an overset grid approach. The code also allows modeling propellers in form of thin actuator disks in which the disk area corresponds to the propeller diameter. The use of uniform or non-uniform thrust distributions are available. A non-uniform case requires a given radial position for maximum thrust force. The loading profile is assumed to be linear with a zero thrust at the inner blade radius and then increases until the radial position of maximum thrust, and then decreases to zero at the rotor tip.

3. Propeller Performance

Rotating propellers have significant influence on an aircraft aerodynamics and its stability and control due to slipstream and propeller wake effects. The installed propeller performance is altered due to wing upwash as well [21]. The installed configuration should therefore achieve maximum propeller efficiency while minimizing the adverse impacts on aircraft aerodynamics [22]. The propeller increases air speed and alters the flow direction behind it. The rise in dynamic pressure will increase the wing lift and drag. The change of flow direction leads to a variation of the wing local angle of attack. The propeller slipstream delays the aircraft stall as well [23]. While this is a favorable effect, the stall behavior from propellers can be unacceptable [24]. For example, advanced propellers used in initial designs of C-130J prevented the inner wing from stalling [24] and therefore stall started at the wing tips causing the loss of roll control. Additionally, highly loaded propellers produce a propeller wake because of strong tip vortices formed at the tips of propeller blades. When these propellers are mounted in the front of the wing, the propeller wake causes a considerable variation in the lift and drag distribution across the wingspan [25]. This can cause an unsteady load distribution over the aircraft as well.

Reference [26] describes several propeller aircraft interference effects. In this reference, the wing section is divided into regions and the propeller performance is detailed for four points of the blade tips. Following the guidelines of Reference [26], Figure 1 shows C-130H with two counter-spinning Hamilton Standard 54H60 propellers. Only the inboard propeller effects are considered. The wing is divided into four regions: (1) region one ("R1") is from fuselage to the propeller tip; (2) region two ("R2") covers the propeller right tip to the hub; (3) region three ("R3") extends from the hub to the left propeller tip; (4) and finally region four ("R4") is from the propeller disk towards the wing tip. In addition, four points are shown on the displayed inboard propeller. These points are at the tip of each blade. Wing regions of 2 and 3 are behind the propeller and are affected by the propeller slipstream. In R2, the lift increases due to an increase in dynamic pressure and local angle of attack. In R3, the angle of attack decreases and it counteracts the tendency of the lift increase due to a rise in dynamic pressure behind the propeller. As reported in Reference [26], the propeller effects are not limited to R2 and R3 and some changes in R1 and R4 can be experienced as well. In terms of propeller performance, the wing upwash causes an asymmetric load on the propeller blades such that angle of attack increases at P2 and decreases at P4. Points 1 and 3 are affected by the wing presence as well. The presence of the nacelle also increases axial velocity in all shown points.

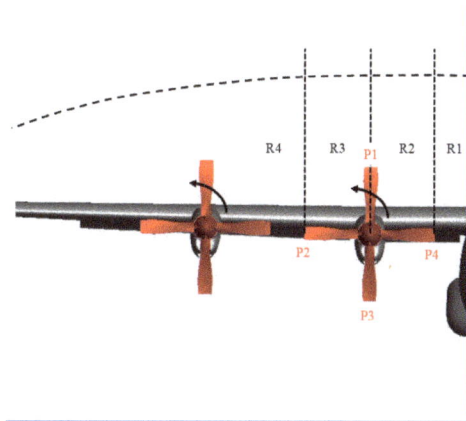

Figure 1. Propeller wing interference effects for counterclockwise spinning propeller of C-130H.

4. Test Cases

The main focus of this work is on the propeller aircraft interference effects of C-130J which uses a Rolls-Royce AE2100 turboprop series with Dowty 391 six-bladed propeller system with a diameter of 162 inches. The blades have a high-speed design with a thin airfoil section and a swept back blade made of composite materials.

Both propellers (inboard and outboard) spin counterclockwise (as viewed from front) at constant rotational speed of 1020 rpm or 6120 deg/s. Different blade angles are tested. The propeller with 20 deg is shown in Figure 2.

Computational grids were generated in Pointwise version 18.0 (Fort Worth, TX, USA). The surface grid cells are mostly structured quadrilateral, but anywhere that these cell types are not possible to make, triangular surface cells are used. The interface between structured and unstructured mesh uses a surface T-rex cells technique that ensures high quality transition between the structured and unstructured surface meshes.

Figure 2. Dowty six-blade R391 propeller with 20-deg blade angle is shown. This propeller has a diameter of a diameter of 162 inches (4.12 m).

The main motivation for using the quadrilateral mesh is to have very good grid resolution on the blade leading and trailing edges and at the blade tips. A part of the hub is covered with patches of structured meshes as well. The volume mesh is fully unstructured with a 50 prism layer on the propeller surface. The growing ratio of the prism layer is 1.25 and the growth is terminated when the transition between the prism layer and the tetrahedral mesh is smooth.

Two set of grids were generated for each propeller at each given blade angle. In the first grid, the free-stream boundary condition was used with an outer diameter of about 25 times of the blade diameter. These grids have approximately 51.1 million cells and are used for simulation of non-installed propellers. In the second set of grids, the outer boundary is an overset with a diameter of about 1.5 times of blade diameter. These girds are used for installed propeller simulations. These grids have approximately 50 million cells consisting of 27 million prismatic cells around blades and hub surfaces. Finally, for the propeller overset grid with a 20-deg blade angle, a new grid was generated with blade surfaces being mirrored in order to have a clockwise spinning propeller.

The wing geometry is extended to a symmetric plane and has inboard and outboard flaps down 50% (or 22.5°) with two engine nacelles mounted under the wing. The engine inlets are modelled as solid walls in this work. No-slip conditions are assumed at all solid walls. The wing grid is also generated in Pointwise version 18 and has about 72.5 million cells consisting of around 41 million prismatic cells in proximity of the wall surfaces and 31 million tetrahedral cells elsewhere. The grid units are in inches and in this system the wing half span measures 783.5 inches as shown in Figure 3. The centerline of inboard and outboard nacelles are at 193 inches and 397 inches from the wing root, respectively.

In the wing and propeller simulations, the wing is defined as the parent body with the propeller as its children. This is a helpful approach as any motion applied to the wing will be applied to all children, i.e., propellers. The propeller bodies use the same grids and are defined with a translation vector to have propellers installed inboard or outboard. Different wing/propeller configurations are then considered; some examples are shown in Figure 4. In the first case, only wing geometry is considered including engine nacelles and the propeller hub geometries. In the second case, a single propeller is mounted at the inboard nacelle. The propeller could spin clock or counterclockwise and could have different blade angles. In the third case, a single propeller is installed at the outboard wing section; the propeller could again spin clockwise or counterclockwise and could have different blade angles. In the final case, two propellers are installed at both inboard and outboard nacelles. They can spin at the same or opposite directions. Note that, in the overset approach of this work, a small gap is needed between wing and propeller grids. In addition, the motion files are only applied to the propellers with hub and blades spinning simultaneously.

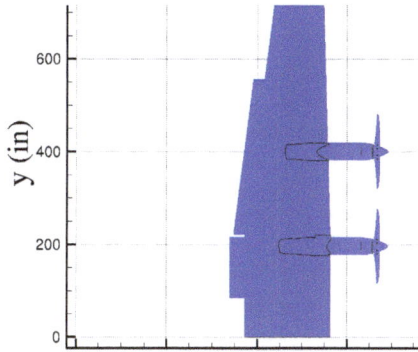

Figure 3. The location of flaps, nacelles, and propeller on the wing.

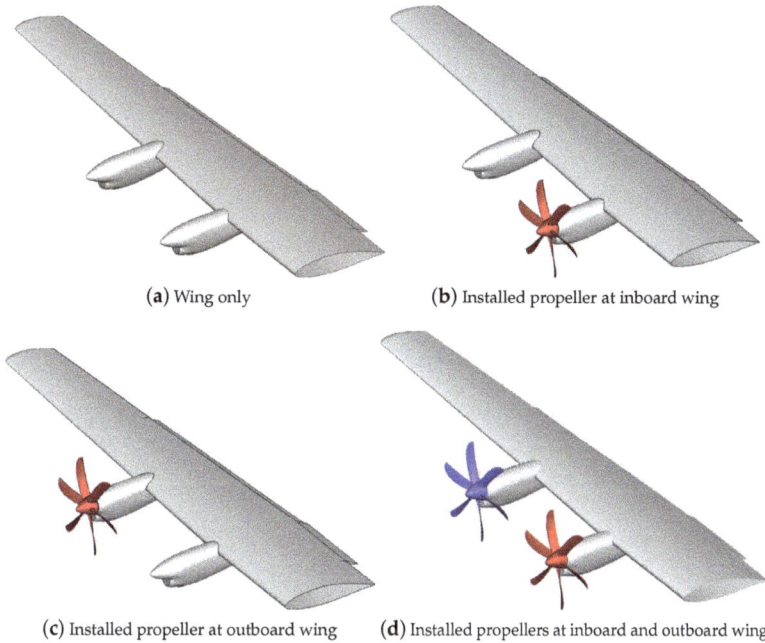

(**a**) Wing only

(**b**) Installed propeller at inboard wing

(**c**) Installed propeller at outboard wing

(**d**) Installed propellers at inboard and outboard wing

Figure 4. Test cases include no propeller; propeller installed inboard; propeller installed outboard; and propellers installed on the inboard and outboard wing.

5. Results and Discussion

In all CFD simulations, the SARC turbulence model DDES simulation is used. Kestrel simulations are run in an unsteady mode in which second order accuracy in time is also used. A time step of 0.001 s, a temporal damping of 0.01, and three Newton sub-iterations are used for non-moving grids. For spinning propellers, eight subiterations are set.

5.1. Propeller Performance

The performance of the Dowty six-bladed R391 propeller is investigated using Kestrel simulation tools. Flow conditions in all simulations correspond to the air speed of 130 KIAS (Knots Indicated Air Speed) at 1000 ft altitude and zero angles of attack and sideslip. The propeller grids with large outer boundaries and free-stream conditions are used to investigate the non-installed propeller performance. The grids (hub and propeller blades) rotate around x-axis at 6120 deg/s speed. The forces and moments of all (noslip wall) surfaces (hub and blades) as well as each blade surface are written in separate files. The simulation results show that blade forces and moments reach steady-state values for constant speed propellers. The results confirm that aerodynamic forces and torques exerted on each blade are symmetric as well.

Figure 5 shows the blade loadings for clockwise and counterclockwise spinning propellers mounted on the inboard wing section at the final time of simulation. Note that propellers spin at a constant speed of 1024 rpm. The solutions are colored by a pressure coefficient. In the computational setup of these simulations, the forces and moments at each blade are written separately. The ratio of thrust force at each blade to total propeller thrust is given in Figure 5. Notice that, for isolated (non installed) propellers, all blades report similar thrust values. Figure 5 shows that installed propeller have different loading depending on the direction of rotation. Figure 5a shows the solution of the propeller spinning counterclockwise. The results show that blades moving downward (opposite of the wing upwash) have more thrust force than those moving upward. The maximum thrust is at the lowest positioned blade. Likewise, Figure 5b shows the solution of the propeller spinning clockwise with constant rotational speed of 1024 rpm. As observed in counterclockwise spinning case, blades moving downward (opposite of the wing upwash) have more thrust force than those moving upward. The maximum thrust again occurs at the lowest positioned blade.

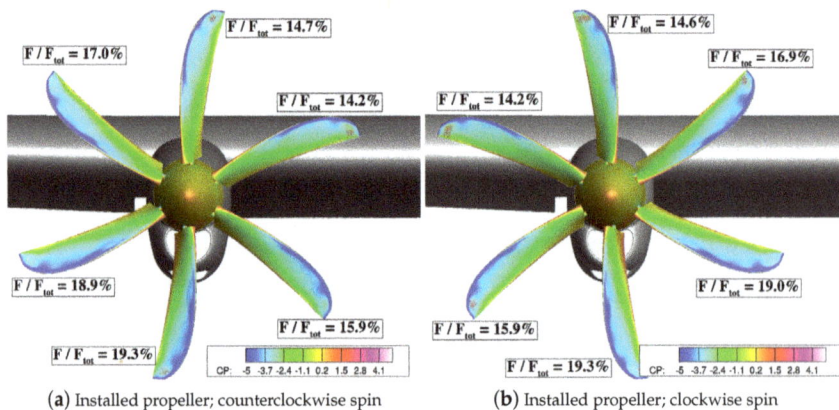

(a) Installed propeller; counterclockwise spin   (b) Installed propeller; clockwise spin

Figure 5. Installed propeller surface pressure data. Propeller installed inboard; propellers have a 20-deg blade angle and spin at 1024 rpm clockwise or counterclockwise. Propeller solutions are at final simulation time.

5.2. Wing/Propeller Aerodynamic Interaction

The number of time steps in all simulations is 6500. Out of these time steps, 500 are used in startup mode that helps to fade away the effects of solid walls, ramp up time, ramp down advective damping effects, and prepare the solution for grid motions or unsteady simulations [27]. However, simulation and therefore physical time will remain zero during these startup time steps. Flow conditions in all simulations again correspond to the air speed of 130 KIAS at 1000 ft altitude and zero angles of attack and sideslip. For stall behavior simulations, the angle of attack varies from zero to 12 degrees. For the wing surfaces and C_p-plots, time-averaged solutions from the last 3000 iterations were used.

All propellers spin counterclockwise unless stated otherwise. The rotational speed is 1020 rpm or 6120 deg/s. Table 1 gives a list of simulations.

Table 1. Simulation runs.

Simulation Cases	Inboard Propeller	Outboard Propeller	Angle of Attack (deg)	Blade Angle (deg)
Case 1	CCW		[0, 9, 10, 11, 12]	[20, 28]
Case 2	CW			20
Case 2		CCW	[0, 9, 10, 11, 12]	[20, 28]
Case 4	CCW	CCW	[0, 9, 10, 11, 12]	20
Case 5	CW	CCW	[0, 9, 10, 11, 12]	20
Case 6	CCW	CW	[0, 9, 10, 11, 12]	20

A number of scripts were written to extract slices at different spanwise locations of the wing. These locations are given in inches and can be visualized in Figure 3. Another script will calculate local lift and drag coefficients from pressure coefficients of each slice to make a local lift distribution over the wing. The first set of results compare pressure coefficient values over the wing for a number of slices ranging from $y = 20$ to $y = 420$ inches for a wing only and a wing with an inboard mounted propeller. The propeller has a blade angle of 20deg and can spin clockwise or counterclockwise; for each setting, a different propeller grid was selected to have a positive thrust force by spinning propellers. Propellers spin at 6120 deg/s (1020 rpm) as well. The pressure data of these configurations are shown and compared in Figure 6. Notice that these data correspond to time-averaged wing solutions for the final three seconds of simulations.

Note that the wing region behind an inboard propeller ranges approximately from $y = 120$ to $y = 280$ inches. Figure 6 shows that inboard propeller effects can be seen at smaller y positions, even at $y = 20$ inches as the C_p-plots do not match with each other at these locations. A counterclockwise propeller mounted on the inboard wing causes the pressure differences between upper and lower surfaces to increase compared with a wing without propeller for $y = 20$ to $y = 120$ inches. A counterclockwise rotation causes an upwash in these region and an increased local angle of attack. Instead, a clockwise spinning propeller causes the pressure differences between upper and lower surfaces decrease compared with a wing without propeller for positions $y = 20$ to $y = 120$ inches. This is due to downwash effects of the propeller over this region of the wing. Notice that the effects of deflected flaps can be seen in C_p-plots of positions at and larger $y = 100$ in. In the range of $y = 120$ to $y = 200$, the counterclockwise spinning propeller causes significant differences between upper and lower surfaces again compared with the wing-only configuration. The reason is due to the combined effects of upwash and increases momentum behind the propeller at this region. The clockwise spinning propeller also shows larger differences because of the momentum increase, but differences are still smaller than the counterclockwise spinning propeller.

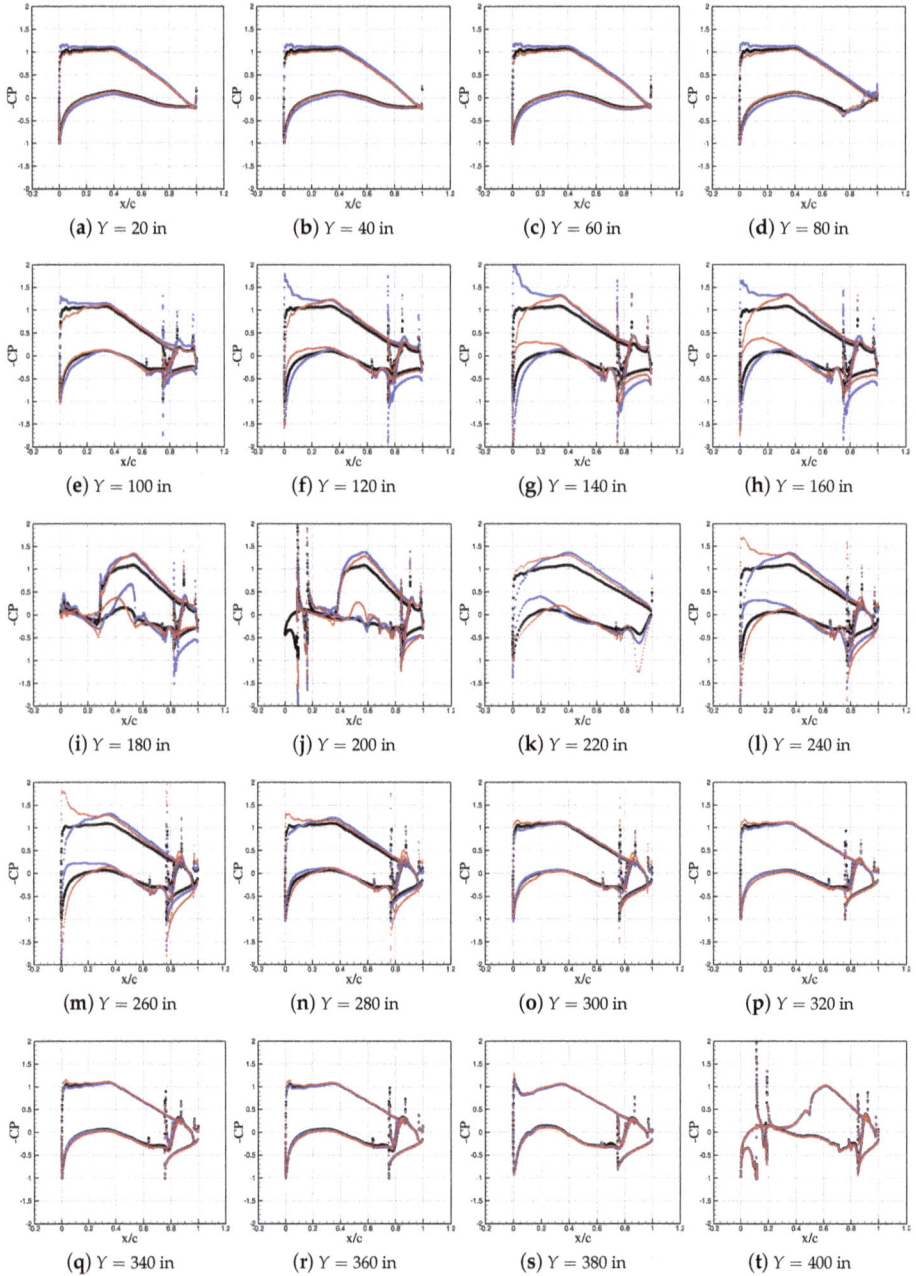

Figure 6. Pressure distribution $(-C_p)$ over the wing for wing only and wing with a prop mounted at inboard nacelle; the propeller spins counterclockwise and clockwise and has a 20-deg blade angle. Black dots show no prop. Blue-colored square markers show a propeller spinning counterclockwise. Red triangles show a propeller spinning clockwise. Pressure data are time-averaged for the final three seconds of simulations.

At around the nacelle center, i.e., $y = 193$ inches, both propellers show nearly the same C_p plots as downwash and upwash velocities are small in this region. From $y = 200$ inches outwards, the propeller effects change to become the opposite, i.e., a clockwise spinning propeller induces upwash over the wing and the counterclockwise spinning propeller induces downwash. The largest effects can be seen from a clockwise spinning propeller for locations between $y = 200$ to $y = 280$ inches, where it shows the largest differences between C_p values at upper and lower surfaces. For $y = 280$ outwards, the counterclockwise spinning propeller effects become small and the pressure data almost matches the wing only data. The clockwise spinning propeller still shows some changes in C_p plots up to $y = 380$ inches due to upwash effects over these regions. These results show that a propeller installed on the front of the wing can significantly change the wing aerodynamics in particular behind the propeller; these effects depend on the propeller direction of rotation and they can even be seen at different wing locations that are not behind the propeller.

The next results compare wing only pressure data with a counterclockwise spinning propeller with 20-deg blade angle and mounted either on the wing inboard or outboard. Figure 7 shows and compares these numerical data for wing slices extracted from $y = 20$ to $y = 520$ inches. The outboard mounted propeller has no significant changes over wing local pressure plots for positions of $y = 20$ to $y = 300$ inches. The inboard mounted propeller, however, creates pressure on the upper surface smaller and on the lower surface larger for positions of $y = 20$ to $y = 300$ inches due to propeller upwash effects. The changes become significant for $y = 120$ to $y = 200$ inches due to combined effects of upwash and increased dynamic pressure behind the propeller. Large gradients of C_p increase and decrease can be seen at the wing leading edge and behind the propeller. For $y = 200$ to $y = 280$ inches, the inboard propeller effects are decreased as the downwash effects opposite from the dynamic pressure increase. For $y = 280$ outwards, the inboard propeller causes smaller differences between upper and lower wing surfaces due to downwash effects.

The outboard propeller shows very similar trends as well; however, the effects over the wing can be seen from $y = 300$ inches outward. In these regions, the wing is subject to propeller upwash. From $y = 320$ to $y = 400$, there are combined effects of upwash and increased dynamic pressure. For $y = 400$ to $y = 480$ inches, the downwash due to propeller opposite from the effects of increased dynamic pressure. Finally, for $y = 480$ outwards, the propeller downwash causes there to be smaller pressure on the lower surface and larger pressure values on the upper surface.

Figure 8 shows vorticity isosurfaces for simulated cases of wing only, propellers installed inboard or outboard wing with 20-deg blade angle. The inboard propellers spin either clockwise or counterclockwise with a rotational speed of 1024 rpm. Isosurfaces correspond to the vorticity magnitude of 100. Figure 8 shows the slipstream generated behind the propellers. A negative pressure region is formed over the upper wing surface behind the propellers. For counterclockwise propellers, the pressure is more negative behind the right side of propeller (viewed from front) than the left side. This is again due to combined effects of upwash and increased dynamic pressure. For the clockwise spinning propeller, the wing pressure is more negative behind the left side propeller than its right side.

Wing tip and flap vortices can be seen in Figure 8. The engine inlet was assumed to be a solid wall. Therefore, the inlet surface experiences stagnation pressure. The flow separates as it makes a 90-deg turn at the inlet edge. The separated flow will roll into two vortices around each nacelle and they will move upwards. The interaction of these vortices with wings will form two vortices near each other on the upper wing surface behind each nacelle. These vortices can be seen in Figure 8a. In the presence of the propeller, these vortices become much larger and are lifted up from surface as shown in Figure 8c,d. There is a vortex shedding at where the propeller slipstream interacts with the wing surface.

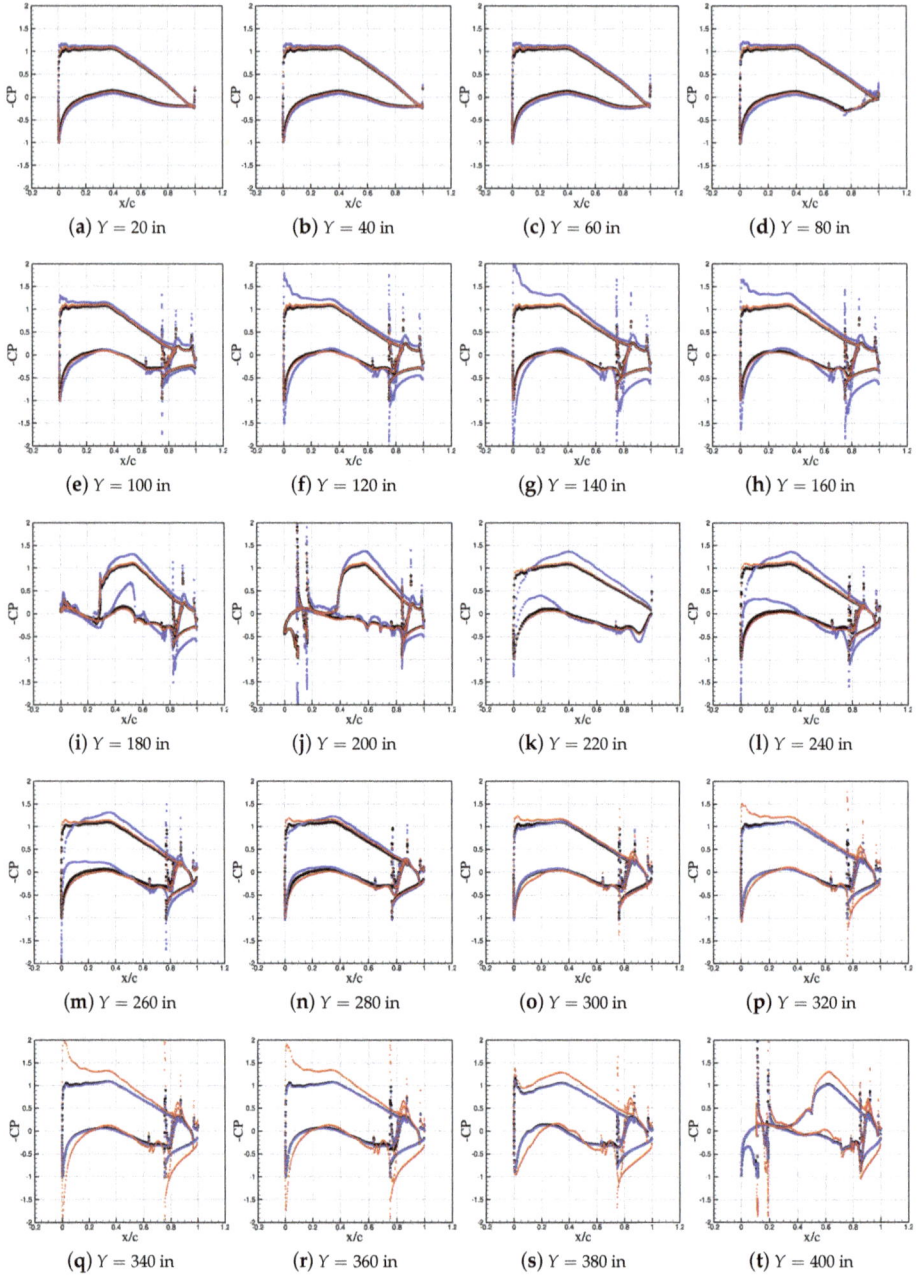

Figure 7. *Cont.*

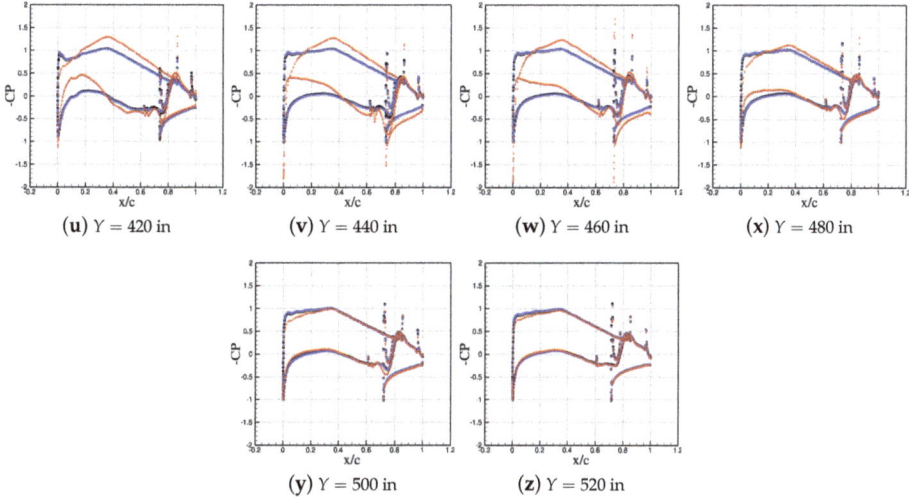

(u) $Y = 420$ in **(v)** $Y = 440$ in **(w)** $Y = 460$ in **(x)** $Y = 480$ in

(y) $Y = 500$ in **(z)** $Y = 520$ in

Figure 7. Pressure distribution ($-C_p$) over the wing for wing only and a wing with a prop mounted either on the inboard or outboard nacelle; the propeller spins counterclockwise and has a 20-deg blade angle. Black dots show no prop. Blue-colored square markers show a propeller mounted on the wing inboard. Red triangles show a propeller mounted wing outboard. Pressure data are time-averaged for the final three seconds of simulations.

Figure 9 compares the wing local lift distribution for the wing only and propellers installed inboard or outboard. The local lift is presented as $C_l.c$ which is local lift times local chord. The propellers have a 20-deg blade angle and spin at 1024 rpm. The inboard propellers can either spin clockwise or counterclockwise. The data calculated correspond to time-averaged data. Figure 9 shows that wing local lift increases behind the propeller. The lift rise in the left and right sections of the propellers are different and will depend on the direction or rotation.

In more detail, Figure 9a compares local lift distribution of a wing only configuration (No Prop) with data of wings and a propeller mounted inboard spinning clockwise (Prop CW) or counterclockwise (Prop CCW). Note that the lift distributions of all wings are affected by the flap deflections (flaps are located approximately at $y = 85$ to $y = 550$ inches). For example, moving towards the wing tip, the local lift of "No Prop" configuration increases, then gradually decreases, and then falls outside the outboard flap. In the "No Prop" case, there are local lift changes behind nacelles due to inlet vortices formed over the upper surface as well. In the "Prop CCW" case, the lift distribution is larger than the "No Prop" case for all spanwise distances from 20 to 280 inches. For further distances, the local lift is very close to "No Prop" data. Figure 9a shows that the local lift of "Prop CCW" suddenly increases, moving towards the left side of the propeller until it reaches a maximum and then drops. The effects of vortex shedding can be seen on the plots, especially near the right tip ($y = 276$ inches) of the propeller spinning CCW and the left tip ($y = 196$ inches) of the propeller spinning CW. The wing of "Prop CW" configuration has smaller lift than "No Prop" for distances from wing root to $y = 80$ inches due to induced upwash from propeller. The maximum lift occurs behind the right side of propeller. Both CW and CCW spinning propellers have the same thrust at the center of hub. Figure 9b compares the wing data with a propeller installed inboard or outboard. The outboard propeller effects can be seen even at the wing root. The propeller causes less lift than the "No Prop" case at a location right of the propeller.

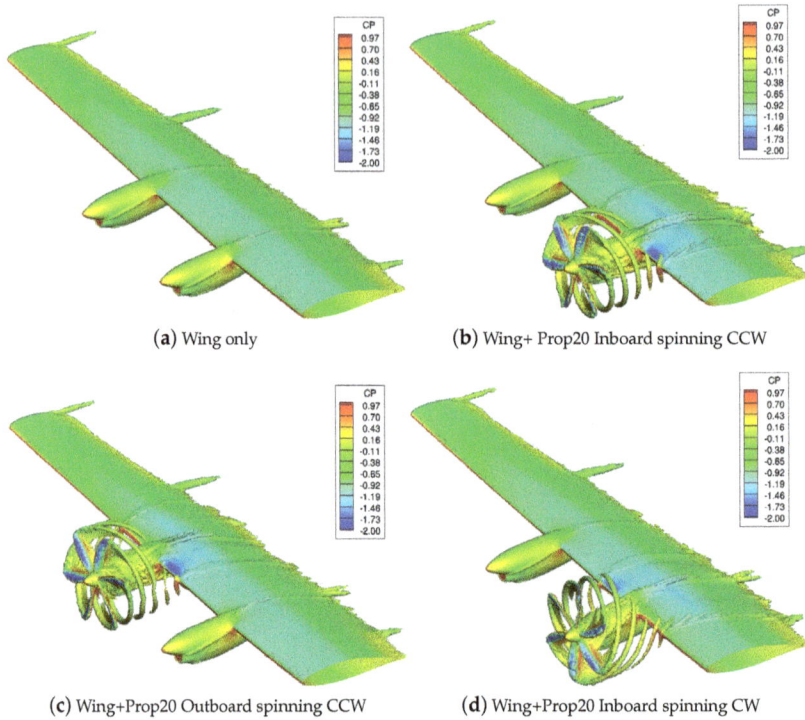

(**a**) Wing only   (**b**) Wing+ Prop20 Inboard spinning CCW

(**c**) Wing+Prop20 Outboard spinning CCW   (**d**) Wing+Prop20 Inboard spinning CW

Figure 8. Propeller installed inboard/outboard; vorticity isosurfaces are colored with pressure coefficients. Propellers have a 20-deg blade angle and spin at 1024 rpm. Wing only solution is time-averaged for the final three seconds of simulations. Wing+Prop solutions are at the final simulation time.

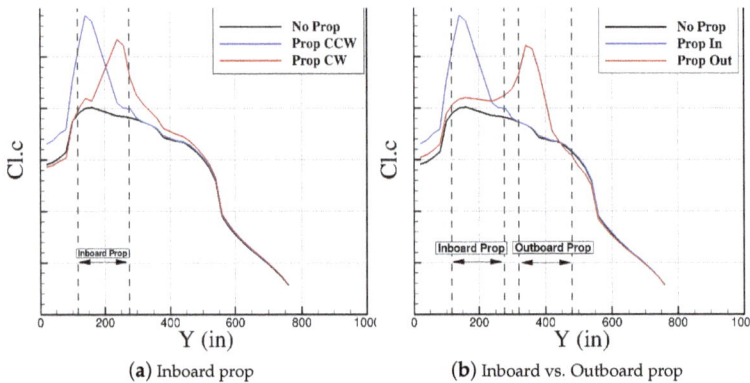

(**a**) Inboard prop   (**b**) Inboard vs. Outboard prop

Figure 9. Local lift distribution for wing and propellers are installed inboard or outboard the wing. Propellers have a 20-deg blade angle. The inboard propellers spin either CW or CCW. In these figures, $C_l.c$ denotes the local lift times the local chord length. Local lift data are found from time-averaged solutions.

The next results compare the effects of blade angle on the wing aerodynamics. Two blade angles of 20 and 28 degrees are considered. A single propeller is installed on either the inboard or outboard section of the wing. All propellers spin counterclockwise at a spinning speed of 1024 rpm. In both (inboard and outboard mounted) cases, the propeller with a 28-deg blade angle have similar trends with the propellers with a 20-deg blade angle, but much larger differences are obtained between pressure data at upper and lower surfaces at 28-deg blade angles. In more detail, Figure 10 presents the iso-surfaces of the vorticity magnitude for these simulations. All visualizations correspond to the final simulation time step. Figure 10 shows that a larger slipstream is formed behind the propeller with a 28-deg blade angle. More negative pressure regions over the wing were formed with propellers having a 28-deg blade angle as well. Vortex shedding at the junctions of the wing and propeller slipstream are stronger for propellers with a 28-deg blade angle. Finally, Figure 11 compares the local wing lift distributions of these configurations. The propeller with a 28-deg blade angle leads to larger lift values over the wing. The vortex shedding effects are more visible in the plots of propellers with a 28-deg angle as well.

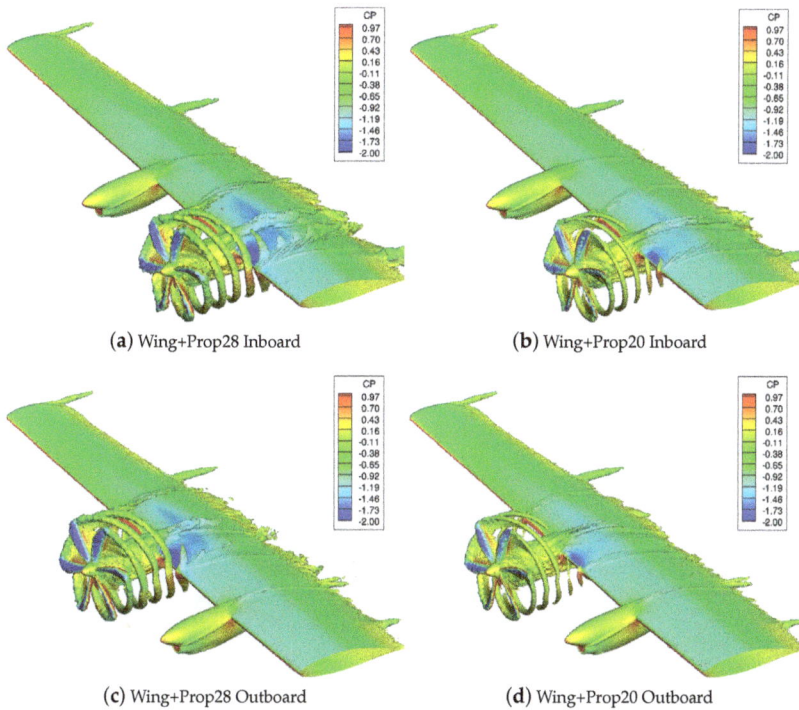

(a) Wing+Prop28 Inboard   (b) Wing+Prop20 Inboard

(c) Wing+Prop28 Outboard   (d) Wing+Prop20 Outboard

Figure 10. Propeller installed inboard/outboard; vorticity isosurfaces are colored with a pressure coefficient. Propellers have a 20-deg blade angle and spin at 1024 rpm counterclockwise. The wing only solution is time-averaged for the final three seconds of simulations. Wing+Prop solutions are at the final simulation time.

Next, results of a single propeller and a wing with both inboard and outboard mounted propellers are compared. In both cases, propellers have a 20-deg blade angle, spin counterclockwise at a rotational speed of 1024 rpm. Figure 12 shows the local lift distribution and vorticity isosurfaces of the wing with both propellers installed. In regions between propellers, the two-propellers increase the local wing lift compared with single propeller cases. In other regions, the two-propeller data follow the trends of the single propeller locally installed. In addition, Figure 13 compares the local lift distributions of the wings with two propellers but different spinning scenarios. Figure 13 shows that very different lift distributions are obtained depending on the spinning directions. These effects will be important in the aircraft design and how to control where the wing will stall first.

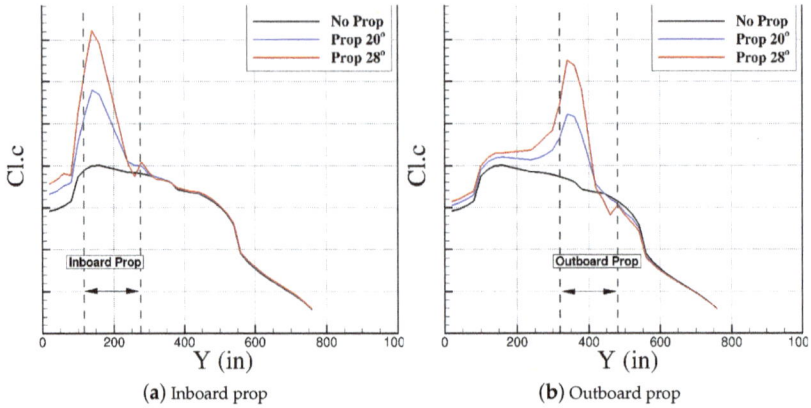

(a) Inboard prop

(b) Outboard prop

Figure 11. Local lift distribution for wing and propellers installed inboard or outboard the wing. Propellers have 20-deg or 28-deg blade angles. The propellers spin counter clockwise. In these figures, $C_l.c$ denotes the local lift times the local chord length. Local lift data are found from time-averaged solutions.

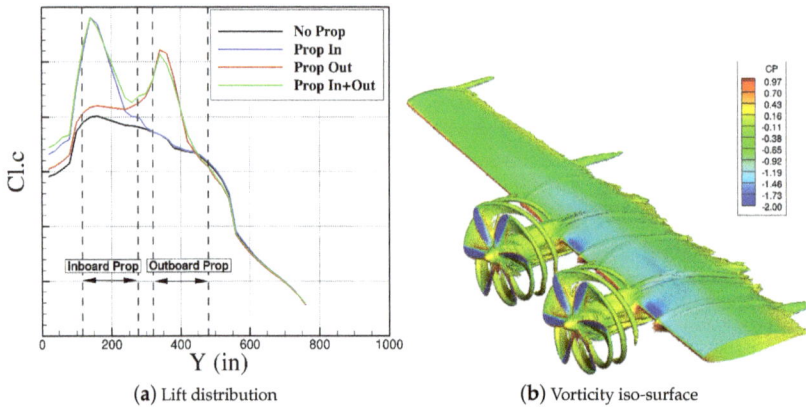

(a) Lift distribution

(b) Vorticity iso-surface

Figure 12. Propeller installed at both inboard and outboard wing; In (a), local lift distribution for wing and propellers installed inboard and outboard the wing are shown. In (b), vorticity isosurfaces are colored with pressure coefficient. Propellers have 20-deg blade angle and spin at 1024 rpm.

Final results present the effects of propeller on the wing stall behavior. Figure 14 shows the lift distribution of four configurations at angles of attack of 9, 10, 11, and 12 degrees. The configurations include wing without propeller, wing with inboard propeller, wing with outboard propeller, and wing with both inboard and outboard propellers. All propellers have a blade angle of 20 degrees and spin counterclockwise. Figure 14a shows that the wing only case has stalled at an 11-deg angle of attack. Increasing the angle of attack to 12 degrees does not increase local lift in most regions; it even falls behind the outboard nacelle. Figure 15 shows that, at an 11-deg angle of attack, flow is separated at the wing roots and behind nacelles. However, the tip has not been stalled yet and the lift increases with increasing angle of attack at the tip. Figure 14b,c show that local wing stalls behind propellers are delayed by mounting propellers at the inboard and outboard wing; however, the single propeller causes flow separation in other regions. The two-propeller case, however, delays stall at most positions. Figure 15 compares the vorticity iso-surfaces of all these configurations for tested angles of attack.

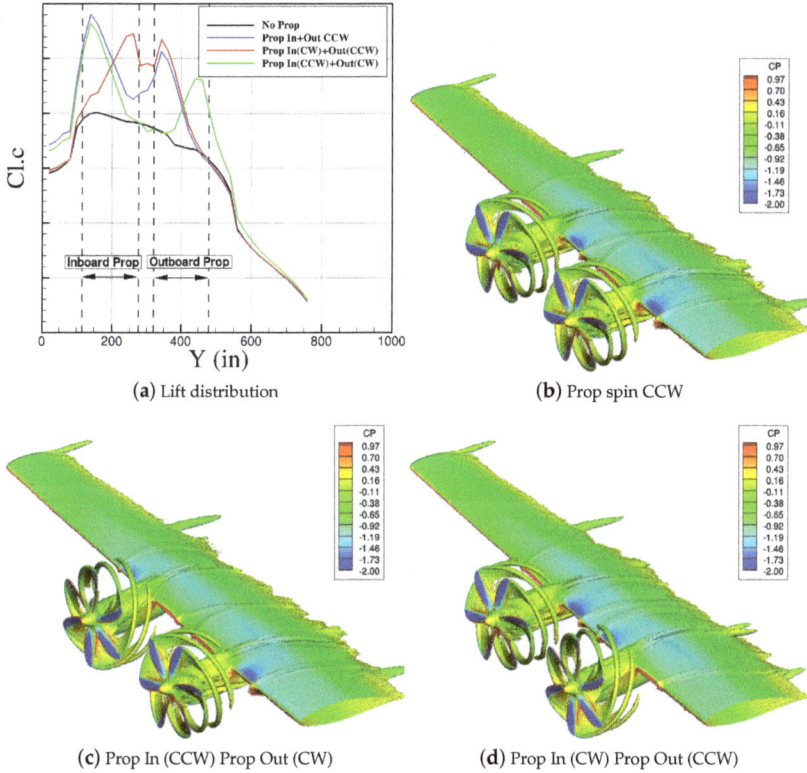

(**a**) Lift distribution

(**b**) Prop spin CCW

(**c**) Prop In (CCW) Prop Out (CW)

(**d**) Prop In (CW) Prop Out (CCW)

Figure 13. Propeller installed on both inboard and outboard wings, but they spin at different directions.

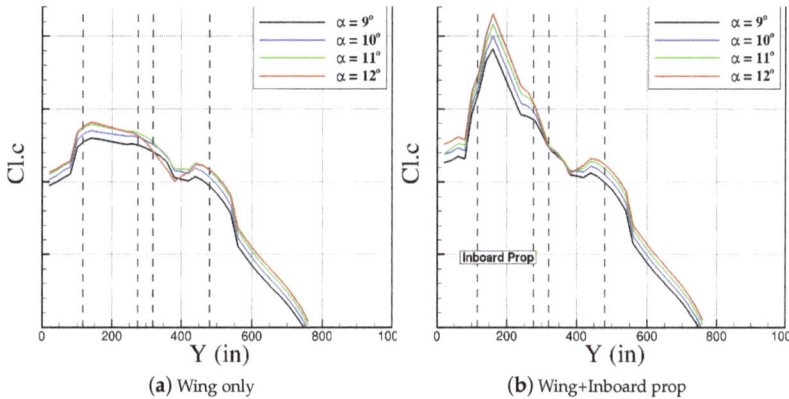

(**a**) Wing only

(**b**) Wing+Inboard prop

Figure 14. *Cont.*

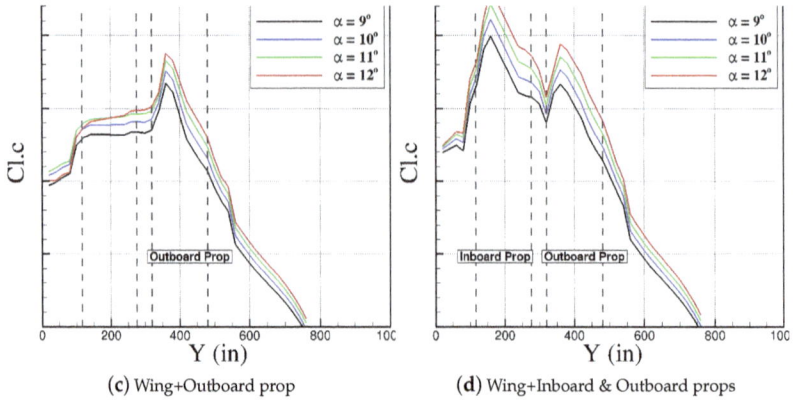

(**c**) Wing+Outboard prop   (**d**) Wing+Inboard & Outboard props

Figure 14. The wing stall behavior with and without propellers. All propellers spin counterclockwise.

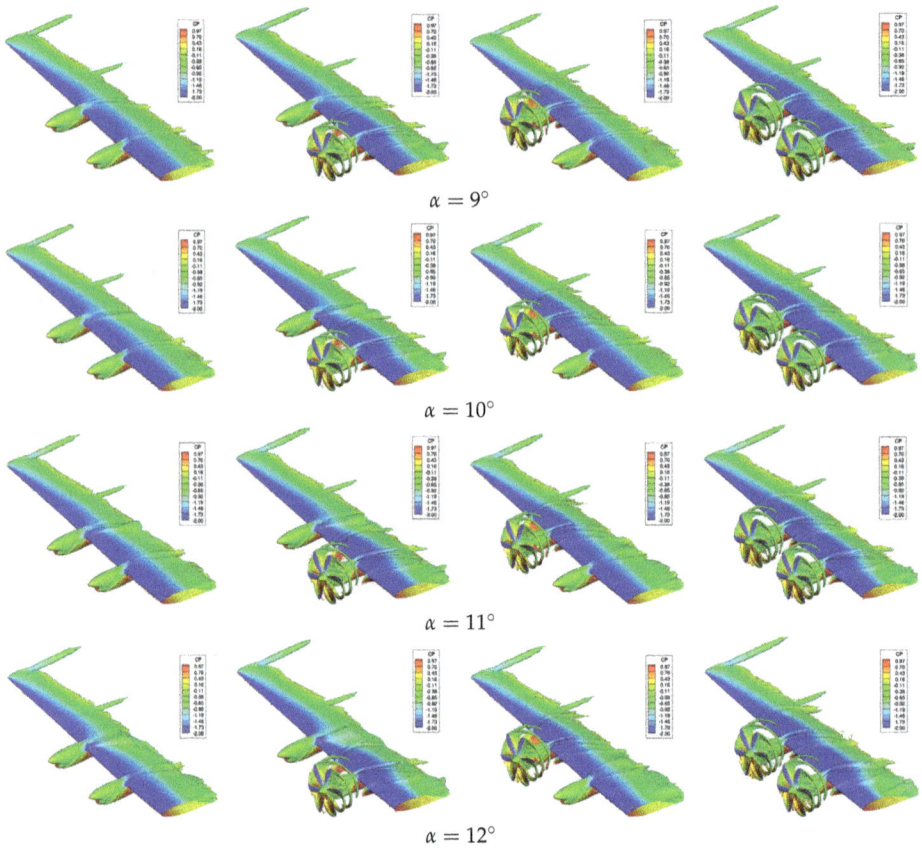

$\alpha = 9°$

$\alpha = 10°$

$\alpha = 11°$

$\alpha = 12°$

Figure 15. Propeller installed inboard/outboard; it has a 20-deg blade angle. Wing solution is time-averaged. Wing+Prop solutions are at a time of six seconds.

6. Conclusions

The propeller significantly changes the wing aerodynamic performance. The effects will depend on the blade angle, direction of rotation, and position of propellers on the wing. The most significant effects were seen beyond the propeller. For a counterclockwise spinning propeller (viewed from front), the upwash on the left side of the propeller caused the local angle of attack to increase and hence the local lift coefficient. The lift increase will be substantial behind the propeller because of combined effects of upwash and rise in the flow momentum. On the right side of the propeller, downwash will reduce the dynamic pressure rise effects. Outside the propeller disk, downwash causes the local lift to decrease. For tested propellers, increasing the blade angle from 20 to 28 degrees increased the local lift as well. In addition, two-propeller configuration was simulated and the results show that this improved wing lift distribution compared with single installed ones. Finally, the propeller presences will delay flow separation and local stall over the wing behind the propeller disk. The results of this work show the capability of Kestrel simulation tools for modeling and design of propellers and investigate their effects over aircraft during a conceptual design in which no experimental or flight test data are available yet. This will lead to reducing the number of tests required later. In addition, these results can be used for teaching purposes.

Author Contributions: All authors contributed to the research conceptualization and methodology. K.B. provided geometries. A.J. developed all computational grids. M.G. and P.A. performed all simulations and prepared the original draft. M.J.S. is the HPCRC director. K.B. funded this research.

Funding: This material is based in part on research sponsored by the US Air Force Academy under the agreement numbers of FA7000-17-2-0007 and FA7000-16-2-0010.

Acknowledgments: The authors acknowledge the support of the HPMCP computing resources and CREATE support team. The U.S. Government is authorized to reproduce and distribute reprints for Governmental purposes notwithstanding any copyright notation thereon. The views and conclusions contained herein are those of the authors and should not be interpreted as necessarily representing the official policies or endorsements, either expressed or implied, of the organizations involved with this research or the U.S. Government.

Conflicts of Interest: The authors declare no conflict of interest.

Nomenclature

a	acoustic speed, m·s^{-1}
CFD	computational fluid dynamics
C_p	pressure coefficient, $(p - p_\infty)/q_\infty$
CREATE	Computational Research and Engineering Acquisition Tools and Environments
CCW	counterclockwise
CW	clockwise
D	propeller diameter, m
DDES	delayed detached eddy simulation
F	thrust force, N
M	Mach number, V/a
p	static pressure, N/m^2
p_∞	free-stream pressure, N/m^2
q_∞	free-stream dynamic pressure, N/m^2
SARC	Spalart–Allmaras with rotational and curvature correction
RANS	Reynolds Averaged Navier–Stokes
t	time, s
VZLU	Czech aerospace research center
V_∞	free-stream velocity, m·s^{-1}
x,y,z	grid coordinates, m

Greek

α	angle of attack, deg
β	blade angle, deg

References

1. Froude, R.E. On the Part Played in Propulsion by Differences of Fluid Pressure. *Trans. Inst. Nav. Archit.* **1889**, *30*, 390.
2. Rankin, W.J. On the Mechanical Principles of the Action of Propellers. *Trans. Inst. Nav. Archit.* **1865**, *6*, 13–39.
3. Westmoreland, W.S.; Tramel, R.W.; Barber, J. Modeling Propeller Flow-Fields Using CFD. In Proceedings of the 46th AIAA Aerospace Sciences Meeting and Exhibit, Reno, NV, USA, 7–10 January 2008; AIAA Paper 2008-0402 .
4. Stuermer, A. Unsteady CFD Simulations of Propeller Installation Effects. In Proceedings of the 42nd AIAA/ASME/SAE/ASEE Joint Propulsion Conference & Exhibit, Sacramento, CA, USA, 9–12 July 2006; AIAA Paper 2006-4969.
5. Shafer, T.; Green, B.; Hallissy, B.; Hine, D. Advanced Navy Applications Using CREATE^{TM}-AV Kestrel. In Proceedings of the 52nd Aerospace Sciences Meeting, National Harbor, MD, USA, 13–17 January 2014; AIAA Paper 2014-0418.
6. McDaniel, D.; Nichols, R.; Klepper, J. Unstructured Sliding Interface Boundaries in Kestrel. In Proceedings of the 54th AIAA Aerospace Sciences Meeting, San Diego, CA, USA, 4–8 January 2016; AIAA Paper 2016-1299.
7. Steij, R.; Barakos, G. Sliding Mesh Algorithm for CFD Analysis of Helicopter Rotor–Fuselage Aerodynamics. *Int. J. Numer. Methods Fluids* **2008**, *58*, 527–549. [CrossRef]
8. Ghoreyshi, M.; Bergeron, K.; Lofthouse, A.J. Numerical Simulation of Wake Flowfield Behind the C-130 with Cargo Ramp Open. *J. Aircr.* **2017**, *55*, 1103–1121 . [CrossRef]
9. Bergeron, K.; Ghoreyshi, M.; Jirasek, A. Simulation of C-130 H/J Troop Doors and Cargo Ramp Flow Fields. *Aerosp. Sci. Technol.* **2018**, *72*, 525–541. [CrossRef]
10. Bergeron, K.; Ghoreyshi, M.; Jirasek, A.; Aref, P.; Lofthouse, A.J. Computational Modeling of C-130 H/J Propellers and Airdrop Configurations. In Proceedings of the 35th AIAA Applied Aerodynamics Conference, Denver, CO, USA, 5–9 June 2017; AIAA Paper 2017-3574.
11. Roth, G.L.; Morton, S.A.; Brooks, G.P. Integrating CREATE-AV Products DaVinci and Kestrel: Experiences and Lessons Learned. In Proceedings of the 50th AIAA Aerospace Sciences Meeting Including the New Horizons Forum and Aerospace Exposition, Nashville, TN, USA, 9–12 January 2012; AIAA Paper 2012-1063.
12. Morton, S.A.; McDaniel, D.R. A Fixed-Wing Aircraft Simulation Tool for Improving DoD Acquisition Efficiency. *Comput. Sci. Eng.* **2016**, *18*, 25–31. [CrossRef]
13. Morton, S.A.; McDaniel, D.R.; Sears, D.R.; Tillman, B.; Tuckey, T.R. Kestrel: A Fixed Wing Virtual Aircraft Product of the CREATE Program. In Proceedings of the 47th AIAA Aerospace Sciences Meeting including The New Horizons Forum and Aerospace Exposition, Orlando, FO, USA, 4–7 January 2009; AIAA Paper 2009-0338.
14. McDaniel, D.; Nichols, R.; Eymann, T.; Starr, R.; Morton, S. Accuracy and Performance Improvements to Kestrel's Near-Body Flow Solver. In Proceedings of the 54th AIAA Aerospace Sciences Meeting, San Diego, CA, USA, 4–8 January 2016; AIAA Paper 2016-1051.
15. Godunov, S.K. A Difference Scheme for Numerical Computation of Discontinuous Solution of Hydrodynamic Equations. *Sbornik Math.* **1959**, *47*, 271–306.
16. Tramel, R.; Nichols, R.; Buning, P. Addition of Improved Shock-Capturing Schemes to OVERFLOW 2.1. In Proceedings of the 19th AIAA Computational Fluid Dynamics Conference, San Antonio, TX, USA, 22–25 June 2009; AIAA Paper 2009-3988.
17. Spalart, P.R.; Allmaras, S.R. A One Equation Turbulence Model for Aerodynamic Flows. In Proceedings of the 30th Aerospace Sciences Meeting and Exhibit, Reno, NV, USA, 6–9 January 1992; AIAA Paper 1992-0439.
18. Spalart, P.R.; Schur, M. On the Sensitisation of Turbulence Models to Rotation and Curvature. *Aerosp. Sci. Technol.* **1997**, *1*, 297–302. [CrossRef]
19. Menter, F. Eddy Viscosity Transport Equations and Their Relation to the k-ε Model. *ASME J. Fluids Eng.* **1997**, *119*, 876–884. [CrossRef]
20. Spalart, P.R.; Jou, W.H.; Strelets, M.; Allmaras, S.R. Comments on the Feasibility of LES for Wings, and on a Hybrid RANS/LES Approach. In Proceedings of the 1st AFSOR International Conference on DNS/LES, Ruston, LA, USA, 4–8 August 1997; Greyden Press: Columbus, OH, USA, 1997; pp. 137–147.

21. Ferraro, G.; Kipouros, T.; Savill, A.M.; Rampurawala, A.; Agostinelli, A. Propeller–Wing Interaction Prediction for Early Design. In Proceedings of the 52nd Aerospace Sciences Meeting, National Harbor, MD, USA, 13–17 January 2014; AIAA Paper 2014-0564.

22. Stuermer, A.; Rakowitz, M. *Unsteady Simulation of a Transport Aircraft Propeller Using MEGAFLOW*; RTO-MP-AVT-123 Technical Report; Meeting Proceedings RTO-MP-AVT-123, Paper 7; RTO: Neuilly-sur-Seine, France, 2015.

23. Swatton, P. *Principles of Flight for Pilots*; Aerospace Series; Wiley: Hoboken, NJ, USA, 2011.

24. Mikolowsky, W. A Short History of the C-130 Hercules. In Proceedings of the AIAA International Air and Space Symposium and Exposition: The Next 100 Years, Dayton, OH, USA, 14–17 July 2003; AIAA Paper 2003-2746.

25. Thom, A.; Duraisamy, K. Computational Investigation of Unsteadiness in Propeller Wake–Wing Interactions. *J. Aircr.* **2013**, *50*, 985–988. [CrossRef]

26. Veldhuis, L.M. Review of Propeller–Wing Aerodynamic Interference. In Proceedings of the 24th International Congress of the Aeronautical Sciences, Yokohama, Japan, 29 August–3 September 2004.

27. Computational Research and Engineering Acquisition Tools And Environments (CREATE), Eglin AFB, FL 32542. In *Kestrel User Guide, Version 6.0*; High Performance Computing Modernization Program: Lorton, VA, USA, 2015.

aerospace

MDPI

Article

CFD Validation and Flow Control of RAE-M2129 S-Duct Diffuser Using CREATE™-AV Kestrel Simulation Tools

Pooneh Aref, Mehdi Ghoreyshi *, Adam Jirasek and Matthew J. Satchell

High Performance Computing Research Center, U.S. Air Force Academy, USAF Academy, El Paso County, CO 80840, USA; Pooneh.Aref@usafa.edu (P.A.); Adam.Jirasek@usafa.edu (A.J.); Matthew.Satchell@usafa.edu (M.J.S.)
* Correspondence: Mehdi.Ghoreyshi@usafa.edu

Received: 17 January 2018; Accepted: 14 March 2018; Published: 16 March 2018

Abstract: The flow physics modeling and validation of the Royal Aircraft Establishment (RAE) subsonic intake Model 2129 (M2129) are presented. This intake has an 18 inches long S duct with a 5.4 inches offset, an external and an internal lip, forward and rear extended ducts, and a center-positioned bullet before the outlet. Steady-state and unsteady experimental data are available for this duct. The measurements include engine face conditions (pressure recovery, static pressure to free-stream total pressure ratio, and distortion coefficient at the worst 60° sector or DC60), as well as wall static pressure data along the duct. The intake has been modeled with HPCMP CREATE™-AV Kestrel simulation tools. The validation results are presented including the effects of turbulence models on predictions. In general, very good agreement (difference errors are less than 6%) was found between predictions and measurements. Secondary flow at the first bend and a region of flow separation are predicted at the starboard wall with an averaged DC60 coefficient of 0.2945 at the engine face. Next, a passive and an active flow control method are computationally investigated. The passive one uses vane-type vortex generators and the active one has synthetic jet actuators. The results show that considered passive and active flow control methods reduce the distortion coefficient at the engine face and the worst 60° sector to 0.1361 and 0.0881, respectively. The flow control performance trends agree with those obtained in experiments as well. These results give confidence to apply the Kestrel simulation tools for the intake design studies of new and unconventional vehicles and hence to reduce the uncertainties during their flight testing.

Keywords: S-duct diffuser; flow distortion; flow control; vortex generators

---

1. Introduction

A key challenge in the development of an aircraft is to integrate the propulsion system with the airframe such that a balance between the integrated propulsion requirements and the overall aircraft design demands are met. Many of the engineering challenges in this issue arise from the fact that the successful integration of airframes and engines involves major compromises between the wishes of the aircraft and engine manufacturers. For a multi-mission aircraft propulsion system, the installation study involves the design of intake, exhaust system, secondary air system (i.e., ejector nozzles), and the integration of those components with the engine and aircraft.

Typically, the intake design is the aircraft manufacturer's responsibility, but in terms of the installed-engine performance, the intake effects need to be understood. The optimal intake design is a trade-off study about desirable requirements, namely high pressure recovery, low installation drag, low radar and noise signatures, as well as minimum weight and cost. In terms of propulsion, the function of the intake is to introduce a sufficient mass of air from the ambient environment flow uniformly and with stability to the engine compressor under all flight conditions.

Turbojet and low bypass turbofan engines are often located within the fuselage of an aircraft. This is especially employed in combat aircraft which should face a compact layout in order to reduce radar cross section. In addition, this type of engine installation results in less installation drag. However, the optimal incorporation of an engine into the aircraft fuselage for reducing the radar cross-section is a complex task because such a configuration requires an S duct with the possibility of thick boundary layer ingestion into the engine compressor under the severe conditions of the adverse pressure gradient inside the duct. Thus, the design of these intakes must ensure that the engine operates effectively in the presence of such boundary-layer ingestion and to avoid significant flow distortion. The intake flow distortion is characterized by the non-uniformity in the flow parameters (such as velocity and pressure) in planes perpendicular to the flow direction [1]. Specifically, flow distortions at the engine face should be minimized. In a subsonic intake diffuser, the flow distortion is caused by the ingestion of fuselage boundary layer or aircraft vortices into intake, flow separation at the cowl lips during maneuvering conditions, or formation of secondary and separated flow regions at the intake bends with small radius of curvature. These airflow distortions will lead to total pressure loss and non-uniformity at the engine face which reduces the compressor surge margin and may eventually cause the compressor to stall, engine instability, and the performance deviation from design conditions [2]. Careful consideration should therefore be given to the design of the shape of the cowl lip and diffuser of these intakes.

The purpose of this article is to determine the flow characteristics of S duct intakes via the extensive use of computational simulations. The Computational Fluid Dynamic (CFD) tools, however, need to be first verified against experimental data. In order to verify the used CFD tools, the Royal Aircraft Establishment (RAE) subsonic intake Model 2129 (M2129) is considered here. This subsonic S duct intake was designed under a joint program between NASA and the UK Ministry of Defense (MOD) and was tested in the DRA/Bedford wind tunnel. The aim of these experiments was to calibrate CFD codes and achieve the inlet distortion control.

The flow inside M2129 S-duct intake has been extensively studied using experimental and numerical methods. However, all CFD codes based on solutions of the Reynolds-Averaged Navier-Stokes (RANS) equations failed to exactly predict measured data because of the complexity of flow-field inside this duct [3,4]. As detailed in Ref. [5], the flow initially accelerates before reaching the throat section of this intake. At the first bend, the flow passing along the starboard side (inside of the bend) is subject to the centrifugal and pressure forces which cause the flow streamlines move towards the port side (outside of the bend). The interaction of the incoming flow from the inner side with the adverse pressure gradient region occurring at the port side leads to forming two swirling secondary flows. The flow continues to decelerate inside the diffuser and the ram pressure will rise as well. At the second bend, the low (kinetic) energy flow at the outside wall does not form any strong secondary flows that balance or cancel those generated by the first bend. In addition, the low energy flow along the starboard side of the second bend is subject to adverse pressure gradient region and will separate. Most CFD codes using RANS solutions failed to predict the observed secondary flows and the exact location of flow separation point. However, hybrid RANS and large eddy simulation methods have improved these predictions. This work in particular focuses on the simulation of the RAE M2129 intake using the HPCMP (High Performance Computing Modernization Program) CREATE™-AV (Air Vehicles) Kestrel simulation tools. These relatively new tools from the U.S. DoD (Department of Defense) HPCMP have extensively been tested and validated for a wide range of external flow applications, however, much less effort has been devoted to the validation of the code for internal flow problems.

It is quite impossible to design an S duct diffuser with small flow distortions at all flight conditions. Therefore, many studies have focused on the reduction and control of intake flow distortion using passive [6,7] and active flow controls [8,9]. The most common methods for passive and active controls are using vanes/plates and synthetic jet actuators, respectively. Passive vortex generators have already been used to control the boundary layer separation in many applications. The mixture of low and high momentum flow regions by these devices can locally control the effects of separation. Vane-type vortex generators have been used in S duct intakes as well to improve pressure recovery, for example see Ref. [10]. The recent applications of these devices attempt to improve both total pressure recovery and uniformity. In these applications, an array of vane type vortex generators are typically placed at or behind the first bend to control the secondary flow formation [11]. Though these devices are very simple to install, they cannot effectively control and reduce distortion at all off-design conditions.

Anderson and Gibb [11] presented the experimental results of applying different vortex generator configurations for the secondary flow control of the M2129 S duct. The experiments were again conducted at the DRA/Bedford 13 ft × 9 ft wind tunnel with the throat Mach number in range of 0.2 to 0.8. The VG170 configuration of Anderson and Gibb's study is considered in this work. This configuration has the best performance among others for all tested conditions. It consists of 22 vanes with chord length ratio of 0.2703 and blade height ratio of 0.07.

Air jet vortex generators have been tested for the flow control of the M2129 S duct as well [12]. These devices are more difficult to install than passive ones and require bleeding high pressure air from engine that might affect its performance. However, these devices lead to lower distortion coefficients and improve engine face conditions at all off-design conditions. In this work, an active control method with 22 jets is considered; each jet has a diameter of 1 mm, an imping angle of 30 degrees relative to the walls, with a bled air total to free-stream pressure ratio of two.

This work uses the HPCMP CREATE^{TM}-AV Kestrel simulation tools to validate and investigate the flow predictions of the RAE M2129 S duct baseline. Computational tools are then investigated for studying the distortion and flow changes of this baseline with vane type and air jet vortex generators. The article is organized as follows. First, CFD solver and test cases are described. The intake performance is briefly presented. Next, the validation results are given for the baseline followed by the results obtained from baseline with control methods. Finally, concluding remarks are provided.

2. CFD Solver

The simulation tools used in this work is the fixed wing computational tool of CREATE^{TM}-AV program, i.e., Kestrel. The code is a DoD-developed solver in the framework of the CREATE^{TM} Program, which is funded by the DoD High Performance Computing Modernization Program (HPCMP). The CREATE^{TM} focuses on addressing the complexity of applying computationally based engineering to improve DoD acquisition processes [13]. CREATE^{TM} consists of three computationally based engineering tool sets for design of air vehicles, ships, and radio-frequency antennae. The fixed wing analysis code, Kestrel, is part of the Air Vehicles Project (CREATE^{TM}-AV) and is a modularized, multidisciplinary, virtual aircraft simulation tool incorporating aerodynamics, jet propulsion integration, structural dynamics, kinematics, and kinetics [13]. The code has a Python-based infrastructure that integrates Python, C, C++, or Fortran-written components [14]. New modules can easily integrated into the code.

Kestrel version 7.2 is used in this work. The flow solver of the code discretizes Reynolds–Averaged Navier Stokes (RANS) equations into a cell-centered finite-volume form. The code then solves unsteady, three-dimensional, compressible RANS equations on hybrid unstructured grids [15]. The Method of Lines (MOL) in the code separates temporal and spatial integration schemes from each other [16]. The spatial residual is computed via a Godunov type scheme [17]. Second-order spatial accuracy is obtained through a least squares reconstruction. The numerical fluxes at each element face are computed using various exact and approximate Riemann schemes with a default method based on HLLE++ scheme [18]. In addition, the code uses a subiterative, point-implicit scheme method (a typical Gauss-Seidel technique) to improve the temporal accuracy.

Kestrel receives an eXtensible Markup Language (XML) input file generated by Kestrel User Interface and stores the solution convergence and volume results in a common data structure for later use by the Output Manager component. Some of the turbulence models available within Kestrel include the one-equation turbulence models of Spalart–Allmaras (SA) [19], Spalart–Allmaras with rotational/curvature correction (SARC) [20], Mentor's SST model [21], and Delayed Detached Eddy Simulation (DDES) with SARC [22] and SST RANS turbulence model [21].

3. Test Case

An S-duct diffuser, named RAE M2129, was designed around 1990 under a joint program between NASA and the UK defense Ministry and was tested in the DRA/Bedford 13 ft × 9 ft wind tunnel. A large number of diffuser geometries have been studied with different lip shapes, cross-sectional changes, and etc. The test case used in this study corresponds to the geometry used in the Aerodynamics Action Group AD/AG-43, "Application of CFD to High Offset Intake Diffusers" [23]. The aim of this project was calibration of CFD codes for an S duct diffuser.

The M2129 geometry is shown in Figure 1. This intake has a circular entry section followed by an S bend diffuser. The model is a side-mounted duct with a horizontal plane of symmetry, i.e., $y = 0$. The diffuser offset is therefore in the horizontal plane. Based on this setup, the duct side with minimum z was named starboard, port was the side with the maximum z, the minimum y side was named bottom, and finally the maximum y side top.

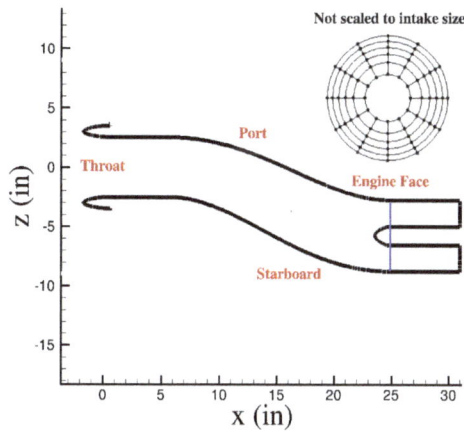

Figure 1. The sketch of the RAE M2129 geometry used in this work. The pressure rakes at the engine face are shown as well. They include 12 equally spaced arms with 30-degree intervals with six pitot pressure probes at each arm.

The inlet throat diameter is approximately 5.06 inches. The diffuser length is 18 inches with 5.4 inches centerline offset. Two constant-area sections were added at the upstream and end of the duct as shown in Figure 1. With this modification the overall built duct length was 29.969 inches. The engine face bullet starts at $x = 23.129$ inches measured from the most forward part of the intake lip and becomes parallel at $x = 24.569$ inches. The engine face, where measurements were made, is at $x = 24.669$ inches as well. In more detail, the locus of the centerline offset curve and the variation of the radius in the diffusing part of the intake can be determined by following equations:

$$z = 0.15L\left[1 - \cos\left(\pi\frac{x}{L}\right)\right] \tag{1}$$

$$\left(\frac{R - R_t}{R_f - R_t}\right) = 3\left(1 - \frac{x}{L}\right)^4 - 4\left(1 - \frac{x}{L}\right)^3 + 1 \tag{2}$$

where L is the diffuser length, R_t and R_f are the throat and engine face radii, respectively. By studying these equations, the 5.4 inches be deduced to be the duct offset between the throat and engine face center.

The RAE M2129 intake experiments were conducted in the DRA/Bedford 13 ft × 9 ft wind tunnel which is a closed-circuit type. The experimental data used in this work corresponds to Data Point (DP) 78. The diffuser of this run has a bullet and a static rake in the compressor entry plane. The free-stream conditions in these experiments correspond to a Mach number of 0.204 and zero angles of attack and side slip. Free-stream total pressure and total temperature were 105,139.5 Pa and 293.7 K, respectively. Table 1 lists all experimental conditions. The static rake at the engine face (see Figure 1) had 12 equally spaced arms with 30 degrees intervals, such that each arm had six pitot pressure probes. The inner probes were at 1.1 inches radius, while the outer probes were positioned at a radius of 2.89 inches. The measurements included pressure recovery, static pressure to total free-stream pressure, and DC60 coefficient at the engine face. In additions, DP78 experiments had four rows of static pressure taps along the duct at starboard, port, top, and bottom sides.

Table 1. Flow conditions of the DP78 run of RAE-M2129 baseline.

Flow Conditions	Experimental Data
Free-stream Mach	0.204
Free-stream total pressure	105,139.5 Pa
Free-stream total temperature	293.7 K
Angle of attack	0 degree
Sideslip angle	0 degree
Mass flow ratio	1.9382

In a different experimental campaign by Gibb and Anderson [12], the RAE M2129 diffuser was tested with different vane type vortex generator designs. The general geometries of these vanes and their locations relative to the diffuser entry plane are given in Ref. [12] and shown in Figure 2. In this work, the VG170 configuration is used. Referring to Figure 2, this configuration has 22 vanes located at X_{VG}/R_i of 2.0. Each vane has a blade height ratio (h/R_i) of 0.070 and chord length ratio (c/R_i) of 0.2703. In addition, the spacing angle and the vane angle of attack were set at $15°$ and $16°$, respectively.

Finally, the M2129 diffuser intake with air jet vortex generators (AJVG) was studied as well. Experimental data of such a flow control method were again reported by Gibb and Anderson [12]. In the present work, similar designs to Ref. [12] are used as well. The flow control consists of the row of high pressure jets located in the position of VG170 vane-equipped configuration described earlier. Unlike the designs used in Ref. [12], the jets of this work are inclined 17 degrees towards the inlet wall and 30 degrees towards the surface normal direction following recommendation of Ref. [24]. The jet diameters are 1 mm as well.

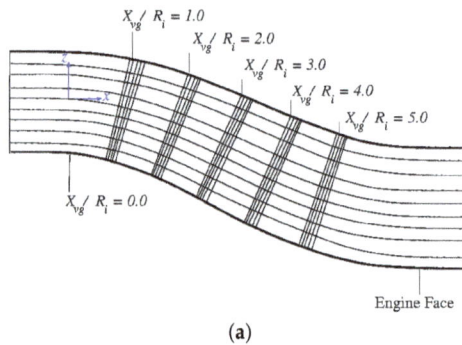

(a)

Figure 2. *Cont.*

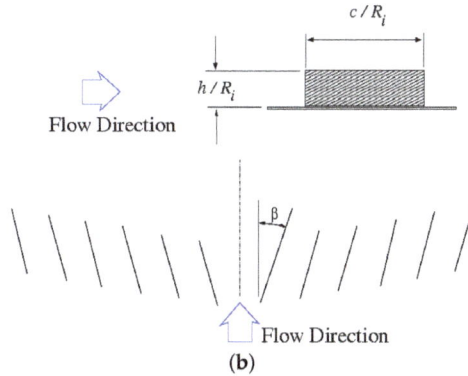

Figure 2. The RAE M2129 vortex generator locations and geometry parameters. These pictures were adapted from Ref. [12]. (**a**) VG locations; (**b**) Geometric parameters.

4. Computational Grids

Four grids are considered in this study. Detail of these grids are given in Table 2. The first grid corresponds to the RAE M2129 wind tunnel geometry without any flow control. This corresponds to the wind tunnel model of DP78 run with a bullet. However, the computational model has slightly longer forward extension than the section of experimental model. The grid for this geometry is a hybrid (structured and unstructured) and was obtained from Bernhard Anderson of NASA Glenn Research Center and has about 31.2 million cells. This grid has structured cells over all wall surfaces. Body-fitted structured grid layers exist at the wall and tetrahedra cells used elsewhere. This model was used for CFD validation of the M2129 diffuser baseline.

Table 2. Detail of computational grids.

Grid	Description	Number of Cells (Millions)
Grid1	Obtained from NASA; baseline intake	31.2
Grid2	generated at USAFA; baseline intake	15
Grid3	generated at USAFA; baseline intake + 22 vanes	19.9
Grid4	generated at USAFA; baseline intake + 22 jets	62.3

A second intake diffuser was modeled at the U.S. Air Force Academy (USAFA) using M2129 intake data and Equations (1) and (2). These two geometries (USAFA and NASA) do not match everywhere as shown in Figure 3. The new geometry has slightly different offset from the original one and the cross-sectional changes are different from experimental models as well.

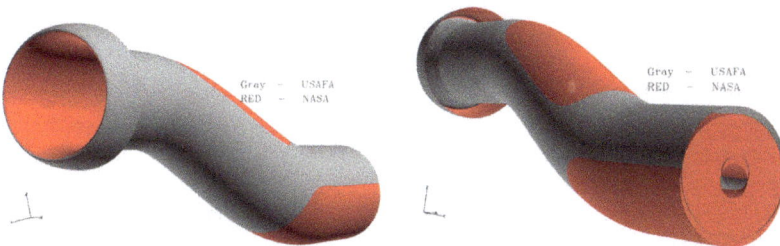

Figure 3. RAE M2129 geometries. NASA model was obtained from Bernhard Anderson. USAFA model was created from given intake data and offset and diameter changes equation with length.

Three grids were then made from this new intake geometry: one without any flow control, one with passive, and one with active flow control. Geometry and grids are shown in Figure 4. The passive flow control consist of one row of vanes which are designed according to Ref. [12] and its detailed VG170 configuration. The grid generator builds a prism layer around the vane surfaces to enable their modeling as viscous wall. The active flow control consists of the row of high pressure jets located in the position of VG170 vanes configuration as well.

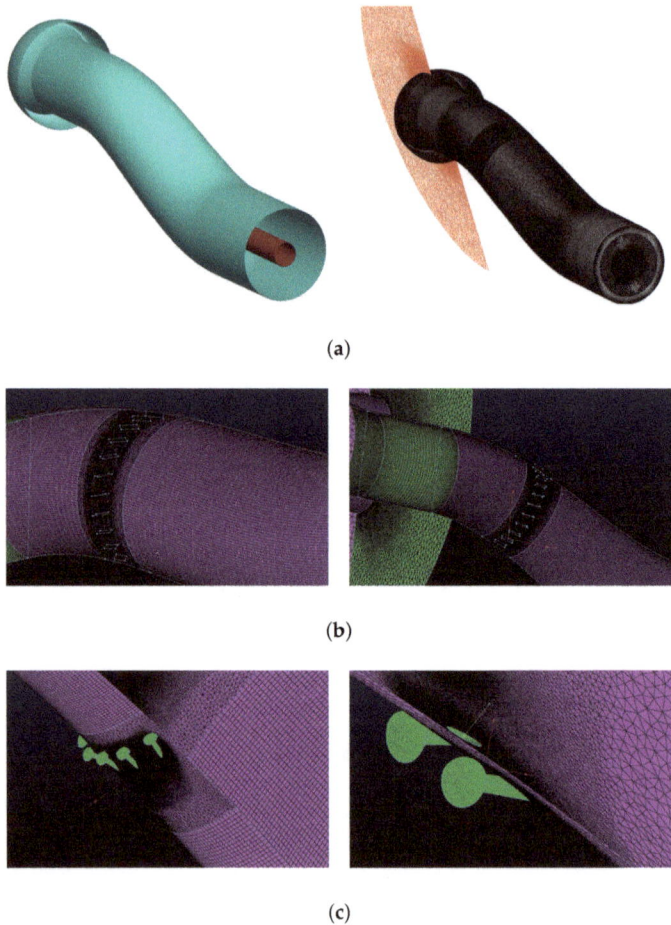

(a)

(b)

(c)

Figure 4. USAFA baseline intake model with vortex generator plates and jets. (**a**) USAFA baseline geometry and symmetry grid; (**b**) Vortex generator plates; (**c**) Jets.

The computational grids were made in Pointwise version 18. Most of the surface mesh is made of patches of structured mesh with prism layers build upon the surface mesh. The remaining parts of interior meshes above the prism layer are made of tetra cells. The thickness of the first layer is determined by the condition of $y+ < 1$ with a cell wall normal growth ratio of 1.25. The baseline grid has approximately 15 million cells. The approximate numbers of cells for intake with passive and active controls are 19.9 and 62.3 million cells, respectively.

5. Intake Performance

In general, the loss of total pressure inside a diffuser intake can be the result of the wall frictions, the flow separation, and formation of shock waves. A term named pressure recovery can represent the level of losses of a diffuser. It is defined as follows:

$$PR = \frac{p_{0f}}{p_{0\infty}} \tag{3}$$

where p_{0f} is the mean total pressure at the engine's face plane and $p_{0\infty}$ denotes free-stream total pressure. If the diffuser flow is assumed to be isentropic, there is no pressure losses and the outlet and inlet total pressures of the intake are the same which gives a pressure recovery of one. Wallin et al. [25] presented a relationship between the pressure recovery and geometric patterns of a not bent intake. Based in this relationship, as the duct length increases, the pressure losses will rise due to the higher skin friction losses. These pressure losses will be higher if the intake has a bend and when the internal cross-section shape changes, e.g., from elliptic to circular. In addition, the pressure recovery values drop if the engine operates at the larger throttle setting. The pressure recovery will fall off with increasing yaw and incidence angle as well. Additionally, the influence of external pressure field (wing, fuselage, etc.) upon the intake pressure recovery needs to be determined. For the intake types that the entry is from fuselage, the run up distance and wetted fuselage area ahead of intake significantly affect the flow-field inside the intake. The prediction of all these effects upon the intake's performance is a challenging task that must be concerned within the intake performance analysis.

The sensitivity of the engine performance on the intake's pressure-recovery depends on the design of the engine. Antonatos et al. [26] have assumed a linear relationship between the total pressure losses and the engine thrust as:

$$\frac{\Delta F}{F} = CR \cdot (1 - PR) \tag{4}$$

where ΔF and F denote thrust drop and thrust force respectively. CR is the correction factor that depends on the engine configuration. The correction factor for turbojet and turbofan engines lies between 1.1 and 1.6 over the Mach number range 0.8 to 2.2.

In addition to the total pressure losses, the diffuser flow is distorted where the flow separates from boundary surfaces or in the presence of secondary flows. The separation in duct can be result of sharp bends, thin lip, shock and boundary-layer interactions and etc. Since the 1950s, the following distortion parameter is used widely to identify distortion level:

$$D_t = \frac{p_{0f,\max} - p_{0f,\min}}{p_{0f}} \tag{5}$$

where the maximum and minimum total pressures are taken from a series of measurements of equally spaced radial probes around the engine face plane. Later, new distortion definitions were used based on the differential total pressure values between the probes average and minimum pressure total pressures. Different distortion coefficients have been defined as well. The one that has been adopted in this study is from Ref. [23] which is defined as:

$$DC60 = \frac{p_{0f} - p_{0f,60}}{\bar{q}_f} \tag{6}$$

where $p_{0f,60}$ is the mean total pressure in the worst 60-degree sector of the engine face. $\bar{q}_f$ is the mean dynamic pressure at the engine face as well.

6. Results and Discussions

Computational resources were provided by the DoD's High Performance Computing Modernization Program. All simulations are run on the Engineering Research Development Center (ERDC) Topaz

System (3456 computing nodes with 36 cores per node running two Intel Xeon E5-2699v3 processors at a base core speed of 2.3 GHz with 117 GBytes of RAM available per node). Each job asks for 2024 computing nodes and 8 h wall-clock time. The convergence criterion is based on tracking data (CFD parameters, coefficients, forces) that Kestrel prints at each time step and whether the averaged values have reached a converged state or not.

Validation results use the NASA baseline grid without any flow control mechanism. Kestrel was used to solve the flow inside this intake diffuser. The boundary conditions for the far-field included the free-stream Mach number, total pressure and total temperature corresponding to DP78 experimental data of Ref. [23]. These data are given in Table 1. The outlet plane was defined as a sink boundary with the given mass flow rate of Table 1.

All CFD simulations are run in unsteady mode with second order accuracy in time and three Newton subiterations as recommended by the Kestrel user guide [27]. A global time step of 5×10^{-4} s is used in all simulations. Notice that the accuracy of predictions of secondary and separated flows inside a S duct will largely depend on the temporal order of accuracy and selected time step. Too large the time step can cause instability, inaccuracy, and not capturing important time-dependent features. Too small the time step will increase the computational cost to simulate flow changes over a given period of time. In order to understand the effects of time step on the solution, pressure recovery and DC60 values of simulations of the NASA baseline grid using SARC–DDES turbulence model are plotted versus time step in Figure 5. Note that all simulations were run for 2.25 s of physical time; the solutions between 2 and 2.25 s were then time-averaged. Figure 5 shows that pressure recovery and DC60 values have nearly converged to their final values and experimental data for a time step of 5×10^{-4} s. Interestingly, the values corresponding to the smallest time step used, i.e., 1×10^{-5} s, show slightly larger differences with experimental data than the selected time step of 5×10^{-4} s. In addition, CFD data using this small time step are ten times more expensive to obtain.

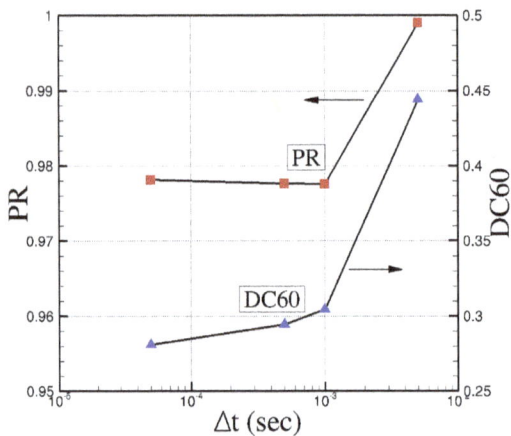

Figure 5. Time step sensitivity study. All cases were run for 2.25 s. Predictions between time 2 to 2.25 s were time-averaged. NASA baseline grid with SARC–DDES turbulence model was used.

All simulations are initially run with 500 startup iterations and then are continued for 4500 regular time steps or 2.25 s of physical time. Note that during startup iterations, simulation time remains zero. Time-averaged data are then written for last 500 time steps or simulation times between 2 and 2.5 s. Two sets of tap positions are defined: In the first, tap points are located at the engine face and correspond to the pitot tube locations of a static rake used in the DP78 experimental run. This rake has six pitot pressure probes at each of 12 equally spaced arms with 30 degrees intervals, i.e., 72 tap positions. These tap positions are shown in Figure 1. The second set of tap data correspond to the intake wall at starboard and port sides. These taps were generated using Carpenter, a mesh manipulation tool of Kestrel, as the result of the symmetry plane cut with the duct wall. Density, static and total pressure, and velocity components at each tap positions are written by the solver for each time step from 4000 to 4500. Finally, simulations are run for different turbulence models.

First set of results compare the time-averaged pressure recovery (total pressure to free-stream total pressure) at the engine face with available experimental data measured at pitot pressure probes. The comparison plots are shown in Figure 6 using the SARC–DDES turbulence model. Time-averaged data are plotted using visualization files with averaged data from simulation times between 2 to 2.5 s. Notice that measurements were taken at shown pitot pressure probes of Figure 6b and therefore the shown experimental plot is a partial representation of the actual engine face, e.g., the bullet is not shown. Additionally, pressure recovery data at regions between arms or measurements were interpolated from the rake measured data. However, CFD data show the full engine face plane generated in TECPLOT.

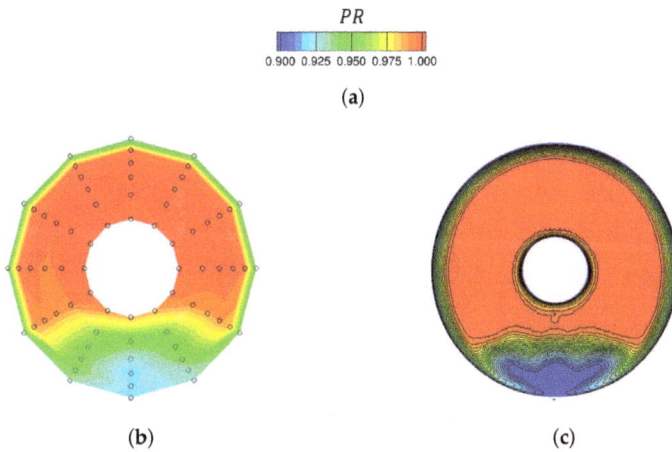

Figure 6. Experimental and simulated engine face pressure ratio. NASA grid was used with SARC-DDES turbulence model. CFD data are time-averaged from 2 to 2.5 s of simulation time. (**a**) Pressure recovery range; (**b**) Experimental data; (**c**) CFD—Baseline.

Figure 6 shows that overall pressure recovery trends predicted by CFD solver are similar to those found in the experiments. Pressure recovery values at and near walls are small because of skin friction effects. There is a separated flow region at the starboard side of the engine face plane with low pressure recovery values which make non-uniform flow entering the engine compressor. At this region (blue-colored region of Figure 6), there are two large counter-rotating vortices near the intake's symmetry plane and at the starboard side of intake. Two smaller secondary vortices are formed outside and above the primary vortices as well (these vortices are not shown in the figure). These vortices were not well captured with the experiments using shown static probes.

In more detail, Table 3 compares Kestrel predictions using SARC-DDES with available measurements of DP78 run. CFD data are again time-averaged values for simulation times between 2 to 2.5 s. These averaged data are calculated in two ways. In the first approach, CFD data (total, static pressure, density, velocities) are written at each rake point shown in Figure 7 and then used to find engine face conditions. These data are available at each time step from time 2 to 2.5 s meaning that the tap data in CFD are reported for last 500 time steps. A script was then written to find averaged pressure recovery, Mach, static pressure ratio and DC60 coefficient at the engine face using these data.

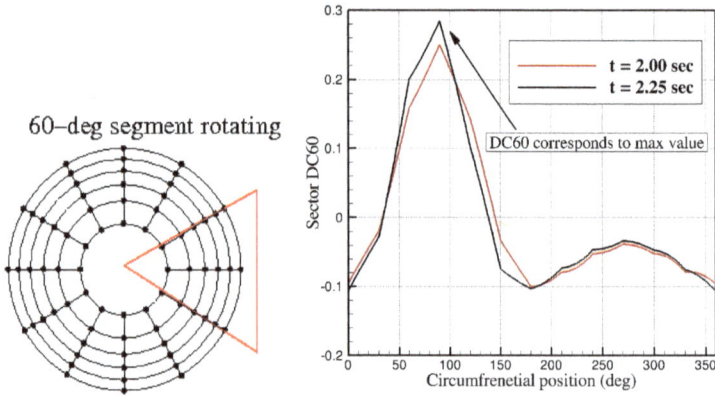

Figure 7. Illustration of the DC60 calculation approach. In the left, a 60-degree segment is shown which is at zero circumferential angle. The sector rotates clockwise one degree at a time. In the right, DC60 values for the 60-degree segment at each angle are shown. These plots correspond to simulation times of 2 and 2.25 s.

Table 3. Validation data of the RAE-M2129 baseline intake. NASA grid was used with SARC–DDES turbulence model. CFD data are time-averaged from 2 to 2.5 s of simulation time.

Engine Face Station	Experimental Data	CFD Data (Rake)	CFD Data (Face)	Error (Rake), ε %
$p_{0f}/p_{0\infty}$	0.9744	0.9776	0.9752	−0.328
Mach No.	0.4193	0.4329	0.4329	−3.243
$p_f/p_{0\infty}$	0.8522	0.8737	0.8497	−2.523
DC60	0.3130	0.2945	0.3294	5.910

The process of calculating DC60 from pressure rakes is illustrated in Figure 7. In more detail, the inputs to the code are flow conditions at locations of the shown rake probes. The number of probes is specified as a number of arms of the rake probe and a number of probes at each arm. DC60 index is then calculated for a 60 degrees sector rotating clockwise, one degree at each time. The swirl index and DC60 then corresponds to the critical sector with maximum DC60 index value. The data can be steady or a sequence of time dependent solutions. In addition, the utility saves flow field data for visualization purposes in Ensight format. In the second method, the engine face conditions are estimated from engine face data input in TECPLOT. Time-averaged solutions are used for this purpose. Table 3 shows that CFD data are about 6% of measured data. This is perhaps one of the best matches seen for this run. Both methods (rake and face data) agree well to each other but DC60 values are slightly overpredicted using the face data compared with rake data.

Additionally, Figure 8 compares static pressure to free-stream total pressure ratio at the port and starboard sides of the intake wall. Note that the origin was set to the most forward lip point in the experiments and the fact that CFD geometry is about 1.7 inches longer than experimental model to allow steady flow through diffuser as recommended in previous studies. Figure 8 shows that CFD data match very well with experimental data at the port side up to 14 inches. At the starboard side, measurements show flow separation at approximately $x = 10.5$ inches, however, CFD predicts flow separation at further upstream distance. The starboard pressure data do not match after separation as well. At the first bend, static pressure increases (flow decelerates) at the port side and decreases (flow accelerates) at the starboard side. The diffuser flow will then decelerate at all wall sides as the diffuser cross-section increases. The flow at the port side accelerates at the second bend as well.

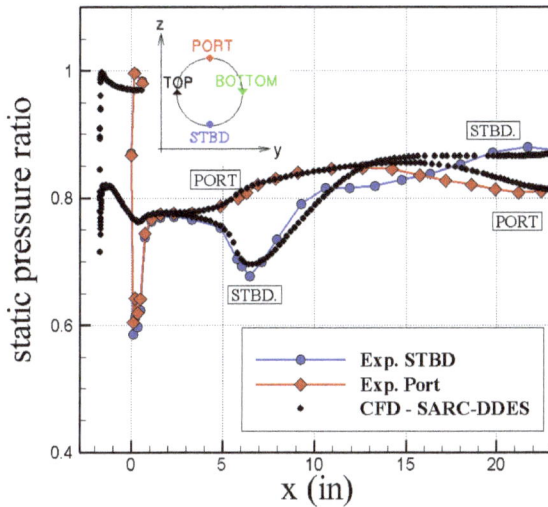

Figure 8. M2129 DP78 wall pressure measurements and simulations using Kestrel and SARC–DDES turbulence model. Static pressure ratio, $p_f / p_{0\infty}$, is the ratio of averaged engine face static pressure to free-stream total pressure. NASA baseline grid was used. CFD data are time-averaged from 2 to 2.5 s of simulation time.

The effects of turbulence models on diffuser predictions are given in Table 4. Hybrid DDES models show better agreement than RANS turbulence models. SARC–DDES, in particular, has the best agreement among tested turbulence models. In more detail, Figures 9 and 10 compare effects of turbulence models on the engine face and wall duct predictions. Figure 9 shows that RANS turbulence models (SA, SARC, and SST) predicted smaller distorted flow regions than hybrid DDES models and experiments. Specifically, the RANS models failed to predict the secondary vortices formed at the engine face. Additionally, SA and SARC models predict smaller primary vortices than other models and hence DC60 differences with experimental data are much larger for these models.

Table 4. Turbulence modeling effects on the RAE-M2129 baseline predictions. NASA grid was used. CFD data are time-averaged from 2 to 2.5 s of simulation time.

	$p_{0f}/p_{0\infty}$	Mach No.	$p_f/p_{0\infty}$	DC60
Experiments	0.9744 (-)	0.4193 (-)	0.8522 (-)	0.3130 (-)
CFD–SARC + DDES	0.9776 (−0.3280%)	0.4329 (−3.243%)	0.8737 (−2.523%)	0.2945 (5.910%)
CFD–SA	0.9809 (−0.6671%)	0.4245 (−1.2402%)	0.8780 (−3.0275%)	0.2233 (28.65%)
CFD–SARC	0.9805 (−0.6260%)	0.4291 (−2.3372%)	0.8778 (−3.004%)	0.2370 (24.28%)
CFD–Menter SST	0.9790 (−0.4721%)	0.4296 (−2.4565%)	0.8770 (−2.9101%)	0.2899 (7.38%)
CFD–Menter SST + DDES	0.9778 (−0.3489%)	0.4380 (−4.4598%)	0.88715 (−4.1011%)	0.2880 (7.98%)

Figure 10 shows that all models have similar predictions up to 12 inches distance and then they become different in the separated flow region. At the port side, all models show similar trends, but hybrid DDES models match better with experimental data than RANS models. Hybrid DDES models have better agreement with experimental data at the starboard region as well. SARC and SST models do not predict any flow separation at the starboard side up to shown distance of $x = 23$ inches. The air pressure at starboard side increases inside the diffuser using these RANS models. However, SARC–DDES and SST-DDES show flow separation at the starboard side at $x = 14.5$ and $x = 12.5$ inches, respectively. The flow is attached at further downstream distance and pressure will increase again.

From these results and predictions shown in Table 4, SARC–DDES found to bring predictions closer to experiments than other models. Notice that DDES models should perform better for unsteady turbulent flows with large flow separation regions. In addition, Figure 11 compares SARC and SARC + DDES model predictions for port and starboard wall sides. The lines shows time-averaged data and the bars denote maximum deviation from averaged values. The results show that DDES model predicts more unsteadiness and larger variations in predictions, particularly at separated flow regions.

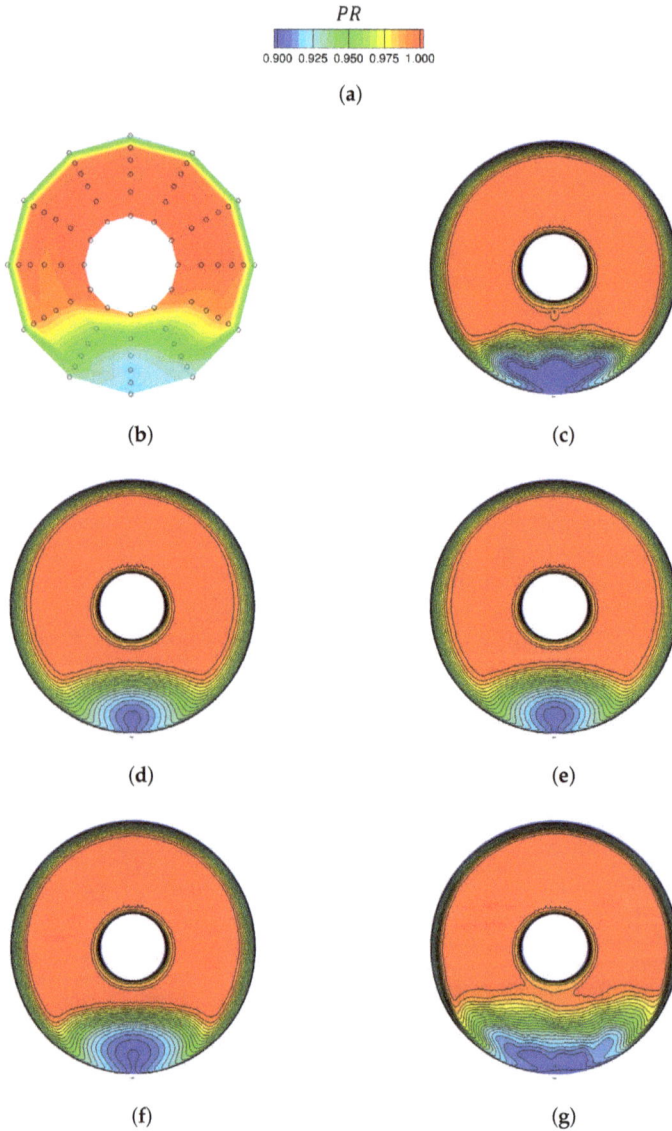

Figure 9. Effects of turbulence model on engine face pressure ratio. Time-averaged solutions are shown for the NASA baseline grid. (**a**) Pressure recovery range; (**b**) Experimental data; (**c**) CFD–SARC + DDES; (**d**) CFD–SA; (**e**) CFD–SARC; (**f**) CFD–Menter SST; (**g**) CFD–Menter SST + DDES.

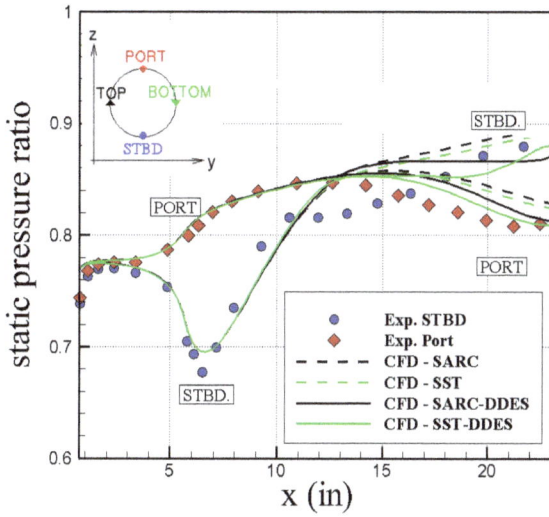

Figure 10. Turbulence model effects on CFD predictions of wall static pressure. Time-averaged solutions are shown for the NASA baseline grid. Static pressure ratio ($p_f / p_{0\infty}$) is the ratio of averaged engine face static pressure to free-stream total pressure.

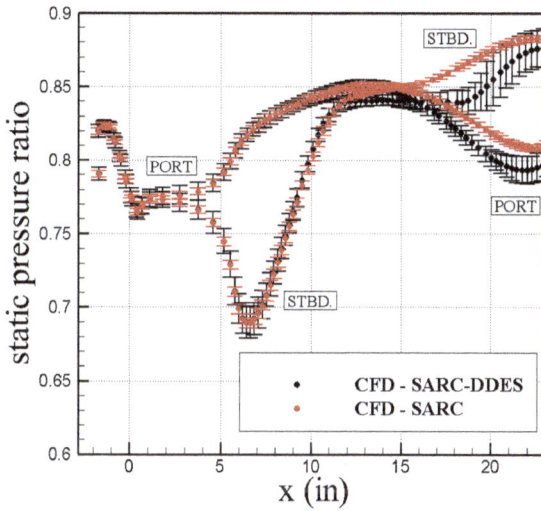

Figure 11. Unsteadiness in solutions of M2129 intake. Dot markers show time-averaged static to free-stream total pressure value of the NASA baseline grid. Bar denotes maximum difference from averaged values. Static pressure ratio ($p_f / p_{0\infty}$) is the ratio of averaged engine face static pressure to free-stream total pressure.

Figures 12 and 13 show scaled Q (Q-criterion normalized by shear strain) isosurfaces and streamlines, respectively. These isosurfaces and streamlines are colored by pressure coefficients. Secondary and separated flow can be seen in these figures. Secondary flows are much stronger at the first bend than those formed at the second bend. Figure 12 shows that SARC and SST models predict no flow separation upstream of the bullet. These models predict primary vortices at the engine face, though SST predicted stronger vortices than SARC model. The hybrid DDES models show a flow separation region and then attachment before the bullet. Primary and secondary vortices are predicted using these turbulence models, though the strength and position of vortices are different from SARC-DDES to SST-DDES model.

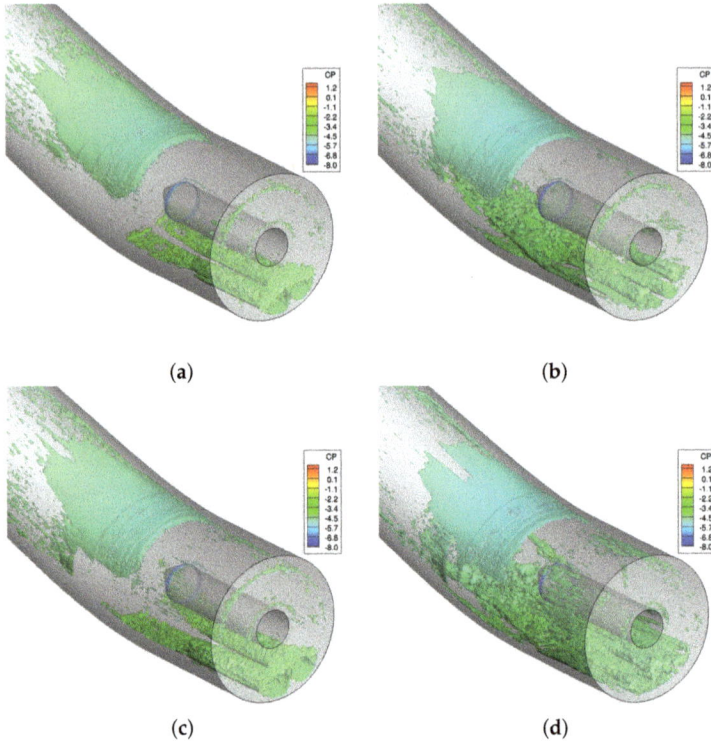

(a)

(b)

(c)

(d)

Figure 12. Isosurfaces of scale Q-criterion (iso value of 0.25) colored by pressure coefficient. NASA baseline grid was used. CFD data are time-averaged from 2 to 2.5 s of simulation time. (**a**) SARC turbulence model; (**b**) SARC + DDES turbulence model; (**c**) SST turbulence model; (**d**) SST + DDES turbulence model.

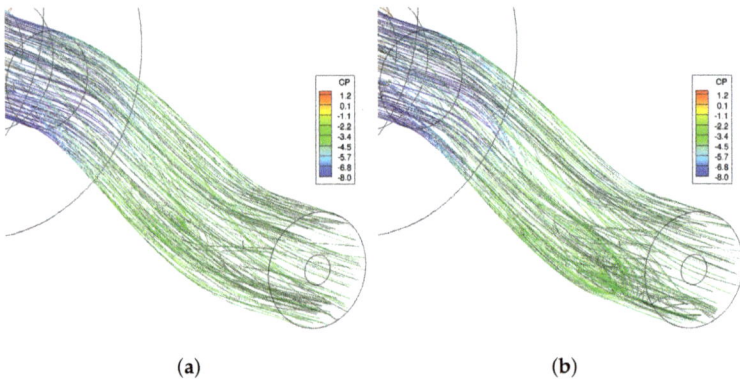

(a)

(b)

Figure 13. *Cont.*

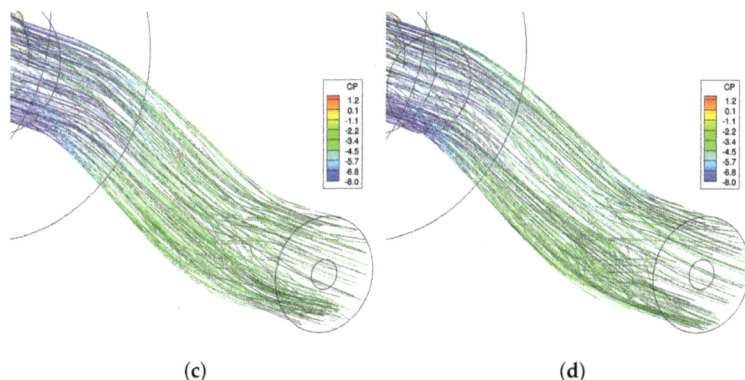

(c) (d)

Figure 13. Streamlines colored by pressure coefficient. NASA baseline grid was used. CFD data are time-averaged from 2 to 2.5 s of simulation time. (**a**) SARC turbulence model; (**b**) SARC + DDES turbulence model; (**c**) SST turbulence model; (**d**) SST + DDES turbulence model.

Final results present the flow control simulations. The USAFA baseline grids with and without flow control are tested. The flow controls include plates and active jet vortex generators. The pressure recovery data at the engine face and symmetry plane of these configurations are shown in Figure 14. Both vortex generators reduce flow separations and first bend secondary flows and hence improve the pressure uniformity. Figure 14 shows small region of low pressure recovery at port side by using vanes. This is due to formation of secondary flows at the second bend and flow separation. Overall, the active jet flow control has better flow uniformity than vane-type method tested. In more detail, Table 5 compares engine face data of these configurations. DC60 drops by 67.7% using vanes and 79.1% using jets. Pressure recovery changes are small: the jets cause the pressure recovery to increase by 1.8%. Figure 15 compares the starboard and port side static pressures of the baseline, and the baseline with vanes and jet vortex generators. Figure 15 shows that both vortex generators eliminate the flow separation region at the starboard side and at the engine face. There is a small pressure drop at about $x = 9$ inches where vanes and jets are installed. The vanes cause some flow separation at the port side as well.

Figure 16 compares the streamlines and scaled Q isosurfaces of three configurations. Figure 16 shows that both vortex generators reduce the secondary flows at the first bend and eliminate the separation flow region at the starboard side. However, secondary flows are formed at the second bend. Vane-type vortex generators show some flow separation at the port side as well.

Table 5. Flow control predictions of the RAE-M2129 intake. USAFA grids were used with SARC-DDES turbulence model. CFD data are time-averaged from 2 to 2.5 s of simulation time.

Configuration	$p_{0f}/p_{0\infty}$	Mach No.	$p_f/p_{0\infty}$	DC60
Baseline	0.97089 (-)	0.4275 (-)	0.8500 (-)	0.4215 (-)
Baseline + Vortex generators	0.97255 (0.171%)	0.4333 (1.357%)	0.8512 (0.1412%)	0.1361 (−67.7%)
Baseline + jets	0.98187 (1.131%)	0.4198 (−1.80%)	0.8675 (2.058%)	0.0881 (−79.1%)

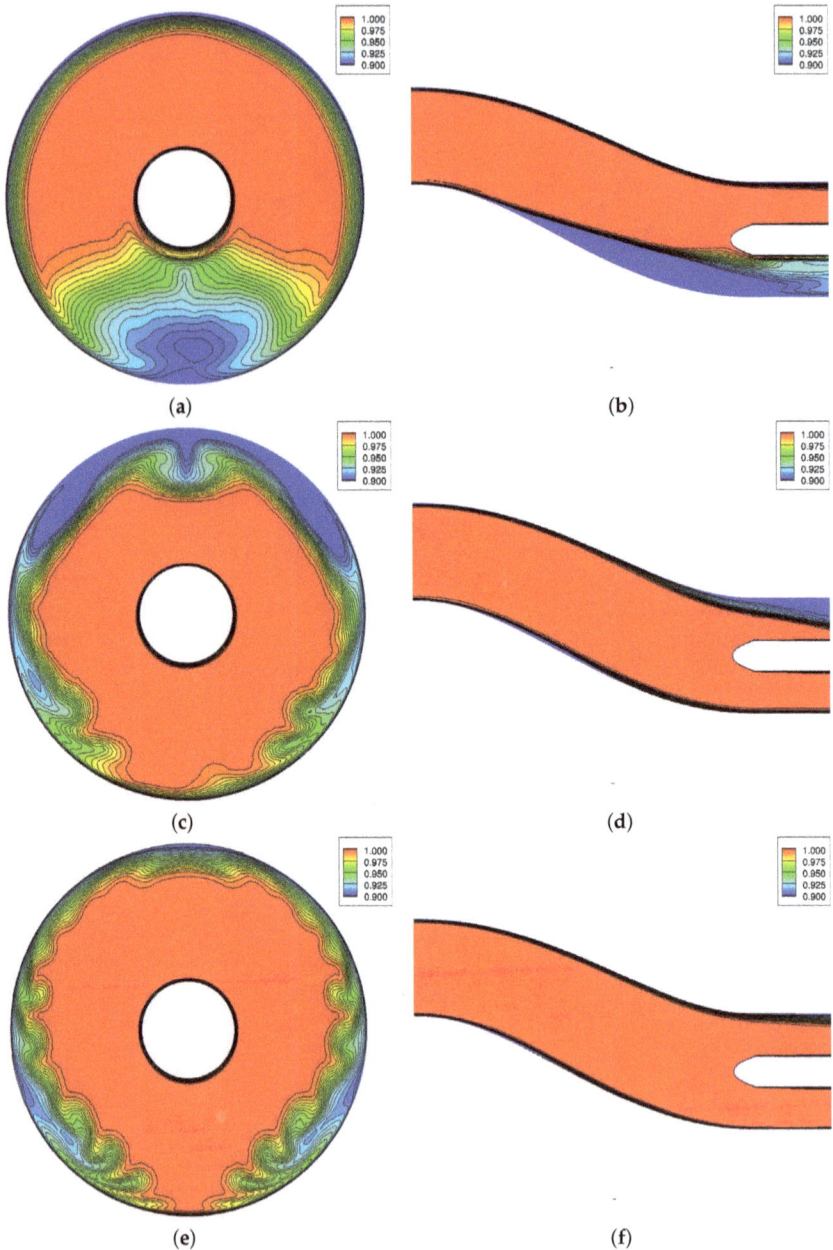

Figure 14. Engine face and symmetry pressure ratio using USAFA grids and SARC-DDES turbulence models. CFD data are time-averaged from 2 to 2.5 s of simulation time. (**a**) Baseline engine face; (**b**) Baseline symmetry; (**c**) Baseline + VG engine face; (**d**) Baseline + VG symmetry; (**e**) Baseline + Jets engine face; (**f**) Baseline + Jets symmetry.

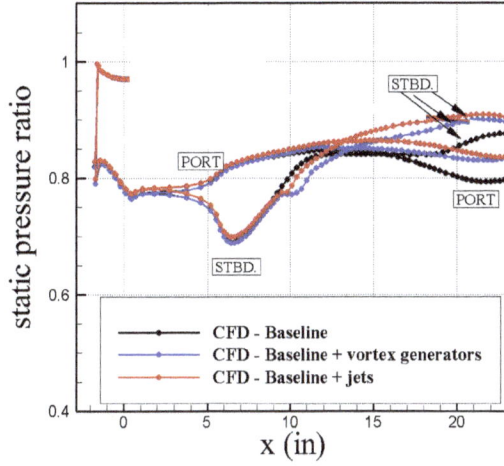

Figure 15. Wall pressure data for the baseline, baseline + vortex generators, and baseline + jets simulations. Static pressure ratio ($p_f/p_{0\infty}$) is the ratio of averaged engine face static pressure to free-stream total pressure. USAFA grids and SARC-DDES turbulence models were used. CFD data are time-averaged from 2 to 2.5 s of simulation time.

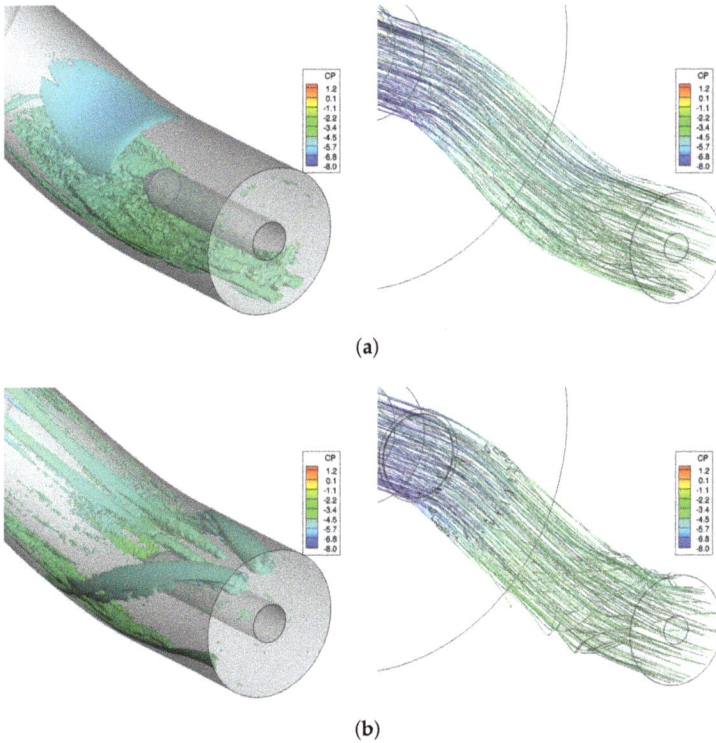

(a)

(b)

Figure 16. *Cont.*

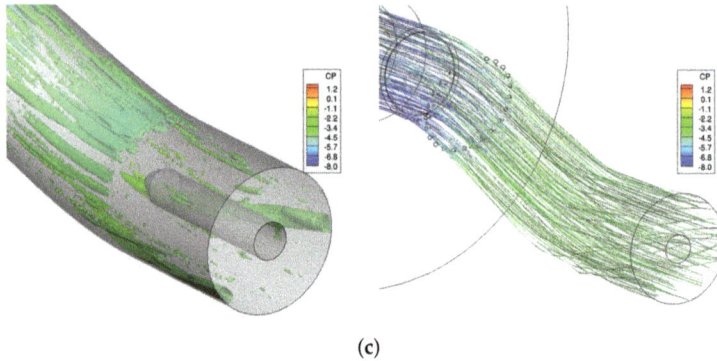

(c)

Figure 16. Isosurfaces of scale Q-criterion (iso value of 0.25) and streamlines colored by pressure coefficient. USAFA grids and SARC-DDES turbulence models were used. CFD data are time-averaged from 2 to 2.5 s of simulation time. (**a**) Baseline; (**b**) Baseline + VG; (**c**) Baseline + Jets.

The mass flow rate through three considered ducts was adjusted such that to vary throat Mach number at the throat from 0.2 to 0.8. The predicted pressure recovery and DC60 of these ducts calculated and compared with experimental data of Ref. [12] in Figure 17. Note that baseline geometries are slightly different; the jet controls have different installation angles as well. Figure 17 shows that CFD and experiments do not match due to geometry differences, however the trend of changes are similar. Experiments shows that jets and vanes have better pressure recovery than the baseline. The pressure recovery drops by increasing throat Mach number or mass flow rate as well. CFD shows similar trends, however, the vanes of this work have lower pressure recovery values than the baseline. In addition, Figure 17b shows that jet has the smallest DC60 compared with the baseline and vanes. DC60 increases for the baseline and the baseline with jets with increasing throat Mach number. Notice that jet pressure ratio was fixed at the experiments and CFD. Predictions show similar trends for the baseline and the diffuser with jets. Both experiments and CFD show that DC60 of the diffuser with vanes slightly change with throat Mach number and even drops at larger Mach numbers.

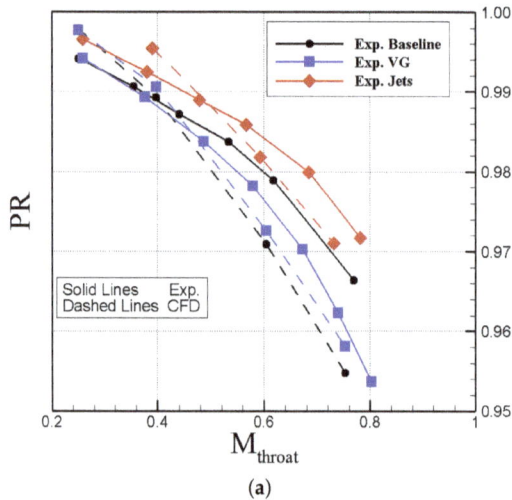

(a)

Figure 17. *Cont.*

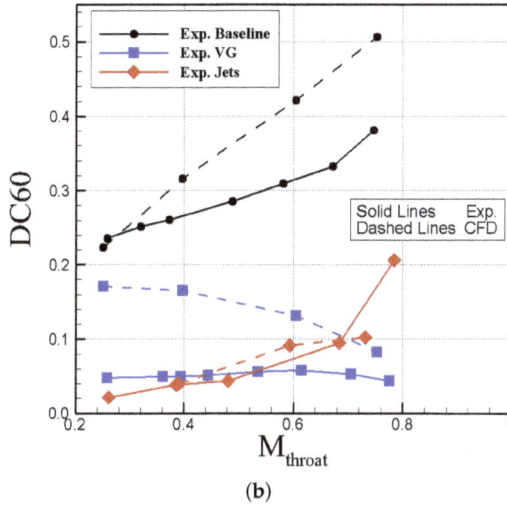

(b)

Figure 17. Validation of CFD data. USAFA grids and SARC-DDES turbulence models were used. CFD data are time-averaged from 2 to 2.5 s of simulation time. (**a**) Pressure Recovery; (**b**) DC60.

7. Conclusions

Kestrel simulation tools were used for the validation and flow control study of RAE-M2129 intake diffuser. Two set of geometries were used: one from NASA which represent actual tested intake model and the second one a USAFA model created from intake given parameters and geometry relationships. CFD predictions of the NASA model matched well with experiments, perhaps one of the best agreements seen for this configuration and used experimental setup. All results show that there is a flow separation region at the starboard side of the intake wall with large distortion coefficients measured at the engine face. It was shown that SARC-DDES turbulence model has a better agreement compared with other tested models.

Two flow control mechanisms were investigated: (a) one using vortex generators (b) and second one using synthetic jet actuators. The results showed that both control methods reduce the distortion coefficient at the engine face. Jet vortex generators cause 79.1% drop in the flow distortion and improve the pressure recovery compared with the baseline diffuser. Vane-type vortex generators also improve the flow distortion but slightly decrease the pressure recovery values. In both control methods, secondary flows at the first bend are reduced, however both show the secondary flow at the second bend. The vanes cause the flow separate at the port side of the diffuser before the engine face. The results show that increasing the diffuser mass flow rate will cause the pressure recovery drop and the distortion coefficients to increase. The diffuser with flow control has smaller pressure recovery at diffuser conditions with larger mass flow rate. The vanes DC60, however, seems to be insensitive to mass flow rate changes and even decreases at larger values. The jets were tested at fixed bled air pressure and therefore DC60 will increase with increasing mass flow rate. However, adjusting the incoming pressure with diffuser mass flow rate can improve pressure recovery and DC60 at all tested conditions. Future work extends these results to include more flow control configuration and installation locations. The results of this work will be applied to the NATO AVT-251 muli-disciplinary configuration named Muldicon. This UCAV has a fuselage buried small turbofan engine and the intake shape design will significantly impact the engine performance.

Acknowledgments: This article was approved for Public Release with Distribution A. This material is based in part on research sponsored by the US Air Force Academy under agreement numbers of FA7000-17-2-0007 and FA7000-16-2-0010. The U.S. Government is authorized to reproduce and distribute reprints for Governmental purposes notwithstanding any copyright notation thereon. The views and conclusions contained herein are those of the authors and should not be interpreted as necessarily representing the official policies or endorsements, either expressed or implied, of the organizations involved with this research or the U.S. Government. The authors would like to thank Bernhard Anderson of NASA Glenn Research Center for passing the M2129 baseline grid to us. Special thanks to aeronautical cadets Jesse Montgomery and David Sargent who took Advanced CFD course (AE472) in the fall semester of 2017. The numerical simulations have been performed on HPCMP's Topaz cluster located at the ERDC DoD Supercomputing Resource Center (DSRC).

Author Contributions: Pooneh Aref and Mehdi Ghoreyshi run simulations, generated all plots, and prepared this article. Adam Jirasek generated the computational grids and provided a script to find engine face conditions from simulations. Matthew Satchell is the program manager of this research at USAFA.

Conflicts of Interest: The authors declare no conflict of interest.

Nomenclature

a	acoustic speed, ms^{-1}
c	vane chord, m
CFD	Computational Fluid Dynamics
Cp	pressure coefficient, $(p - p_\infty)/q_\infty$
CREATE	Computational Research and Engineering Acquisition Tools and Environments
DC	Distortion Coefficient
DDES	Delayed Detached Eddy Simulation
F	thrust force, N
h	vane height, m
L	diffuser length, m
M	Mach number, V/a
PR	pressure recovery, p_t/p_{t0}
p	static pressure, N/m^2
p_f	averaged engine face static pressure, N/m^2
p_∞	free-stream static pressure, N/m^2
p_{0f}	averaged engine face total pressure, N/m^2
$p_{0\infty}$	free-stream total pressure, N/m^2
q	dynamic pressure, N/m^2
q_∞	free-stream dynamic pressure, N/m^2
R	Radius, m
R_t	Radius at throat, m
R_f	Radius at engine face, m
RAE	Royal Aircraft Establishment
SARC	Spalart–Allmaras with rotational and curvature correction
RANS	Reynolds Averaged Navier Stokes
t	time, s
V	free-stream velocity, ms^{-1}
VG	Vortex Generator
x,y,z	grid coordinates, m
X_{VG}	position of vortex generator vanes, m
$y+$	non-dimensional wall normal distance

Subscripts

f	engine face
t	throat
∞	free-stream

Aerospace **2018**, *5*, 31

References

1. Ward, T. *Aerospace Propulsion Systems*; John Wiley & Sons: Hoboken, NJ, USA, 2010; pp. 330–331.
2. Gil-Prieto, D.; MacManus, D.G.; Zachos, P.K.; Bautista, A. Assessment Methods for Unsteady Flow Distortion in Aero-Engine Intakes. *Aerosp. Sci. Technol.* **2018**, *72*, 292–304.
3. May, N.E. *The Prediction of Intake/S-Bend Diffuser Flow Using Various Two-Equation Turbulence Model Variants, Including Non-Linear Eddy Viscosity Formulations*; Technical Report; Aircraft Research Association Contractor Report; Aircraft Research Association: Beford, UK, 1997.
4. Menzies, R.D.; Badcock, K.J.; Barakos, G.N.; Richards, B.E. *Validation of the Simulation of Flow in an S-Duct*; AIAA Paper 2002–2808; AIAA: Reston, VA, USA, 2002.
5. Menzies, R. Computational Investigation of Flows in Diffusing S-shaped Intakes. *Acta Polytech.* **2001**, *41*, 61–67.
6. Reichert, B.; Wendt, B. Improving Curved Subsonic Diffuser Performance with Vortex Generators. *AIAA J.* **1996**, *34*, 65–72.
7. Jirasek, A. Design of Vortex Generator Flow Control in Inlets. *J. Aircr.* **2006**, *43*, 1886–1892.
8. Hamstra, J.; Miller, D.; Truax, P.; Anderson, B.; Wendt, B. Active Inlet Flow Control Technology Demonstration. *Aeronaut. J.* **2000**, *104*, 473–479.
9. Amitay, M.; Pitt, D.; Glezer, A. Separation Control in Duct Flows. *J. Aircr.* **2002**, *39*, 616–620.
10. Brown, A.; Nawrocki, H.; Paley, P. Subsonic Diffusers Designed Integrally with Vortex Generators. *J. Aircr.* **1968**, *5*, 221–229.
11. Anderson, B.H.; Gibb, J. Vortex-Generator Installation Studies on Steady-State and Dynamic Distortion. *J. Aircr.* **1998**, *35*, 513–520.
12. Gibb, J.; Anderson, B. Vortex Flow Control Applied to Aircraft Intake Ducts. In *High Lift and Separation Control, Proceedings of the Royal Aeronautical Society Conference, Bath, UK, 29–31 March 1995*; Royal Aeronautical Society: London, UK, 1995; p. 14.
13. Roth, G.L.; Morton, S.A.; Brooks, G.P. Integrating CREATE-AV Products DaVinci and Kestrel: Experiences and Lessons Learned. In Proceedings of the 50th AIAA Aerospace Sciences Meeting Including the New Horizons Forum and Aerospace Exposition, Nashville, TN, USA, 9–12 January 2012.
14. Morton, S.A.; McDaniel, D.R. A Fixed-Wing Aircraft Simulation Tool for Improving DoD Acquisition Efficiency. *Comput. Sci. Eng.* **2016**, *18*, 25–31.
15. Morton, S.A.; McDaniel, D.R.; Sears, D.R.; Tillman, B.; Tuckey, T.R. Kestrel: A Fixed Wing Virtual Aircraft Product of the CREATE Program. In Proceedings of the 47th AIAA Aerospace Sciences Meeting Including the New Horizons Forum and Aerospace Exposition, Orlando, FL, USA, 5–8 January 2009.
16. McDaniel, D.; Nichols, R.; Eymann, T.; Starr, R.; Morton, S. Accuracy and Performance Improvements to Kestrel's Near-Body Flow Solver. In Proceedings of the 54th AIAA Aerospace Sciences Meeting, San Diego, CA, USA, 4–8 January 2016.
17. Godunov, S.K. A Difference Scheme for Numerical Computation of Discontinuous Solution of Hydrodynamic Equations. *Sb. Math.* **1959**, *47*, 271–306.
18. Tramel, R.; Nichols, R.; Buning, P. Addition of Improved Shock-Capturing Schemes to OVERFLOW 2.1. In Proceedings of the 19th AIAA Computational Fluid Dynamics, San Antonio, TX, USA, 22–25 June 2009.
19. Spalart, P.R.; Allmaras, S.R. A One Equation Turbulence Model for Aerodynamic Flows. In Proceedings of the 30th Aerospace Sciences Meeting and Exhibit, Reno, NV, USA, 6–9 January 1992.
20. Spalart, P.R.; Schur, M. On the Sensitisation of Turbulence Models to Rotation and Curvature. *Aerosp. Sci. Technol.* **1997**, *1*, 297–302.
21. Menter, F. Eddy Viscosity Transport Equations and Their Relation to the k-ε Model. *ASME J. Fluids Eng.* **1997**, *119*, 876–884, doi:10.1115/1.2819511.
22. Spalart, P.R.; Jou, W.H.; Strelets, M.; Allmaras, S.R. Comments on the Feasibility of LES for Wings, and on a Hybrid RANS/LES Approach. In Proceedings of the 1st AFSOR International Conference on DNS/LES, Ruston, LA, USA, 4–8 August 1997; Greyden Press: Columbus, OH, USA, 1997; pp. 137–147.
23. Berens, T.M.; Delot, A.L.; Chevalier, M.; Van Muijden, J. Numerical Simulations for High Offset Intake Diffuser Flows. In Proceedings of the 52nd Aerospace Sciences Meeting, National Harbor, MD, USA, 13 January 2014.

24. Samuelsson, I. Transonic Wind Tunnel Test at High Reynolds Numbers with an Air Inlet with a S-shaped Diffuser Duct with Flow Control Devices (Vortex Generators and Micro-Jets). In Proceedings of the Conference Flygteknik 2010, Stockholm, Sweden, 18–19 October 2010.

25. Wallin, M.; Håll, U.; Grönstedt, T. Effects of Engine Intake and Nozzle Design the Performance of a Low-Signature Reconnaissance UAV. In Proceedings of the 40th AIAA/ASME/SAE/ASEE Joint Propulsion Conference and Exhibit, Fort Lauderdale, FL, USA, 11–14 July 2004.

26. Antonatos, P.; Surber, L.; Laughrey, J.; Stava, D. *Assessment of the Influence of Inlet and Aftbody/Nozzle Performance on Total Aircraft Drag*; AGARD Aerodyn. Drag 28 p (SEE N 74-14709 06-01); AGARD: Izmir, Turkey, 1973.

27. *Computational Research and Engineering Acquisition Tools And Environments (CREATE)*; Version 6.0; Kestrel User Guide; U.S. DoD HPCMP: Eglin AFB, FL, USA, 2015.

aerospace

MDPI

Article

Assessing the Ability of the DDES Turbulence Modeling Approach to Simulate the Wake of a Bluff Body

Matthieu Boudreau [1,*], **Guy Dumas** [1] and **Jean-Christophe Veilleux** [2]

[1] CFD Laboratory LMFN, Département de Génie Mécanique, Université Laval, 1065 Avenue de la Médecine, Québec, QC G1V 0A6, Canada; gdumas@gmc.ulaval.ca

[2] Graduate Aerospace Laboratories, California Institute of Technology, Pasadena, CA 91125, USA; jveilleu@caltech.edu

* Correspondence: matthieu.boudreau.1@ulaval.ca

Received: 26 June 2017; Accepted: 28 July 2017; Published: 1 August 2017

Abstract: A detailed numerical investigation of the flow behind a square cylinder at a Reynolds number of 21,400 is conducted to assess the ability of the delayed detached-eddy simulation (DDES) modeling approach to accurately predict the velocity recovery in the wake of a bluff body. Three-dimensional unsteady Reynolds-averaged Navier–Stokes (URANS) and DDES simulations making use of the Spalart–Allmaras turbulence model are carried out using the open-source computational fluid dynamics (CFD) toolbox OpenFOAM-2.1.x, and are compared with available experimental velocity measurements. It is found that the DDES simulation tends to overestimate the averaged streamwise velocity component, especially in the near wake, but a better agreement with the experimental data is observed further downstream of the body. The velocity fluctuations also match reasonably well with the experimental data. Moreover, it is found that the spanwise domain length has a significant impact on the flow, especially regarding the fluctuations of the drag coefficient. Nonetheless, for both the averaged and fluctuating velocity components, the DDES approach is shown to be superior to the URANS approach. Therefore, for engineering purposes, it is found that the DDES approach is a suitable choice to simulate and characterize the velocity recovery in a wake.

Keywords: wake; bluff body; square cylinder; DDES; URANS; turbulence model

---

1. Introduction

Numerical simulations of turbulent flows involving multiple interacting bodies are of great interest in a large variety of disciplines. Studies on wind farms, the flow around buildings in a city, heat exchangers, or vehicles in close proximity, come to mind. For such studies, an accurate modeling of the wakes is crucial but challenging. It should be noted that one is often not only interested in the mean quantities of the flow in order to characterize a wake, but also in obtaining accurate information regarding the unsteadiness and the turbulence of the flow field. As a result, steady Reynolds-averaged Navier–Stokes (RANS) models are inadequate for the task as they solely provide the averaged quantities. Moreover, even the averaged quantities in a wake predicted with steady RANS modeling can prove to be unreliable [1–3]. Further, the unsteady RANS alternative (URANS) often predicts a flow field which is almost periodic in time without significant amplitude modulations in the temporal signals of the physical quantities [2,4]. This often turns out not to be representative of the reality, especially when separation occurs [5]. The large eddy simulation (LES) approach would resolve these issues, but its computation cost makes it impractical for simulating complete turbines operating at high Reynolds number under various operating conditions. One possible solution is the use of a hybrid approach such as the delayed detached-eddy simulations (DDES) technique [3,6], which uses a RANS approach in the attached boundary layers and a LES approach in the separated regions of the flow.

Initially, the detached-eddy simulation methodology has been developed to obtain accurate force predictions on bodies with massively separated flow. A list of successful examples of the application of the detached-eddy simulation (DES) approach for several different geometries is given in the review paper of Spalart [3]. More recently, some researchers have started to show some interest in this turbulence modeling approach for the simulation of turbulent wakes. Among such studies, Paik et al. [7] compared the performances of the URANS approach to different DES methodologies for the flow around two wall-mounted cubes in tandem, Nasif et al. [8] investigated the wake characteristics of sharp-edged bluff body in a shallow flow, Muld et al. [9] observed the flow structures in the wake of a high-speed train and Muscari et al. [1] used this approach to study the wake of a marine propeller and observed a good agreement with the experimental results of Felli et al. [10]. Lastly, the authors of the current work have also used this turbulence modeling approach to study the vortex dynamics and the wake recovery of two different types of hydrokinetic turbines, namely the horizontal axis and the vertical axis turbines [11].

While the capacity of the DDES approach for providing accurate force predictions for bodies with massively separated flows has been largely investigated in the literature, its performances in modeling turbulent wakes have attracted much less attention. In this context, a benchmark case is revisited with the current state-of-the-art numerical methodology making the use of the innovative DDES approach. As the ability of the RANS technique to model attached boundary layers has already been addressed and is well documented [12], this study mainly focus on the performances of the DDES approach in the separated regions of the flow. The sharp-edged square-cylinder case studied experimentally by Lyn et al. [13] at a Reynolds number of 21,400 has been chosen here to achieve this task because its wake dynamics are not dependent on the RANS modeling inherent to a DDES simulation. This is due to the fact that the boundary layers on the upstream face of the square cylinder are laminar and because the separation occurs at fixed locations, namely the upstream sharp edges.

Although the case of Lyn et al. [13] has already been investigated during two LES workshops held in 1994 [14] and in 1995 [15], the results at the time did not show a good match with the experimental data and the numerical results also did not agree well with each other. The available computational resources led to an insufficient sampling period, a too-coarse resolution and a too-short spanwise length of the computational domain in most of the simulations. This could partly explain the unsatisfactory results that were obtained. Better results have been obtained more recently using the LES [16–19] and the DES approaches [20,21]. The current work revisits this benchmark case with fine spatial and temporal resolutions and an innovative turbulence modeling technique, namely the DDES. While previous studies used computational domains with a spanwise length of about four cylinder widths, the simulations of the current work have been conducted with different domain sizes in this direction up to a spanwise length of seven cylinder widths, which allows one to better evaluate the effects of this parameter on the flow.

2. Methodology

2.1. Turbulence Modeling

The unsteady Reynolds-averaged Navier–Stokes (URANS) equations are given below [22]:

$$\frac{\partial \langle u_i \rangle}{\partial x_i} = 0 \, , \tag{1}$$

$$\frac{\partial \langle u_i \rangle}{\partial t} + \langle u_j \rangle \frac{\partial \langle u_i \rangle}{\partial x_j} = -\frac{1}{\rho} \frac{\partial \langle p \rangle}{\partial x_i} + \frac{\partial}{\partial x_j} \left(\nu \frac{\partial \langle u_i \rangle}{\partial x_j} - \langle u_i' u_j' \rangle \right) , \tag{2}$$

where u_i is the ith component of the velocity vector, p is the pressure, t is time, ν is the kinematic viscosity, $\langle\rangle$ denotes an ensemble average and $\langle u'_i u'_j \rangle$ are the Reynolds stresses.

A common way to deal with the six unknowns introduced with the Reynolds stress tensor $\langle u'_i u'_j \rangle$ is to make use of the eddy viscosity concept. The original Spalart–Allmaras turbulence model [23] in fully-turbulent mode has been chosen here to achieve this task. This model only involves one additional transport equation, which is for the modified viscosity ($\tilde{\nu}$). This modified viscosity ($\tilde{\nu}$) is related to the eddy viscosity (ν_t) through an empirical relation that accounts for the near-wall viscous effects.

In general, RANS models perform well when the boundary layers are attached. Conversely, they generally have some difficulties when separated flows are encountered [12]. A possible alternative to solve this issue is the use of the LES approach, which consists of resolving the largest scales of the turbulence spectrum and of modeling only the scales smaller than a threshold related to the local grid size [22]. While grid refinement does not extend the resolved part of the energy cascade in the case of URANS simulations [24], it results, in the case of LES simulations, in a wider range of turbulent scales being resolved, thus weakening the role of modeling [2]. Moreover, the smallest scales tend to become more and more isotropic as we go down the energy cascade [22], which makes them easier to model. A relatively simple subgrid-scale model is thus adequate to account for their effect on the largest resolved scales [25]. However, the high computation cost of a LES simulation for complex flows at a high Reynolds number often makes this approach impractical, as previously mentioned. This issue is partly due to the presence of very small turbulent-length scales near solid surfaces resulting in the need for very fine spatial resolution. The use of a hybrid approach, such as detached-eddy simulation (DES), appears to be an interesting alternative with an acceptable computational cost when compared with complete LES simulations. The key idea behind this hybrid methodology is to use a more cost-efficient RANS approach near the walls because of the less restrictive grid spacing requirements, and to use a more complete LES approach away from the walls.

In order to obtain a DES formulation, a RANS model is modified in a way that allows the model to function either in RANS mode in attached boundary layers, or in LES mode in separated regions of the flow. The original DES formulation [26,27] is based on the Spalart–Allmaras turbulence model. In order to switch from a RANS to a LES formulation, the destruction term ($\sim(\tilde{\nu}/d^2)$) in the modified viscosity transport equation is modified: the distance between a point in the domain and the nearest solid surface (d) is replaced with the parameter $\tilde{d}$ defined as:

$$\tilde{d} = \min(d, \, C_{DES} \cdot \Delta) \, , \tag{3}$$

where C_{DES} is a constant equal to 0.65 and Δ is a length scale related to the local grid spacing:

$$\Delta = \max(\Delta x, \, \Delta y, \, \Delta z) \, . \tag{4}$$

To summarize, a DES simulation remains in RANS mode as long as the distance between a point in the domain and the nearest solid surface (d) is smaller than the DES length scale (Δ) times the C_{DES} constant.

A modified version of DES, called delayed detached-eddy simulation (DDES), has been suggested to overcome the possible issue of "grid-induced separation" (GIS) which is dependent on the grid geometry [3,6]. The purpose of this new version is to ensure that the turbulence modeling remains in RANS mode throughout the boundary layers. To do so, the definition of the parameter $\tilde{d}$ is modified as follows:

$$\tilde{d} = d - f_d \max(0, \, d - C_{DES} \cdot \Delta) \, , \tag{5}$$

where f_d is a filter function designed to take a value of 0 in attached boundary layers (RANS region) and a value of 1 in zones where the flow is separated (LES region). The location where the modeling switches between the RANS and the LES modes therefore depends on the flow characteristics, which is not the case in the original DES formulation. As recommended by Spalart [3] in his 2009 Annual

Review paper, the DDES formulation should be the new standard version of DES and it has therefore been chosen to conduct this study.

Several versions of the DDES approach exist with a variety of underlying RANS models. The one chosen in the current study makes use of the Spalart-Allmaras turbulence model. This allows a straightforward comparison with the URANS simulations. The reader is referred to the following papers for a more complete description of the DES and DDES modeling approaches [2,3,6,27].

2.2. Case Description and Numerics

As mentioned in the introduction, the experimental results from Lyn et al. [13] for the flow past a square cylinder of width D were obtained at a Reynolds number of 21,400. The experiment was conducted in a closed water channel measuring 9.75D in the lateral direction and 14D in the transversal direction with the square cylinder going through both lateral walls of the channel. Laser Doppler velocimetry (LDV) measurements of the streamwise and transversal velocity components were made in a plane located at midspan.

To reproduce the results of this experiment with CFD, only a fraction of the experimental lateral extent (9.75D) is considered with the use of periodic boundary conditions in order to reduce the computational cost. DDES simulations have been performed with three different computational domains with a spanwise length of 3D, 5D and 7D. Since it has been observed that the level of velocity fluctuations in the near wake is greatly sensitive to the aspect ratio of the square cylinder up to a value of 7, the largest computational domain has been used as the nominal one and all the results presented in this paper have been obtained with this domain unless otherwise indicated. Note that the square cylinder is located in the center of the domain, as shown in Figure 1, and that the origin of the coordinate system is located at the center of the square cylinder. The distances that separate the square cylinder's center from the inlet and from the outlet have been chosen based on two-dimensional URANS simulations making use of different domain sizes.

It is worth mentioning that a smaller domain size in the spanwise direction had been used for most of the simulations reported in the literature [14–16,18–21]. Two-point auto-correlations of the lateral velocity fluctuations along the lateral direction have been computed with the three different spanwise domain lengths (3D, 5D and 7D). These auto-correlations showed that this specific velocity component is decorrelated over half the span only for the largest domain. The same procedure has been conducted by Garbaruk et al. [28] to validate their choice of four chord lengths in the spanwise direction for their DDES simulation of the flow around an airfoil at an angle of attack of 60°. In order to further reduce the computational cost of the current simulations, symmetry boundary conditions (free-slip walls) are chosen to model the channel walls in the transversal direction. A uniform velocity and a turbulent viscosity ratio (v_t/v) of 0.01 are set at the inlet along with a uniform static pressure at the outlet.

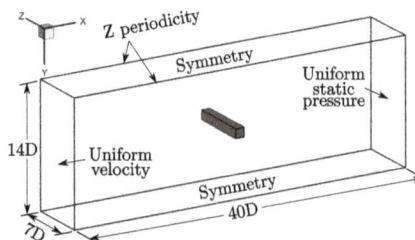

Figure 1. Domain and boundary conditions.

The finite-volume open-source CFD toolbox *OpenFOAM-2.1.x* [29] is used to carry out the simulations. Both the URANS and DDES simulations are three-dimensional and make use of the same spatial and temporal resolutions for the sake of comparison. The choice of a three-dimensional domain for the URANS simulation stems from the fact that three-dimensional URANS simulations generally prove to be more reliable than their two-dimensional counterparts, even for the flow around two-dimensional geometries [4,24]. The convective fluxes are discretized with a second-order-accurate upwind scheme and all other discretization schemes are also second-order-accurate. The PISO algorithm handles the pressure-velocity coupling of the segregated solver that has been used [30]. Further, the residuals convergence criteria have been set to 1×10^{-5} for pressure, momentum and turbulent quantities. Simulations using a less restrictive criteria by an order of magnitude have showed only negligible differences.

The calculation grid corresponds to a two-dimensional unstructured mesh of 28,384 quad elements, as shown in Figure 2, which has been extruded in the spanwise direction resulting in a total of 3,973,760 cells. The grid spacing in the spanwise direction corresponds to the same spacing as the one used in the wake, where an almost constant spacing (Δ_0) of $0.05D$ in all three directions is used. The near-wall resolution gives rise to a dimensionless normal wall distance around one, while the streamwise and spanwise dimensionless wall distances are of the order of 100.

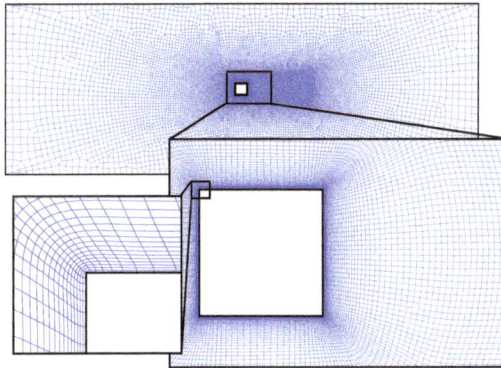

Figure 2. Two-dimensional mesh (28,384 quad elements, $\Delta_0 = 0.05D$) which has been extruded in the spanwise direction to obtain the three-dimensional grid (3,973,760 cells).

A time step of $0.01D/U_\infty$ has been chosen in order to obtain a Courant number of about 0.2 based on the upstream velocity in the wake's refined region of the grid. This value is smaller than the one recommended by Spalart, who suggests that the Courant number should be around unity [31]. Mockett et al. [32] also demonstrated that a local Courant number below or equal to one is necessary in the region of the flow that is resolved in LES mode in order to obtain accurate results. This conclusion has been drawn from their study of the flow around a circular cylinder using a DDES approach [32]. A smaller Courant number of 0.2 based on the upstream velocity has been used in this study to account for the fact that the Courant number can locally reach higher values and to ensure the stability of the PISO algorithm. It is worth noting that the current time step is three times smaller than the finest one used by Mockett [32], which provides confidence that the numerical error associated with the temporal discretization should not be an issue. The time step used in the current study corresponds to roughly 750 time steps per shedding cycle. Simulations on finer grids ($\Delta_0 = 0.0333D$ and $\Delta_0 = 0.025D$) with smaller time steps ($\Delta t = 0.00667D/U_\infty$ and $\Delta t = 0.005D/U_\infty$) have been carried out by the authors in order to make sure that the conclusions of this study were independent of the resolution level that has been used to perform the simulations.

It is worth recalling that the boundary layers on the upstream face of the square cylinder are laminar and that the separation occurs at the two upstream sharp edges. The results of the DDES simulations should therefore not be affected by the RANS modeling in the attached boundary layers since no modeling is in fact needed in these laminar boundary layers. This allows focusing on the performances of the LES region of the DDES approach, as desired.

Because of the relatively low Reynolds number of this flow, some simulations have been carried out with a low-Reynolds-number correction as suggested by Spalart [6]. The correction has been implemented in the open-source computational fluid dynamics (CFD) toolbox OpenFOAM-2.1.x by the authors. With this correction, the ratio $(\tilde{v}/v)$ is replaced with $\max(\tilde{v}/v, 20f_d)$ in the relation responsible for the near-wall viscous effects. The differences observed were negligible and the results presented in this paper have therefore been obtained without using this correction. In order to postprocess the results, every time signal has been decomposed into the sum of a time-averaged and a time-varying component. Applied to the streamwise velocity component, this results in the following decomposition:

$$u(t) \;=\; \bar{u} \,+\, u'(t)\,, \tag{6}$$

where $u(t)$ is the instantaneous signal, $\bar{u}$ is the time-averaged component and $u'(t)$ is the time-varying component.

Lastly, all simulations have been initialized with a flow field obtained from a two-dimensional URANS simulation, and a minimum of 40 convective time units $(40D/U_\infty)$ was calculated in each simulation before recording any temporal signal for further statistical analysis. This time period corresponds to that required for the convection of one domain length based on the upstream velocity. Waiting 200 convective time units $(200D/U_\infty)$ before recording the temporal signals has also been tested, and it resulted in negligible differences. Sufficiently long time samples have been collected to ensure the statistical convergence of all the physical quantities of interest. The values of the time step (Δt) and the grid spacing in the wake region (Δ_0) along with the duration of the recorded time samples that have been chosen for the current study (T_{avg}) are reported in Table 1.

Table 1. Values of the grid spacing in the wake region (Δ_0), the time step (Δt), and the duration of the time samples recorded for statistical analysis (T_{avg}).

Case	Δ_0 [D]	Δt [D/U_∞]	T_{avg} [D/U_∞]
URANS	0.05	0.01	209.5
DDES	0.05	0.01	1507.5

3. Results and Discussion

3.1. Vortex Shedding

A qualitative comparison between the URANS and the DDES simulations is first conducted with respect to their vortex dynamics. The vortices are identified using the λ_2 criterion proposed by Jeong and Hussain [33] for incompressible flows. One can observe in Figure 3 that the DDES simulation allows for the resolution of small-scale three-dimensional vortical structures while the URANS simulation provides in an essentially two-dimensional flow field without any visible three-dimensional instabilities in the shear layers, which is not representative of the reality of bluff-body wakes such as the circular cylinder case in the same range of Reynolds number [34]. This unrealistic behavior can be observed even if the URANS simulation is initialized with the flow field resulting from a DDES simulation. From this qualitative comparison, one can already expect that the level of velocity fluctuations should be smaller in the case of the URANS simulation compared to the DDES simulation, as will be demonstrated in the following sections.

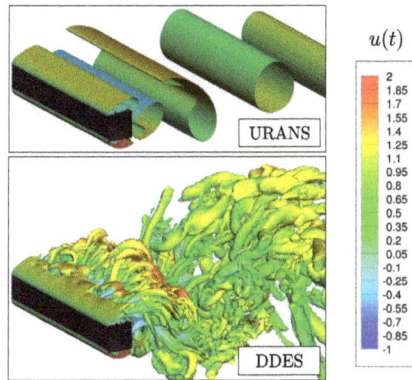

Figure 3. Isosurfaces of the λ_2 criterion for the unsteady Reynolds-averaged Navier–Stokes (URANS) (**top**) and the delayed detached-eddy simulation (DDES) simulations (**bottom**) colored by the instantaneous streamwise velocity component.

3.2. Time-Averaged Velocity Component

For practical purposes, the averaged streamwise velocity component ($\bar{u}$), shown in Figure 4, is probably the most useful physical quantity to analyze in order to characterize the velocity recovery in a wake. The results shown in Figure 4 illustrate the good performance of the DDES approach in comparison to the URANS approach, which largely overestimates the velocity recovery. Indeed, the URANS simulation predicts an almost complete recovery of the upstream velocity only $6D$ downstream of the square cylinder's center. One possible explanation of such a behavior is the very high eddy viscosity values that are predicted by the URANS simulation. Actually, the URANS simulation predicts a maximum value of the turbulent viscosity ratio (ν_t/ν) around 800 in the wake compared with a value of 20 for the DDES simulation. Breuer [25] has made similar observations with results obtained from simulations of the flow around an inclined flat plate in the same range of Reynolds numbers. High effective viscosity ($\nu + \nu_t$) values give rise to an increased transport of momentum which could explain the overestimation of the rate at which velocity is recovered in the URANS wake. Regarding the DDES simulation, the discrepancies observed in the near wake could be partly attributed to the overestimation of the transversal velocity fluctuations in this region as will further be discussed in Section 3.3. It is also interesting to note that the current results are in better agreement with the experimental data [13] than previous DES [21] and LES [14–16,19] simulations performed on this case. Indeed, most of these studies predicted a higher velocity recovery rate than the current DDES simulation. In the authors' opinion, the main cause explaining this might be the use of an insufficient spatial resolution. In fact, a DDES simulation using a coarser resolution (not shown in this paper) has been carried out by the authors, and the results are found to be very similar to those of the aforementioned studies, i.e., showing a more pronounced overestimation of the velocity recovery rate.

Numerical predictions of the averaged streamwise and transversal velocity profiles at several locations downstream of the square cylinder's center are presented in Figures 5 and 6, respectively. The URANS and the DDES simulations provide very similar predictions of the averaged velocity profiles at $x/D = 1$ for both velocity components. At this location, it is observed in Figure 5 that both simulations overestimate the averaged streamwise velocity component in the wake's center, as previously noted, and up to a distance of approximatively $0.7D$ in the transversal direction. The slightly underestimated averaged streamwise velocity component in the region between $0.7D$ and $1.3D$ is consistent with the overestimated velocity in the wake's central region. Indeed, a smaller velocity deficit in the wake's center is necessarily associated with an overestimation of the momentum transport across the wake, which is also responsible for a higher wake spreading rate. This is also consistent with the higher transversal velocities directed from outside the wake toward the wake's central region (higher negative values) compared with the experimental data at this location. This can be observed on the left plot in Figure 6.

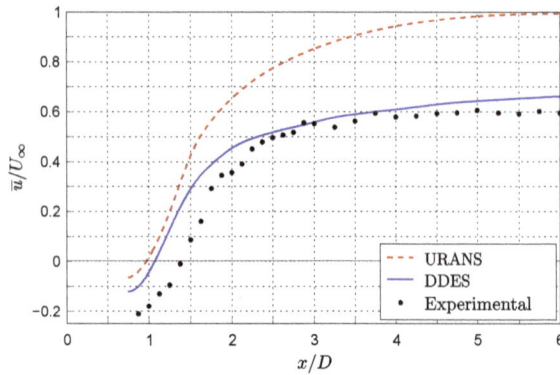

Figure 4. Evolution of the averaged streamwise velocity component normalized with the upstream velocity ($\overline{u}/U_\infty$) in the center of a square cylinder's wake ($y = 0$) compared with the experimental data [13].

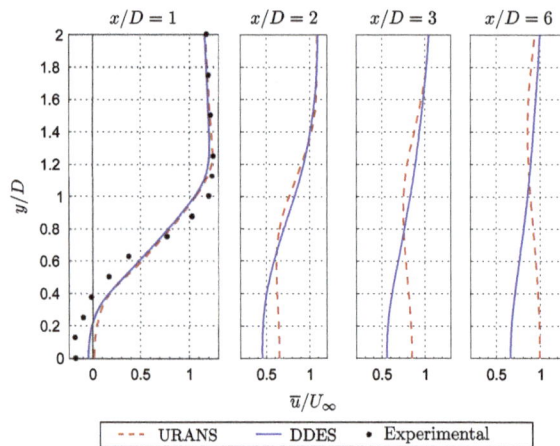

Figure 5. Profiles of the averaged streamwise velocity component along the y-axis normalized with the upstream velocity ($\overline{u}/U_\infty$) at several locations downstream of the square cylinder's center compared with the experimental data [13].

Nonetheless, even if the URANS and the DDES predictions of the averaged velocity profiles are very close to each other $1D$ downstream of the square cylinder's center, large discrepancies are observed further downstream in Figures 5 and 6. In the case of the averaged streamwise velocity profiles obtained with the URANS simulation, shown in Figure 5, a velocity deficit located away from the wake's center is still observed even after the velocity in the wake's center has been completely recovered. This is certainly not representative of a real wake's behavior. Regarding the averaged profiles of the transversal velocity component shown in Figure 6, one can observe that the URANS simulation predicts higher negative values than the DDES simulation, except for the profile taken at $1D$ downstream. This observation is consistent with the higher recovery rate of the averaged streamwise velocity component obtained with the URANS simulation, as observed in Figure 4.

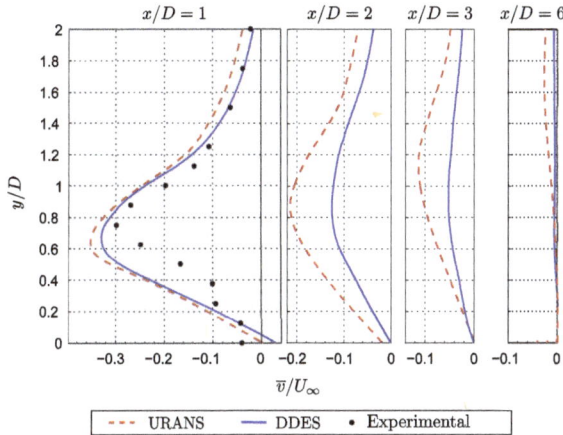

Figure 6. Profiles of the averaged transversal velocity component along the *y*-axis normalized with the upstream velocity ($\overline{v}/U_\infty$) at several locations downstream of the square cylinder's center compared with the experimental data [13].

3.3. Time-Varying Velocity Component

It can be useful for engineering purposes, such as for determining the fatigue loads experienced by an object located in a wake, to know about the time-varying component of velocity. Furthermore, the time-varying component of velocity also provides an insight into the physics at play. It is therefore important to accurately predict the fluctuating components of velocity.

Fluctuations of the streamwise and transversal velocity components are compared with the experimental data [13] in Figure 7. As can be seen in Figure 7a, the level of streamwise velocity fluctuations predicted by the URANS simulation are underestimated while the DDES predictions are slightly overestimated over most of the studied area, the latter still being in good agreement with the experimental data. Regarding the transversal fluctuations, the URANS and DDES results are closer to each other than they are for the streamwise component. However, it is observed that DDES results are still more accurate, especially in the region of the wake located beyond 3*D* downstream of the square cylinder's center.

It is interesting to note that both the URANS and the DDES simulations predict that the location where the highest transversal fluctuations are observed is closer behind the square cylinder than what is reported with the experimental data [13]. This same behavior has also been reported by previous DES simulations [21] and LES simulations [14,18,19] performed on the same case and is consistent with the smaller mean recirculation lengths (l_r/D) [34] predicted by these simulations and by the ones presented in the current paper. As pointed out by Celik et al. [35] in the case of LES simulations of free shear layers, this observation could be related to the difficulty of the subgrid-scale turbulence model to accurately predict the location of the laminar-turbulent transition in the free shear layers emerging from the two upstream sharp edges of the square cylinder. Also, it is interesting to note that the regions in the wake where the transversal fluctuations are overestimated correspond to the ones where the averaged streamwise velocity recovery rate is overestimated, and vice versa.

The lower level of fluctuations associated with the URANS modeling is related to the fact that the turbulence spectrum is essentially entirely modeled. Regarding the DDES simulation, the observed overestimation of the level of fluctuations is more surprising since only a part of the turbulence spectrum is resolved with a part of it being modeled. This same behavior has already been observed by other groups performing DES simulations of the same square cylinder case [20,21] and of a circular cylinder case [24]. However, these studies might have suffered from a too-narrow domain in the spanwise direction (4D and 2D respectively). Indeed, reducing the spanwise length of the domain from 7D to 5D and from 5D to 3D in previous simulations carried out by the authors has resulted in a continuous increase of the velocity fluctuation levels.

It is worth mentioning that there are also some results that have been reported in the literature [35–38] for which the resolved fraction of the turbulent kinetic energy that is predicted through a LES simulation is greater than the total amount of kinetic energy that is obtained with a direct numerical simulation (DNS), or even greater than the experimentally measured values. Among the possible causes, Celik et al. [35] have suggested that it might be attributed to a near-wall resolution that is too coarse. This leads to a deficiency in the amount of resolved eddies, which can contribute to a deterioration in the prediction of the strain rate. This, in turn, could possibly result in an underestimation of the dissipation, which would explain the overestimation of the turbulent kinetic energy for these cases.

A similar phenomenon can arise with the use of DES or DDES. When the attached boundary layers are modeled with a RANS approach, the modeled fraction of the turbulence spectrum in the RANS region does not allow the natural development of the instabilities that are required in the LES region. This gives rise to the existence of a transition zone, called the "gray area" [3,24], where the instabilities have not yet grown enough to compensate for the decrease in the amount of modeled eddies. Therefore, there is an analogy between a too coarse near-wall resolution in a LES simulation and the transition from a RANS to a LES modeling in a DES or a DDES simulation since both are characterized by a deficiency in the amount of resolved eddies.

However, the case chosen for the sake of the current study is not prone to being affected by this so-called "gray area" since the boundary layers on the upstream face of the square cylinder that separate at both upstream sharp edges are laminar. Consequently, the laminar-turbulent transition occurs in the separated shear layers where the LES approach of the DDES simulations is active.

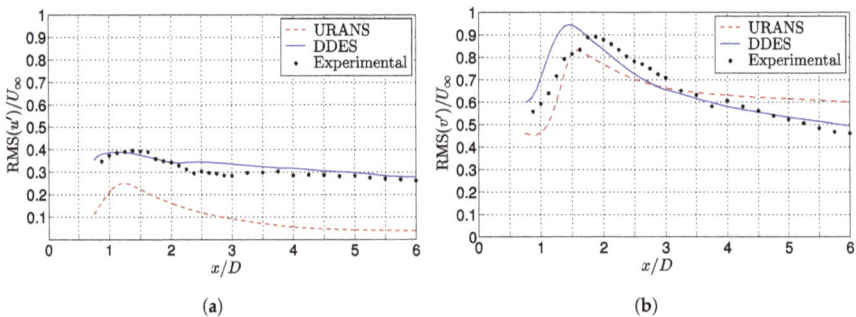

Figure 7. Evolution of the root mean square (rms) of the velocity fluctuations normalized with the upstream velocity (U_∞) in the center of a square cylinder's wake ($y = 0$) compared with the experimental data [13]. (**a**) Streamwise velocity fluctuations; (**b**) Transversal velocity fluctuations.

3.4. Integral Flow Quantities

Samples of the drag coefficient temporal signal obtained with the URANS and the DDES simulations are shown in Figure 8. It is observed that the URANS drag signal is very similar to a sinusoidal wave while the DDES drag signal consists in the superposition of several modes. The same type of signal as the one obtained with the DDES simulation has already been observed in similar simulations around a circular cylinder [24] and a rectangular cylinder [39], and is more representative of real drag signals observed in the case of bluff-body flows [5].

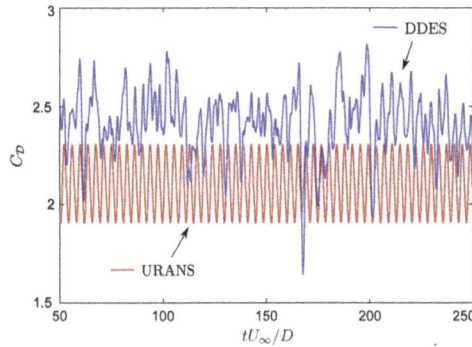

Figure 8. Sample of the drag coefficient (C_D) temporal signals obtained with the URANS simulation (red) and the DDES simulation (blue).

The value of five different integral flow quantities obtained with the current study are reported in Table 2. The first quantity is the time-averaged drag coefficient ($\overline{C_D}$):

$$\overline{C_D} = \frac{1}{T_{avg}} \int_0^{T_{avg}} C_D \, dt = \frac{1}{T_{avg}} \int_0^{T_{avg}} \frac{\mathcal{D}}{0.5 \, \rho \, U_\infty^2 \, D} \, dt \, , \tag{7}$$

where $\mathcal{D}$ is the drag force acting on the square cylinder, T_{avg} is the duration of the signal used to compute the average value (see Table 1), ρ is the fluid density, U_∞ is the freestream velocity and D is the cylinder's width. The other integral quantities considered are the Strouhal number ($St = fD/U_\infty$), the mean recirculation length (l_r) defined as the distance in the wake where a zero mean streamwise velocity is reached, the root mean square (rms) of the drag coefficient fluctuations (C_D'):

$$C_D' = \sqrt{\frac{1}{T_{avg}} \int_0^{T_{avg}} \left(C_D - \overline{C_D}\right)^2 \, dt} \, , \tag{8}$$

and the rms of the lift coefficient fluctuations (C_L'), which is defined in a way similar to C_D'. Note that all the force coefficients are given in units per span length. For comparison purposes, values obtained from experimental measurements [13,40,41] and from DES [20,21] and LES [14–19] simulations performed by other research groups are also reported.

It is important to note that the procedure used by Lyn et al. [13] to determine $\overline{C_D}$ is questionable, as mentioned by Sohankar et al. [19]. Indeed, their value has been obtained from the streamwise momentum flux at $8D$ downstream of the square cylinder's center without taking into account the pressure field. In fact, a negative contribution from the streamwise momentum flux to the time-averaged drag coefficient has been observed with the simulations performed by Sohankar et al. [19] as well as with the simulations of the current study.

Moreover, Bearman and Obasaju [40] and Norberg [41] have performed experiments in similar cases in the same range of Reynolds numbers. Based on the results of these experiments, Sohankar et al. [19] reported a value of $\overline{C_D}$ equal to 2.1 obtained by applying a correction in order to eliminate the blockage effects of these experiments. Regarding the experiment performed by Lyn et al. [13], the blockage effects should result in an increase in $\overline{C_D}$ of approximatively 12%, according to Sohankar et al. [19]. This allows for modification of the corrected $\overline{C_D}$ value reported by Sohankar et al. [19] in order to take into account the blockage effects corresponding to the channel size used in the experiment of Lyn et al. [13] for a straightforward comparison with the results of the current study. The resulting value of $\overline{C_D}$ is equal to 2.35, as reported in Table 2, and is in good agreement with the $\overline{C_D}$ value of 2.4 reported in Blevins' handbook [42] for the same blockage and in the same range of Reynolds numbers.

Table 2. Comparison of the time-averaged drag coefficient ($\overline{C_D}$), the Strouhal number (*St*), the mean recirculation length (l_r/D), the rms of the drag coefficient fluctuations (C_D') and the rms of the lift coefficient fluctuations (C_L'). The results of the DDES simulations performed with three different aspect ratios (AR) are presented.

Source	AR	$\overline{C_D}$	St	l_r/D	C_D'	C_L'
Current study						
URANS	7	2.11	0.133	0.97	0.14	1.56
DDES	3	2.36	0.123	1.20	0.26	1.50
DDES	5	2.40	0.126	1.15	0.21	1.51
DDES	7	2.41	0.126	1.07	0.17	1.47
DES simulations						
Barone/Roy [20]	4	2.36 *	0.131 *	1.42	0.29 *	1.30 *
Schmidt/Thiele [21]	4	2.42	0.13	1.16	0.28	1.55
LES simulations						
Fureby et al. [16]	8	2.0–2.2	0.129–0.135	1.23–1.37	0.17–0.20	1.30–1.34
Moussaed et al. [17]	π	2.06	0.128	≈1.3	0.24	1.28
Schmidt [18]	4	2.18	0.13	1.07	0.19	1.47
Sohankar et al. [19]	4	2.03–2.32	0.126–0.132	≈1	0.16–0.20	1.23–1.54
Rodi et al. [15]	4	1.86–2.77	0.09–0.15	0.89–2.96	0.10–0.27	0.38–1.79
Voke [14]	4	2.03–2.79	0.13–0.16	1.02–1.61	0.12–0.36	1.01–1.68
Experiments						
Bearman/Obasaju [40]	17	2.35 *	0.135 *	-	-	1.34
Lyn et al. [13]	9.75	2.1 †	0.132	1.38	-	-
Norberg [41]	51	2.35 *	0.135 *	-	-	-

* Values corrected for the blockage effects according to the method proposed by Sohankar et al. [19];
† The procedure used by Lyn et al. [13] to obtain this value raises some questions as previously discussed and as mentioned in Sohankar et al. [19].

It is observed that the $\overline{C_D}$ value obtained with the current DDES simulation is closer to the value obtained in the experiments performed by Bearman and Obasaju [40] and by Norberg [41], after being corrected for the blockage effects of the current study, than the one obtained with the URANS simulation. Also, it is interesting to note the large variation in the values of $\overline{C_D}$ predicted by other DES and LES simulations, thus suggesting that this physical quantity is very sensitive to the various flow parameters, as pointed out by Rodi et al. [15].

Regarding the Strouhal number, the value obtained with the URANS simulation is slightly closer to the experimental values than the one obtained with the DDES simulation. However, it is worth noting that the characteristics of the flow field predicted with the current DDES simulation are physically consistent. Indeed, a higher value of the time-averaged drag coefficient is generally associated with a smaller Strouhal number, a smaller mean recirculation length and larger velocity fluctuations in the near wake, as pointed out by Sohankar et al. [19] for the same case, and by Travin et al. [24] and Williamson [34] for the case of a circular cylinder. Lastly, the rms of the drag fluctuations and the lift fluctuations obtained from the DDES simulation agree with the values reported in the various studies that used a LES turbulence model. Moreover, it is observed that the size of the domain in the spanwise direction has a significant effect on the results, especially regarding C_D'. The fact that the spanwise domain length of the other simulations performed using DES in the literature was 4*D* probably explains why the simulation with the smallest spanwise domain length (3*D*) yields the best agreement with these results in terms of C_D'.

4. Conclusions and Outlook

Three-dimensional URANS and DDES simulations of the flow past a square cylinder (Re = 21,400) have been carried out with the open-source CFD toolbox OpenFOAM-2.1.x. The results have been compared with available experimental data [13] in order to assess the ability of DDES modeling at

Aerospace **2017**, *4*, 41

simulating a wake adequately. The Spalart–Allamaras turbulence model has been chosen for both modeling approaches.

The URANS simulation has yielded an essentially two-dimensional behavior, even if a three-dimensional grid with a spanwise length of $7D$ was used, and even after being initialized with the results of a DDES simulation. Large discrepancies have been observed between the URANS results and the experimental measurements [13], especially regarding the time-average and fluctuations of the streamwise velocity component. Unlike the URANS case, the DDES simulation exhibited a more realistic three-dimensional behavior. The agreement of the DDES results with the experimental data [13], regarding both the time-averaged and the time-varying components of velocity, has been similar or far superior to the URANS results, depending on the physical quantity that is considered.

A corrected $\overline{C_D}$ value reported by Sohankar et al. [19], obtained from the experiments performed by Bearman and Obasaju [40] and Norberg [41], has been modified to take into account the blockage effects corresponding to the domain size used in the current study, according to the method proposed by Sohankar et al. [19]. The $\overline{C_D}$ value obtained with the DDES simulation of the current study is in better agreement with this modified experimental value than the one obtained with the URANS simulation. On the other hand, a closer match with the experimental value of the Strouhal number has been observed with the URANS simulation.

The effect of the spanwise domain length has been found to have a considerable impact on the results, especially regarding the fluctuations of the drag coefficient. The fact that most of the previous studies on this benchmark case used a smaller domain size might partly explain the scatter observed between the results of the different numerical and experimental studies.

Based on the current study, it is concluded that DDES modeling appears as a good approach to reliably simulate wakes, especially far wakes of bluff bodies. Some unanswered questions remain regarding the effect of the "gray area" for flows involving turbulent boundary layers modeled in RANS mode as well as for flows for which the separations points are not dictated by the body geometry, namely flows around a body with no sharp edges. These important aspects of DDES simulations should be addressed in future studies.

Acknowledgments: Financial support from the Natural Sciences and Engineering Research Council of Canada (NSERC) and the Fonds de Recherche du Québec-Nature et Technologies (FRQNT) is gratefully acknowledged by the authors. Computations were performed on the Guillimin and Colosse supercomputers at the CLUMEQ HPC Consortium under the auspices of Compute Canada.

Author Contributions: Matthieu Boudreau and Jean-Christophe Veilleux developed the numerical methodology, Matthieu Boudreau and Guy Dumas analyzed the data, and Matthieu Boudreau wrote the paper.

Conflicts of Interest: The authors declare no conflict of interest. The founding sponsors had no role in the design of the study; in the collection, analyses, or interpretation of data; in the writing of the manuscript, and in the decision to publish the results.

Abbreviations

The following abbreviations are used in this manuscript:

CFD	computational fluid dynamics
DDES	delayed detached-eddy simulation
DES	detached-eddy simulation
LES	large eddy simulation
RANS	Reynolds-averaged Navier–Stokes
rms	root mean square
URANS	unsteady Reynolds-averaged Navier–Stokes

References

1. Muscari, R.; Di Mascio, A.; Verzicco, R. Modeling of vortex dynamics in the wake of a marine propeller. *Comput. Fluids* **2013**, *73*, 65–79.
2. Spalart, P.R. Strategies for turbulence modelling and simulations. *Int. J. Heat Fluid Flow* **2000**, *21*, 252–263.

3.   Spalart, P.R. Detached-Eddy Simulation. *Annu. Rev. Fluid Mech.* **2009**, *41*, 181–202.

4.   Shur, M.; Spalart, P.R.; Squires, K.D.; Strelets, M.; Travin, A. Three Dimensionality in Reynolds-Averaged Navier-Stokes Solutions Around Two-Dimensional Geometries. *AIAA J.* **2005**, *43*, 1230–1242.

5.   Najjar, F.; Balachandar, S. Low-frequency unsteadiness in the wake of a normal flat plate. *J. Fluid Mech.* **1998**, *370*, 101–147.

6.   Spalart, P.R.; Deck, S.; Shur, M.L.; Squires, K.D.; Strelets, M.; Travin, A. A new version of detached-eddy simulation, resistant to ambiguous grid densities. *Theor. Comput. Fluid Dyn.* **2006**, *20*, 181–195.

7.   Paik, J.; Sotiropoulos, F.; Porté-Agel, F. Detached eddy simulation of flow around two wall-mounted cubes in tandem. *Int. J. Heat Fluid Flow* **2009**, *30*, 286–305.

8.   Nasif, G.; Barron, R.; Balachandar, R. DES evaluation of near-wake characteristics in a shallow flow. *J. Fluids Struct.* **2014**, *45*, 153–163.

9.   Muld, T.W.; Efraimsson, G.; Henningson, D.S.; Herbst, A.H. Analysis of flow structures in the wake of a high-speed train. In *The Aerodynamics of Heavy Vehicles III. Lecture Notes in Applied and Computational Mechanics*; Springer: Cham, Switzerland, 2016; Volume 79, pp. 3–19.

10.   Felli, M.; Camussi, R.; Di Felice, F. Mechanisms of evolution of the propeller wake in the transition and far fields. *J. Fluid Mech.* **2011**, *682*, 5–53.

11.   Boudreau, M.; Dumas, G. Comparison of the wake recovery of the axial-flow and cross-flow turbine concepts. *J. Wind Eng. Ind. Aerodyn.* **2017**, *165*, 137–152.

12.   Wilcox, D. *Turbulence Modeling for CFD*, 2nd ed.; DCW Industries, Inc.: Flintridge, CA, USA, 1998.

13.   Lyn, D.A.; Einav, S.; Rodi, W.; Park, J.H. A laser-Doppler velocimetry study of ensemble-averaged characteristics of the turbulent near wake of a square cylinder. *J. Fluid Mech.* **1995**, *304*, 285–319.

14.   Voke, P.R. Flow Past a Square Cylinder: Test Case LES2. In *Direct and Large-Eddy Simulation II*; ERCOFTAC Series; Chollet, J.P., Voke, P., Kleiser, L., Eds.; Springer: Dordrecht, The Netherlands, 1997; Volume 5, pp. 355–373.

15.   Rodi, W.; Ferziger, J.H.; Breuer, M.; Pourquiée, M. Status of Large Eddy Simulation: Results of a Workshop. *J. Fluids Eng.* **1997**, *119*, 248–262.

16.   Fureby, C.; Tabor, G.; Weller, H.; Gosman, A. Large Eddy Simulations of the Flow Around a Square Prism. *AIAA J.* **2000**, *38*, 442–452.

17.   Moussaed, C.; Wornom, S.; Salvetti, M.V.; Koobus, B.; Dervieux, A. Impact of dynamic subgrid-scale modeling in variational multiscale large-eddy simulation of bluff-body flows. *Acta Mech.* **2014**, *225*, 3309–3323.

18.   Schmidt, S. *Grobstruktursimulation Turbulenter Strömungen in Komplexen Geometrien und bei Hohen Reynoldszahlen*; Mensch Mensch & Buch-Verlag: Berlin, Germany, 2000.

19.   Sohankar, A.; Norberg, C.; Davidson, L. Large Eddy Simulation of Flow Past a Square Cylinder: Comparison of Different Subgrid Scale Models. *J. Fluids Eng.* **2000**, *122*, 39–47.

20.   Barone, M.; Roy, C. Evaluation of Detached Eddy Simulation for Turbulent Wake Applications. *AIAA J.* **2006**, *44*, 3062–3071.

21.   Schmidt, S.; Thiele, F. Comparison of numerical methods applied to the flow over wall-mounted cubes. *Int. J. Heat Fluid Flow* **2002**, *23*, 330–339.

22.   Pope, S. *Turbulent Flows*; Cambridge University Press: Cambridge, UK, 2000.

23.   Spalart, P.R.; Allmaras, S.R. One-Equation Turbulence Model for Aerodynamic Flows. *La Recherche Aérospatiale* **1994**, *1*, 5–21.

24.   Travin, A.; Shur, M.; Strelets, M.; Spalart, P.R. Detached-Eddy Simulations Past a Circular Cylinder. *Flow Turbul. Combust.* **2000**, *63*, 293–313.

25.   Breuer, M.; Jovicic, N.; Mazaev, K. Comparison of DES, RANS and LES for the separated flow around a flat plate at high incidence. *Int. J. Numer. Methods Fluids* **2003**, *41*, 357–388.

26.   Shur, M.; Spalart, P.R.; Strelets, M.; Travin, A. Detached-eddy simulation of an airfoil at high angle of attack. *Eng. Turbul. Model. Exp.* **1999**, *4*, 669–678.

27.   Spalart, P.R.; Jou, W.; Strelets, M.; Allmaras, S. Comments on the Feasibility of LES for Wings, and on a Hybrid RANS/LES Approach. In Proceedings of the International Conference on DNS/LES, Ruston, LA, USA, 4–8 August 1997.

28. Garbaruk, A.; Leicher, S.; Mockett, C.; Spalart, P.R.; Strelets, M.; Thiele, F. Evaluation of time sample and span size effects in DES of nominally 2D airfoils beyond stall. Notes on Numerical Fluid Mechanics and Multidisciplinary Design. In *Progress in Hybrid RANS-LES Modelling*; Springer: Berlin/Heidelberg, Germany, 2010; Volume 111, pp. 87–99.

29. OpenCFD. *OpenFOAM-The Open Source CFD Toolbox—User's Guide*, 2.1 ed.; OpenCFD Ltd.: London, UK, 2012.

30. Ferziger, J.l H.; Perić, M. *Computational Methods for Fluid Dynamics*; Springer: Berlin/Heidelberg, Germany, 2002.

31. Spalart, P.R. *Young-Person's Guide to Detached-Eddy Simulation Grids*; Technical Report; NASA Technical Reports Server: Hampton, VA, USA, 2001.

32. Mockett, C.; Perrin, R.; Reimann, T.; Braza, M.; Thiele, F. Analysis of Detached-Eddy Simulation for the Flow Around a Circular Cylinder with Reference to PIV Data. *Flow Turbul. Combust.* **2010**, *85*, 167–180.

33. Jeong, J.; Hussain, F. On the identification of a vortex. *J. Fluid Mech.* **1995**, *285*, 69–94.

34. Williamson, C.H.K. Vortex Dynamics in the Cylinder Wake. *Annu. Rev. Fluid Mech.* **1996**, *28*, 477–539.

35. Celik, I.; Cehreli, Z.; Yavuz, I. Index of Resolution Quality for Large Eddy Simulations. *J. Fluids Eng. Trans. ASME* **2005**, *127*, 949–958.

36. Gousseau, P.; Blocken, B.; Van Heijst, G.J.F. Quality assessment of Large-Eddy Simulation of wind flow around a high-rise building: Validation and solution verification. *Comput. Fluids* **2013**, *79*, 120–133.

37. Klein, M. An Attempt to Assess the Quality of Large Eddy Simulations in the Context of Implicit Filtering. *Flow Turbul. Combust.* **2005**, *75*, 131–147.

38. Meyers, J.; Geurts, B.J.; Baelmans, M. Database analysis of errors in large-eddy simulation. *Phys. Fluids* **2003**, *15*, 2740–2755.

39. Mannini, C.; Šoda, A.; Schewe, G. Numerical investigation on the three-dimensional unsteady flow past a 5:1 rectangular cylinder. *J. Wind Eng. Ind. Aerodyn.* **2011**, *99*, 469–482.

40. Bearman, P.; Obasaju, E. An Experimental Study of Pressure Fluctuations on Fixed and Oscillating Square-Section Cylinders. *J. Fluid Mech.* **1982**, *119*, 297–321.

41. Norberg, C. Flow around rectangular cylinders: Pressure forces and wake frequencies. *J. Wind Eng. Ind. Aerodyn.* **1993**, *49*, 187–196.

42. Blevins, R. *Applied Fluid Dynamics Handbook*; Krieger Publishing Company: Malabar, FL, USA, 1992.

MDPI
St. Alban-Anlage 66
4052 Basel
Switzerland
Tel. +41 61 683 77 34
Fax +41 61 302 89 18
www.mdpi.com

Aerospace Editorial Office
E-mail: aerospace@mdpi.com
www.mdpi.com/journal/aerospace